Modern Hydrology

Volume I

Modern Hydrology
Volume I

Edited by **Stacy Keach**

R CALLISTO REFERENCE

New York

Published by Callisto Reference,
106 Park Avenue, Suite 200,
New York, NY 10016, USA
www.callistoreference.com

Modern Hydrology: Volume I
Edited by Stacy Keach

International Standard Book Number: 978-1-63239-464-4 (Hardback)

Printed in the United States of America.

Contents

Preface

The term 'Hydrology' has been derived from two words, 'Hydro' meaning 'water' and 'logy' meaning 'study of'. Thus, etymologically, Hydrology refers to the study of water. The origins of this discipline can be traced to 4000 BC, when in order to improve agricultural productivity of previously barren lands, the Nile was dammed. In a similar fashion, aqueducts were built by the Greeks and Ancient Romans.

Water is one of our most important natural resources. Thus, today, Hydrology has evolved as a science to understand the complex water system of the earth. It encompasses the occurrence, distribution, movement, quality and properties of water. The relationship of each phase of the hydrologic cycle with the environment also happens to be a crucial area of focus in Hydrology. Hydrology is an interdisciplinary field and includes subjects like civil and environmental engineering, physical geography and environmental science among others. The subject is subdivided into surface hydrology and marine hydrology.

For research, masses of data are organized, summarised and analysed, after which simplified, hydrological models are arrived at. This data is primarily used for hydrological prediction such as consequences of reservoir releases, possibility of flooding among others. These models help in quantifying the movement of water between its various states, in a specific region.

I'd like to thank all the people who've contributed into making this book a source of rich knowledge.

Editor

Application of Synthetic Meteorological Time Series in BROOK90: A Case Study for the Tharandt Forest in Saxony, Germany

Kronenberg Rico[1,2*], Güttler Tino[1], Franke Johannes[1], Bernhofer Christian[1]

[1]Meteorology, Technische Universität Dresden, Dresden, Germany; [2]Mathematical Modelling, Bauman Moscow State Technical University, Moscow, Russia.

ABSTRACT

This study presents an extended version of a single site daily weather generator after Richardson. The model is driven by daily precipitation series derived by a first-order two-state Markov chain and considers the annual cycle of each meteorological variable. The evaluation of its performance was done by deploying its synthetic time series into the physical based hydrological model BROOK90. The weather generator was applied and tested for data from the Anchor Station at the Tharandt Forest, Germany. Additionally its results were compared to the output of another weather generator with spell-length approach for the precipitation series (LARS-WG). The comparison was distinguished into a meteorological and a hydrological part in terms of extremes, monthly and annual sums and averages. Extreme events could be preserved adequately by both models. Nevertheless a general underestimation of rare events was observed. Natural correlations between vapour pressure and minimum temperature could be conserved as well as annual cycles of the hydrological and meteorological regime. But the simulated spectrums of extremes, especially, of precipitation and temperature, are more limited than the observed spectrums. While LARS-WG already finds application in practice, the results show that the data derived from the presented weather generator is as useful and reliable as those from the established model for the simulation of the water balance.

Keywords: Richardson Model; Weather Generator; BROOK90; Synthetic Time Series; LARS-WG; Forest Water Balance; Taylor Diagram; Cumulative Periodogram

1. Introduction

The planning, construction and management of precipitation related infrastructure like sewer systems, retention areas or dams highly depend on the occurrence and statistical return period of extreme rainfall events [1,2]. In practise, robust and long time series for simulation and extrapolation of these events are needed to identify and consider such extremes. Hence, the lag of satisfactory long term observations leads to the development of stochastic models which are able to simulate rainfall without the recognition of atmospheric driven processes [3,4]. Their outcomes, long synthetic rainfall series, fulfil the requirements of the engineers. Thus, these models were extended with other meteorological variables and soon the first weather generators were presented [5]. These can be classified into Markov chain, spell length and non-parametric models [6,7]. The main disadvantage of these approaches is their limited capacity to model unobserved states as well as the incomplete preservation of statistical properties. They all depend on historical time series, which by definition can not include unobserved extremes of weather variables. In particular, they are not able to model non-stationary processes. But the transition from stochastic weather generators to stochastic nesting approaches or weather state models is possible. Therefore even changes of the climate can be recognized by integrating different scenarios [8,9] or by including a certain expected trend [10].

The great advantage of these generators is their speed unmatched by any other tool to simulate locally consistent future time series. They are very fast algorithms to produce long time series, which find a vast extent of applications as input in hydrological, hydrodynamic and other climate variable driven models [11-16].

*Corresponding author.

The main focus in this study lies in the derivation and extension of the Richardson model and its application to observed data under the strict preservation of diurnal variations. More particularly not just the preservation of temporal properties was a major goal but also the physical consistencies between the considered meteorological variables. Here from arise the following motivational questions of this paper:

1. Do the simulated empirical distributions of meteorological elements fit the observed ones in terms of rare events (*i.e.* extremes)?

2. Are physical correlations of the water balance preserved by the weather generators?

3. Are the approaches able to retain the annual cycles of the considered hydrological and meteorological elements?

In the tradition of weather generators two philosophies were compared for single site time series in terms of a certain location, a Markov chain with a spell length model. The physical consistencies were investigated by deploying the simulated time series into the hydrological model BROOK90 [17].

2. Data and Study Region

2.1. Tharandt Forest

The Tharandt Forest is situated 20 km south west of Dresden and forms a part of the north eastern boundary of the eastern Ore Mountains. The forest spans a territory of 6000 ha [18], whose extent is shown in **Figure 1**. It is the largest contiguous area of forest dominated land use in Saxony, Germany. In average the territory is located 350 to 400 m asl and therefore features just a small relief intensity [19]. The highest point is the Tännicht with 461 m asl and the lowest is located in Coßmannsdorf with 197 m asl. The area forms a distinct cuesta in the north, which is a boundary to the Mulde loess hills and in the east a boundary to the eastern foothills of the Ore Mountains [20-22].

Due to its location near the north-eastern ridges of the Ore Mountains the study area shows an increased continental character of the climate in contrary to the western parts. Hence, the observed climate is representative for the climatic conditions of the foot hills of the eastern Ore Mountains. Though, the dominance of the forested land in combination with the local topography lead to significant differences from the regional climate conditions [23]. However, the bimodal distribution of the mean annual precipitation also can be observed, which is distinctive for this region.

2.2. Meteorological Data

The meteorological time series were observed at the An-

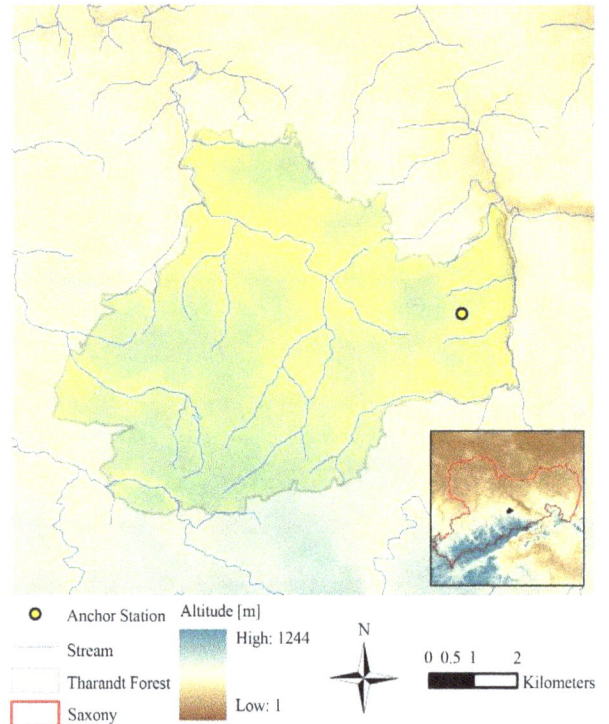

Figure 1. Tharandt Forest with Anchor Station (50°58'N, 13°34'E, 385 m asl) and altitude in m asl; the Wildacker Station is nearby located.

chor Station (50°58'N, 13°34'E) in the Tharandt Forest, Germany. The elements include precipitation, solar radiation, vapour pressure, average, minimum and maximum temperature. They are measured by the Chair of Meteorology at Technische Universität Dresden since the late 1950s. The time series used here is from 01/01/1997 until 09/30/2009 in daily resolution; it reflects the start of continuous flux measurements of, e.g. evapotranspiration at the site [24]. The time series were checked for stationarity, homogeneity and data gaps. The quality assessment was done considering the measurements at the nearby meteorological station Wildacker, which served as reference station. Hence, non-stationarities and heterogeneity could be excluded and no data gaps were found [25]. Further corrections of precipitation measurements were omitted [26,27]. The aforementioned bimodal distribution of the monthly mean precipitation of the Wildacker Station can be seen in **Figure 2**. It summarizes the climatic conditions from 1971 to 2010 at the site.

3. Methods

Stochastic models often find application in hydrology and meteorology as they offer the possibility to generate long and persistent time series.

The main reason for this frequent usage is the short availability of sufficient long and complete observed

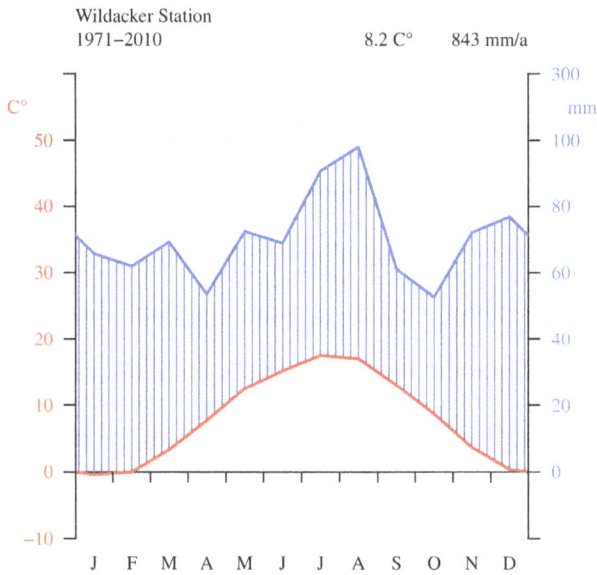

Wildacker Station
1971–2010   8.2 C°   843 mm/a

Figure 2. Climate diagram after Walter and Lieth [28] of the Wildacker Station nearby the Anchor Station in the Tharandt Forest for the period from 1971 until 2010.

time series.

Stochastic models enable to fill gaps in observed series of meteorological variables and to generate synthetic time series without actual time reference but unlimited length. These kinds of models are commonly called "weather generators".

On the one hand a type of model, often deployed is the aforementioned "spell-length" approach. Time series are simulated by the addition of dry and wet periods according to the distribution of such periods in the observed data. The daily amount of precipitation is often derived by parametric distribution function of precipitation (e.g. gamma, log-Gaussian or mixture distributions). On the other hand so called Markov chain models are approaches which are often applied. It is possible based on observation to describe precipitation as stochastic process. These processes are of such a kind that occurrence probabilities of precipitation depend on a finite number of former days. After the generation of wet or dry days the amount of precipitation is similar to the spell-length models derived by aforementioned distribution functions. Richardson introduced a model which extended the synthetic precipitation series with other meteorological elements. Further the applied extended Richardson model is presented for the single site simulation of meteorological variables. All used symbols are summarized in **Table 1**.

3.1. Extended Richardson Model-ERM-Precipitation Estimation

Precipitation was modelled through a first-order two-

state Markov chain for a single station. The stochastic process model is called first-order because the probability if today is a wet day just depends on yesterday. More time lags would increase the order. Likewise it is a two state model which means that just wet or dry states are considered. The conditional probabilities for wet after dry and wet after wet days can be calculated according to Equations (1) and (2), which are based on binary coded series x of wet and dry days. Therefore the shown summations of all j observed days with the defined restriction have to be calculated.

$$p_{01} = \frac{\sum \left(x_j = 1, x_{j-1} = 0\right)}{\sum \left(x_{j-1} = 0\right)} \qquad (1)$$

$$p_{11} = \frac{\sum \left(x_j = 1, x_{j-1} = 1\right)}{\sum \left(x_{j-1} = 1\right)} \qquad (2)$$

Random numbers u and v are normally distributed with zero mean und standard deviation one.

$$u, v \sim N\left(0,1\right) \qquad (3)$$

After the determination of the conditional probabilities they were used in combination with the generated random numbers to simulate new binary series $\tilde{x}$ of wet and dry states. This is done in the following manner. First an initial state has to be announced for the first day which would be the day's probable condition p_c. Second, according to Equations (4) and (5), p_c changes from day to day in dependence of the generated random number u. Hence the daily conditions change according to the estimate transition probabilities.

$$u \le p_c \rightarrow \text{precipitation} \rightarrow \tilde{x} = 1 \rightarrow p_c = p_{11} \qquad (4)$$

$$u > p_c \rightarrow \text{no precipitation} \rightarrow \tilde{x} = 0 \rightarrow p_c = p_{01} \qquad (5)$$

Reference [29] used Fourier's series to simulate day of the year (DOY) dependent transition probabilities (i.e. $\tilde{p}_{01}\left(t\right)$ and $\tilde{p}_{11}\left(t\right)$). This recommendation seemed reasonable. Since, like shown in **Figure 3**, also for the presented case an annual cycle of the transition probabilities can be observed. The Fourier's series and their coefficients are defined in Equations (6) and (7), and in **Table 2**. The equations are already concrete solutions of the presented case study. For, the periods are 365 and 183 days as it can be seen in **Figure 3**.

$$\tilde{p}_{01}\left(t\right) = a_{0,01} + a_{1,01} \cos\left(\omega_{01}t\right) + b_{1,01} \sin\left(\omega_{01}t\right) \qquad (6)$$

$$\tilde{p}_{11}\left(t\right) = a_{0,11} + a_{1,11} \cos\left(\omega_{11}t\right) + b_{1,11} \sin\left(\omega_{11}t\right) \qquad (7)$$

To fit Fourier's series Equations (1) and (2) have to be solved in dependence of DOY. The resulting conditional probabilities are defined as $p_{01}\left(t\right)$ and $p_{11}\left(t\right)$. These are shown in **Figure 3** as observed series.

Table 1. Symbolism and nomenclature.

Symbol	Unite	Description
u, v	[-]	Normally distributed random numbers
$p_{01}, p_{11}, p_{01}(t), p_{11}(t)$ $\tilde{p}_{01}(t), \tilde{p}_{11}(t)$	[-]	conditional probabilities
$x, \tilde{x}$	[-]	Binary observed precipitation series, binary generated precipitation series
p_c	[-]	Probability of today's condition
ω_{01}, ω_{11}	[-]	Frequency
t, j	[d]	Day of year (DOY), day
$a_{0,01}, a_{1,01}, b_{1,01}$ $a_{0,11}, a_{1,11}, b_{1,11}$	[-]	Parameter of Fourier function
μ_s, μ	Element dependent	Maximum-likelihood estimates of shape, of mean
v_{max}	[-]	Maximum probability for precipitation
σ	Element dependent	Maximum-likelihood estimate of standard deviation
k, K	[-]	Meteorological element, number
Z, z	[-]	Observed, estimated standardized series
$X, \tilde{X}$	Element dependent	Observed, estimated series
$\varepsilon(t), \varepsilon^*(t)$	[-]	Residual random error term
M_0, M_1	[-]	Correlation matrix
ρ_0, ρ_1	[-]	Element of correlation matrix
B, A	[-]	Multivariate matrix
e_S	[hPa]	Vapour pressure
h	[°C]	Minimum daily temperature
C_1, C_2, C_3	Coefficient dependent	Coefficients of Magnus formula
P	[mm/time]	Precipitation
E	[mm/time]	Evapotranspiration
Q	[mm/time]	Discharge
dS	[mm/time]	Storage change
$IRVP$	[mm/time]	Canopy evaporation of rain
$ISVP$	[mm/time]	Canopy evaporation of snow
$SNVP$	[mm/time]	General evaporation from snow masses
$SLVP$	[mm/time]	Surface evaporation
$TRAN$	[mm/time]	Transpiration from plants

The coefficients $a_{0,01}$ and $a_{0,11}$ can be seen there as well, for they simply form the arithmetic means of the observed series. Thus, they represent the constant series. Finally the Fourier's series resulting from Equations (6) and (7) were deployed for the modelling.

The daily precipitation amount is calculated in dependence of the binary-coded wet dry series (*i.e.* $\tilde{x}$). As in Equation (8) defined an inverse gamma distribution was deployed. Its shape parameter μ_s was derived as maximum likelihood estimate for each month. The re-

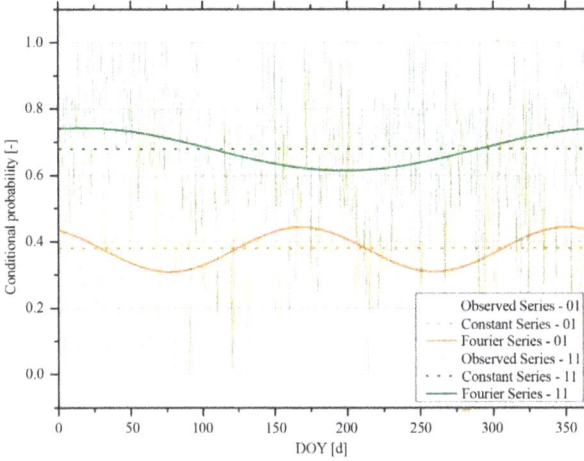

Figure 3. Different series of conditional probabilities in dependence of DOY for the deployed data sets of the Anchor Station, Tharandt.

Table 2. Frequencies and coefficients of the deployed Fourier's series to model in dependence of DOY conditional probabilities of wet and dry days.

Coefficients for $\tilde{p}_{01}(t)$	Coefficients for $\tilde{p}_{11}(t)$
$\omega_{01} = \dfrac{2\pi}{183}$	$\omega_{11} = \dfrac{2\pi}{365}$
$a_{0,01} = \dfrac{1}{365}\sum_{t=1}^{365} p_{01}(t)$	$a_{0,11} = \dfrac{1}{365}\sum_{t=1}^{365} p_{11}(t)$
$a_{1,01} = \dfrac{2}{183}\sum_{t=1}^{365}\cos\left(\omega_{01}p_{01}(t)\right)$	$a_{1,11} = \dfrac{2}{365}\sum_{t=1}^{365}\cos\left(\omega_{11}p_{11}(t)\right)$
$b_{1,01} = \dfrac{2}{183}\sum_{t=1}^{365}\sin\left(\omega_{01}p_{01}(t)\right)$	$b_{1,11} = \dfrac{2}{365}\sum_{t=1}^{365}\sin\left(\omega_{11}p_{11}(t)\right)$

sults are concluded in **Figure 4** of all estimated monthly distributions.

$$P = -\ln\left(v \cdot v_{max} \cdot \mu_s\right) \cdot \mu_s \cdot \tilde{x} \; (\text{mm}) \qquad (8)$$

3.2. Extended Richardson Model-ERM-Estimation of Other Meteorological Elements

The other meteorological elements depend on the generated precipitation series, precisely on the binary coded wet and dry state series $\tilde{x}$. First the observed series X_k were standardized according to Equation (9) over all variables k.

$$Z_k = \frac{\left[X_k - \mu_k\right]}{\sigma_k} \qquad k = 1, 2, \cdots, K \qquad (9)$$

The new estimated meteorological series $\tilde{X}_k$ then simply depend on the daily conditions (*i.e.* wet = 1 or dry = 0) and the estimated DOY (*i.e.* t = DOY) dependent mean and standard deviation like defined in Equation (10).

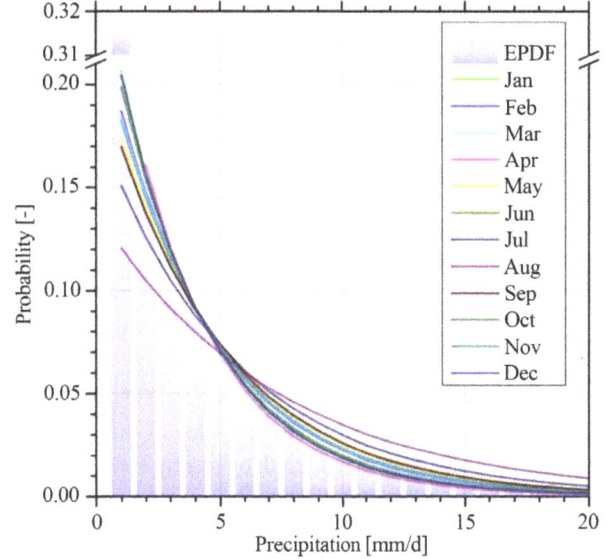

Figure 4. Empirical and theoretical distribution functions (*i.e.* 1 mm bins) of daily precipitation for the Anchor Station, in bars depicted are the annual empirical distribution function (EPDF) of the Anchor Station for the observed period.

$$\tilde{X}_k(t) = \begin{cases} \mu_{k,0}(t) + \sigma_{k,0}(t) \cdot z_k(t) \\ \mu_{k,1}(t) + \sigma_{k,1}(t) \cdot z_k(t) \end{cases} \qquad (10)$$

The standardized series $z_k(t)$ depend on the state the previous day, two multivariate matrices and a residual random error term, like shown in Equation (11).

$$z_k(t) = A \cdot z_k(t-1) + B \cdot \varepsilon(t) \qquad (11)$$

The multivariate matrices A and B are defined in Equations (12) and (13). B has to be calculated through Cholesky or Q/R decomposition [30].

$$BB^{\mathrm{T}} = M_0 - A \cdot M_1^{\mathrm{T}} \qquad (12)$$

$$A = M_1 \cdot M_0^{-1} \qquad (13)$$

M_1 and M_0 are lag-zero and lag-one cross correlation matrices, with elements defined in Equations (14) and (15):

$$\rho_0 = Corr\left[Z_k(t), Z_l(t)\right] \quad k,l = 1,2,\cdots,K \qquad (14)$$

$$\rho_1 = Corr\left[Z_k(t), Z_l(t-1)\right] \quad k,l = 1,2,\cdots,K \qquad (15)$$

The residual random error term is defined in Equations (16) and (17).

$$\varepsilon(t) = B \cdot \varepsilon^*(t) \qquad (16)$$

$$\varepsilon^*(t) \sim MVN(0,1) \qquad (17)$$

3.3. LARS-WG

The stochastic weather generator LARS-WG was developed to simulate daily synthetic meteorological time series for single site [31]. Its latest version 5.0 enables the user to model the actual as well as the future climate. All simulations depend on the observed series from which the necessary model parameters for probabilities and correlations are derived. These were used to simulated synthetic series which are simultaneously randomly distributed. In contrast to the aforementioned extended Richardson approach LARS-WG uses a spell-length approach to derive precipitation series [6,12]. Than the daily precipitation amount is calculated according to semi empiric distribution. Annual cycles of the meteorological elements are considered by using Fourier functions [32]. Unfortunately LARS-WG in its used version is only able to model precipitation, evapotranspiration, minimum and maximum temperature as well as solar radiation. Hence, in contrary to the ERM wind speed and vapour pressure are missing variables.

3.4. BROOK90

The hydrological model BROOK90 was developed to simulate the vertical soil water movement and the evapotranspiration for a certain land surface at a daily resolution. The model is process-orientated and its parameters hold a physical meaning [17].

The model is a complex lumped-parameter model and follows a 'less is more' philosophy, which is characterized by a strong generalization of stream flow generation pathways but enough to compare it to observed time series. This generalization even goes further by ignoring aspects like hill slope hydrology and spatial distribution to focus on factors determining evapotranspiration.

For these reasons its design may serve the purpose of sensitivity analysis by the possibility to include or exclude certain soil water sub-models, which can be necessary to simulate plant growth, biogeography or global hydrology.

In consideration that LARS-WG does not estimate wind speed and vapour pressure the authors of BROOK90 suggest to work with a constant wind speed of 3 m/s and to calculate vapour pressure according to saturated vapour pressure. This was done using Magnus formula, which is defined in Equation (18).

$$e_S = C_1 \cdot \exp\left(\frac{C_2 \cdot h}{C_3 + h}\right)(hPa) \qquad (18)$$

with
over water: $C_1 = 6.1078$ hPa; $C_2 = 17.08085$;
$C_3 = 234.175°C$

over ice: $C_1 = 6.1078$ hPa; $C_2 = 17.84362$;
$C_3 = 245.425°C$

The necessary parameters were chosen according to the minimum temperature (i.e. <0°C the surface was considered as ice), which was suggested by the BROOK90 authors.

The authors state that BROOK90 can fill a wide range of needs. It finds application in teaching and study water budget, water movement on small plots, evapotranspiration and soil water process. In addition it might answer questions related to land management and for the prediction of climate change effects. A further one was added to these tasks by deploying the model to validate the performances of weather generators.

The necessary model parameters of BROOK90 are taken from reference [25] (i.e. B2 configuration) which were determined for the considered period.

4. Results and Discussion

This study focuses on the extension and application of weather generators. Therefore the introductorily asked questions are of significant importance. The results where investigated from two perspectives to answer these questions. On the one hand meteorological properties are analyzed element wise by considering their correlations, periodicity and positive extremes (0.95 quantiles). On the other hand their hydrological properties are surveyed according to the long-term water balance defined in Equation (19). The precipitation (P) is defined as the sum of the discharge (Q), the evapotranspiration (E) and the storage change (dS). It is the water balance in its most common and most simple form [33].

$$P = Q + E + dS \,(\text{mm}) \qquad (19)$$

The results are four data sets, which are named and defined in the following manner: the observed (OBS), the extended Richardson model (ERM), the extended Richardson model without wind speed and vapour pressure (ERMw) and the LARS-WG data set (LARS). 1000 years long synthetic time series were simulated by each approach.

4.1. Meteorological Synthetic Time Series

First the correlations between minimum temperature and vapour pressure were investigated. Hence, the differences of simulated and calculated vapour pressure are outlined. The scatter plots are summarized in **Figure 5**. **Figure 5(a)** shows the natural correlations between minimum temperature and vapour pressure mainly following the generalization of Equation (18), with a not neglected scattering through natural fluctuation. The scattering becomes even broader between the simulated variables in **Figure 5(b)**. But the physical correlations

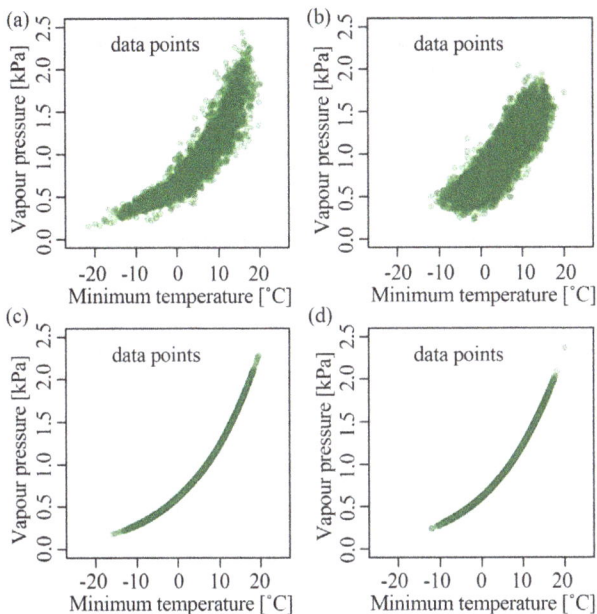

Figure 5. Correlations between minimum temperature and vapour pressure for 5000 randomly selected data points; (a) OBS; (b) ERM; (c) LARS; (d) ERMw.

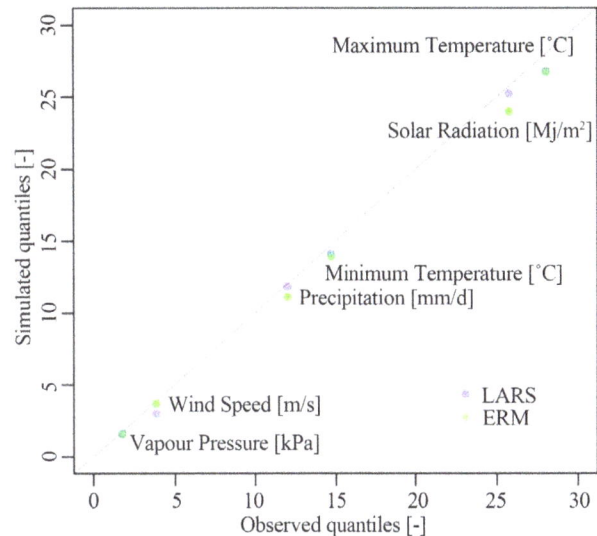

Figure 6. 0.95 Quantiles of all modelled variables for LARS and ERM simulations of 1000 years time series.

and natural fluctuations could be preserved. Whereas **Figures 5(c)** and **(d)** show the curve of Equation (18) for LARS and ERMw, obviously these curves are the same.

The results of the investigations of the 0.95 quantiles are depicted in **Figure 6**. Compared are the Quantiles of simulated and observed variables. They give an insight on how the empirical distributions are located, especially the right tails of these distributions and how they match with the observed distributions. For all simulations, the quantiles of LARS as well as of ERM show a slight under estimation even independent of the season. With no regard on the approach the best fit can be observed in the vapour pressure. The wind speed is slightly better simulated by ERM than by LARS. Tough it must be stated that for LARS only a series of 3 m/s wind speed was generated. This unorthodox procedure arises from the recommendations of the BROOK90 authors. Nevertheless this value nearly represents the stations mean value.

The advantage of the applications of semi empiric distributions becomes obvious looking at the precipitation. The extremes of this meteorological element are better simulated by LARS than by ERM which just uses a gamma distribution.

The largest difference can be found between the simulated solar radiations, which so far cannot be explained.

No differences are observable between minimum and maximum temperature. The underestimations of these variables confirm the results in **Figure 5**, where the spectrum of possible values from minimum to maximum temperature is more limited, in contrast to the observed

spectrum. Thus the tails of the simulated distributions are more limited.

Reference [34] stated that apart from the test of white noise a cumulative periodogram also can be used to examine hidden and suspected periodicities. In this context the monthly synthetic and partly observed time series of the meteorological variables were plotted in **Figure 7**. So a validation of their periodicity is qualitatively possible.

All variables of the observed and simulated series show a jump at a frequency of 0.0833 which is equivalent to 12 month (*i.e.* red highlighted in **Figure 7**). While all variables are characterized by a large significant jump at this point, precipitation follows more or less the 0.5:1 line. The simulated series of precipitation even shows signs of higher periodicity. These jumps just can be explained with the periodicity of months. As obvious also simple series of returning months (*i.e.* {1, 2, 3, …, 12, 1, ..}) are plotted and show significant steps.

These steps are also obvious in the monthly series of precipitation. There it can be explained by the fitting of distributions functions depending on the month (cp. **Figure 4**).

Figure 7 proves that all simulated variables of ERM follow an annual cycle. Even rather artificial inner annual periodicities caused by the model parameterisation of precipitation are preserved.

4.2. Long-Term Water Balance

The Taylor diagram was developed to illustrate the relation between correlations, standard deviation and root mean square error [36]. It is commonly applied for GCM validation. In this case instead of different models the water balance components are drawn in **Figure 8**. The

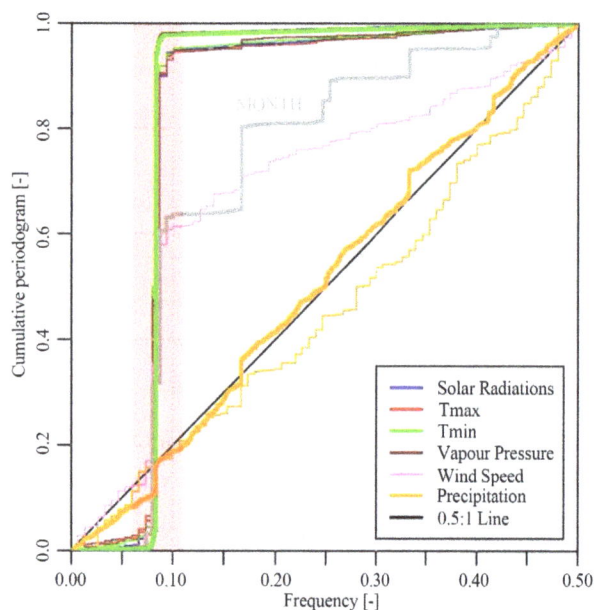

Figure 7. Cumulative periodogram [35] of monthly meteorological variables of OBS and ERM; depicted in bold lines is ERM and in thin lines OBS data.

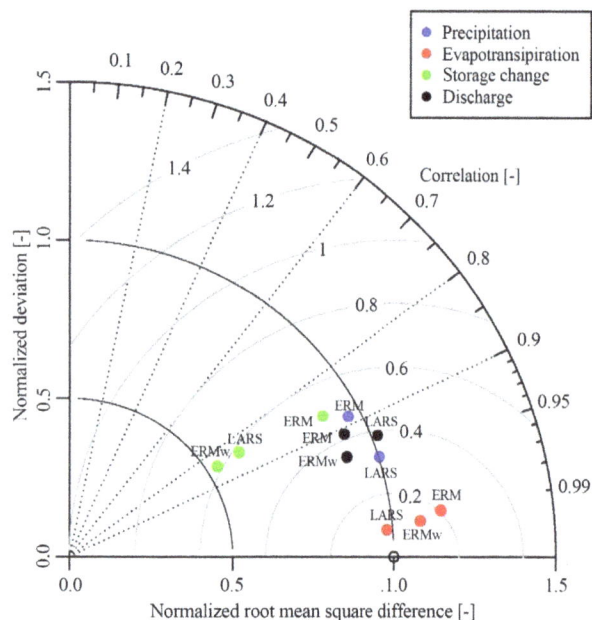

Figure 8. Taylor diagram of the monthly mean standardized values of the long-term water balance elements estimated for 1000 years periods for each model.

meaning remains the same as in its former usage; the closer a point lies to the reference at one on the abscissa the better or closer the simulated component behaves to the reference state (*i.e.* observed state).

The evapotranspiration could achieve the best results, consistently gaining a correlation higher than 0.99 with no dependence on the model. The LARS lies closest to

the reference point followed by ERMw and ERM. But the differences are very small. Generally it can be stated that all models simulated the evapotranspiration really close to the observed state.

These results for ERM could only be achieved because of a post processing of the BROOK90 calculations. Negative E values in winter could be observed in the model runs driven by ERM data. BROOK90 calculates E according to Equation (20). The negative E resulted from negative values caused by the canopy evaporation of snow term (ISVP).

$$E = \mathrm{IRVP} + \mathrm{ISVP} + \mathrm{SNVP} + \mathrm{SLVP} + \mathrm{TRAN}\ (\mathrm{mm}) \quad (20)$$

To solve this contradiction of negative E values, ISVP was excluded from its calculation to obtain realistic results. Therefore the water balance had to be closed by the summation of the excluded term (*i.e.* ISVP) with Q. This post processing occurred to be necessary just for the ERM run. Unfortunately plausible reasons for this behaviour could not be figured out, since the physical correlations between minimum temperature and vapour pressure are preserved, as depicted in **Figure 5(b)**. The authors of BROOK90 give a vague explanation of this particular behaviour by arguing that the canopy evaporation of snow is still a rather unknown process. Hence, its consideration in a hydrological model may lead to the observed uncertainties. Nevertheless, this post processing lead to reasonable results for ERM as **Figures 8** and **9(b)** conclude.

The differences at a monthly scale for the precipitation are larger. While LARS reaches a correlation of 0.95 ERM achieved fairly 0.89. However both results are outstanding compared to other investigations [37] where precipitation, because of its supposed randomness in time and space at this scale, is the most defficile element to simulate. The significant better performance of LARS can be explained by the usage of a semi empiric distribution for the modelling of precipitation, which obviously estimates the daily amount more precise.

Referring to the system's output, the discharges are not wide spread in **Figure 8**. They lie between a correlation of 0.9 and 0.95. The best performance to the references is shown by ERMw followed by LARS and ERM.

The results of the residual storage change term, in contrast, scatter quite within the diagram, while both elements without estimated vapour pressure and wind lie closer to each other (*i.e.* LARS and ERMw) and their correlations are about 0.85. ERM lies apart from them, but nearer to the reference state. Thus the standard deviation of ERM is closer to the reference than those of the others. ERM is considered as best estimated of the storage change, even if its percentage of the overall water balance is marginal.

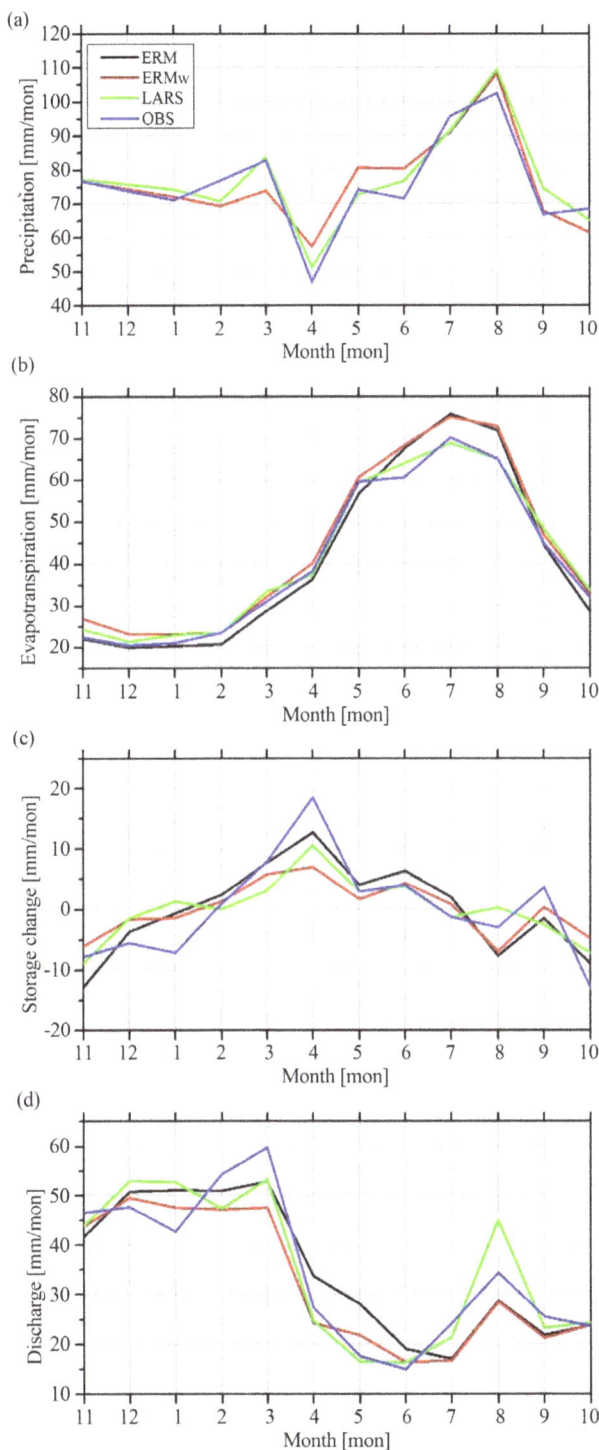

Figure 9. Monthly mean water balance components of 1000 years simulated time series of ERM, ERMw, LARS and OBS, (a) Precipitation; (b) Evapotranspiration; (c) Storage change; (d) Discharge.

The normalized standard deviations for all water components do not differ much from the reference states, except the peculiarities observed within the store change

terms.

Hence, for each component of the water balance a certain model performs best, but in conclusion through the help of the diagram the best model could not be identified.

In **Figure 9** the mean monthly diurnal variations of each component of the water balance are depicted. The precipitation shows no significant differences comparing the curves of the models. ERM and ERMw have the same precipitation curves as they just differ in their wind speed and vapour pressure series. LARS seems to be closer to the observed precipitation, as **Figure 8** already indicates. The overestimation of precipitation by each model in March and August must be mentioned. In context of **Figure 2**, the month with the largest precipitation amount is always overestimated by the weather generators. The main reason for this may be the occurrence of the aforementioned rare extreme event in August 2002, which is included in the observed time series [25]. This event influences the weather generators to much, which lead to the consequent over estimation of summer precipitation.

The curves of the evapotranspiration are very close to the observed results, only ERM and ERMw overestimate evapotranspiration in the summer months about 5 mm, An explanation might be the oversupply of water and energy simulated by ERM in these months.

But the results are still satisfactory taking into account the aforementioned post processing for the ERM driven BROOK90 run.

The curves of the storage change are more distinguishable. The simulated results preserve the annual cycle as clearly depicted in **Figure 9(c)**. The investigation of the curves in detail indicates that they have more in common among themselves than to the observed curve. This curve is characterized by a certain peak in April, most likely caused by the annual snow melting, which is indicating a certain limitation of the weather generators to reproduce these atmospheric conditions. This peak arises, with one month latency, from the melting in March, which clearly can be seen in **Figure 9(d)**. Of further importance is the peak simulated by LARS in August, which results from the larger precipitation and less evapotranspiration in this month. Mainly it is caused by the aforementioned disproportional consideration of the known rare event in this month.

The annual water balance shows less significant differences as depicted in **Figure 10**. LARS and ERM in average overestimate all components of the water balance, whereby the differences are slightly higher for the LARS results. Despite ERMw behaves also as coherent as the other results it shows more distinguishable differences. The lowest uptake of water trough evapotranspira-

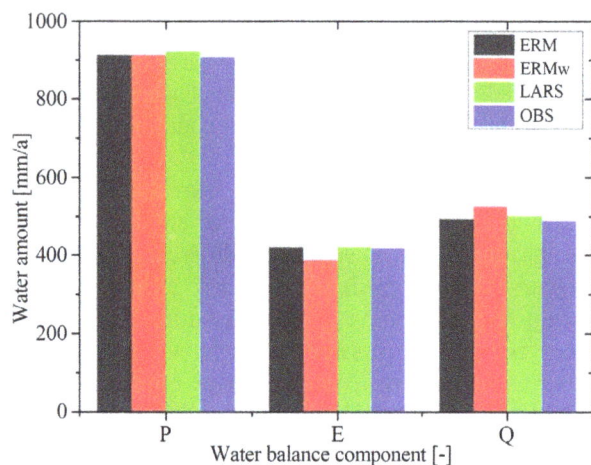

Figure 10. Annual mean water balance components of 1000 years simulated time series of ERM, ERMw, LARS and 12 years OBS.

Table 3. Annual mean sums and differences of the water balances components of 1000 years long synthetic and 12 years long observed time series.

	P	*Q*	*E*	*dS*
OBS [mm/a]	907.70	489.37	418.36	0.03
ERM [mm/a]	913.22	493.52	419.81	0.11
Difference [%]	0.61	0.85	0.35	266.67
ERMw [mm/a]	913.22	525.54	387.94	0.26
Difference [%]	0.61	7.39	-7.27	766.67
LARS [mm/a]	922.83	501.85	421.32	0.34
Difference [%]	1.67	2.55	0.71	1033.33

tion leads to a larger system output by discharge in the catchment.

The differences are even clearer looking at **Table 3**. As already mentioned the deviations of ERM and LARS are neglectable small of <3%. ERM even reaches results of <1% deviation from the observed water balance. The aforementioned differences of ERMw can also be seen clearly. Additionally also the residual storage change term is given in **Table 3**. The observed value is rather small as expected, almost zero. According to this small absolute value are the differences large.

5. Conclusions and Outlook

In this study another extension of a Richardson based weather generator is presented. Its application to the hydrological model BROOK90 for the Anchor Station in the Tharandt Forest, Germany is discussed. To contextualize its results, the performance of the model was compared to another weather generator (*i.e.* LARS-WG).

Rare events of any considered meteorological element are well maintained. Though, LARS better performed in terms of precipitation due to the usage of a semi empiric distribution function. For this reason the application of mixture distributions or non-parametric distribution functions most likely will improve the presented weather generator [38].

Generally, only underestimations could be observed. Through the application of a cumulative periodogram it is proven that also the annual cycles are preserved. The analysis even shows a subtle visible influence of the chosen parameterization of monthly fitted distributions functions for precipitation. The natural correlations of minimum temperature and vapour pressure are sufficiently considered. Thus, through the underestimation of

temperature extremes, the natural spectrum could not be covered completely neither by ERM nor by LARS.

The application of the weather generators in a hydrological context showed temporal and element wise dependences of the performance. While the simulated data set of LARS-WG shows better results for precipitation and evapotranspiration on a monthly basis, ERM performed better on the annual scale for these elements.

The application of ERM in BROOK90 resulted in a post processing due to implausible negative evapotranspiration values in winter, which were caused by a "canopy evaporation of the snow" term. The presented approach is a practical solution, but the authors would always prefer a better description and parameterisation of the responsible processes.

Overall the hydrological perspective emphasises the preservation of annual meteorological and hydrological regimes. Both applied models are useful and reliable for modelling the water balance.

The presented weather generator (*i.e.* ERM) could be extended from a single site to a raster-based multi site weather generator, which might be coupled with a cascade model for the downscaling from daily to 5 min time series [39]. However, this would require a change for the hydrological modelling to a raster based model like WaSim-ETH [37].

As a result of the related demands of information considering the future climate at a regional scale likewise the intention arose to simulate climate scenarios by the recognition of other future atmospheric properties like GCM outputs [40].

6. Acknowledgements

This work was supported by the German Academic Exchange Service (DAAD). Special thanks go to Dr. Uwe Spank and Dr. Klemens Barfus for helpful discussions, and Uwe Eichelmann and Heiko Prasse from the Chair of Meteorology at Technische Universität Dresden for their technical assistance.

REFERENCES

[1]   DIN, "Drain and Sewer Systems outside Buildings," Beuth-Verlag, Berlin, 2008.

[2]   DWA, "Hydraulic Dimensioning and Verification of Drain and Sewer Systems," Beuth-Verlag, Berlin, 2006.

[3]   L. LeCam, "A Stochastic Theory of Precipitation. Proceedings of the 4th Berkley Symposium," Mathematical Statistics and Probability, University of California Press, Berkley, 1961.

[4]   K. R. Gabriel and J. Neumann, "A Markov Chain Model for Daily Rainfall Occurrences at Tel-Aviv," *Quarterly Journal of the Royal Meteorological Society*, Vol. 88, No. 375, 1962, pp. 85-90.

[5]   C. W. Richardson, "Stochastic Simulation of Daily Precipitation, Temperature and Solar Radiation," *Water Resources Research*, Vol. 17, No. 1, 1981, pp. 182-190.

[6]   D. Wilks and R. Wilby, "The Weather Generation Game: A Review of Stochastic Weather Models," *Progress in Physical Geography*, Vol. 23, No. 3, 1999, pp. 329-357.

[7]   J. P. Hughes, P. Guttorp and S. T. Charles, "A Non-Homogeneous Hidden Markov Model for precipitation Occurrence," *Applied Statistics*, Vol. 48, No. 1, 1999, pp. 15-30.

[8]   H. Chen, J. Guo, Z. Zhang and C.-Y. Xu, "Prediction of Temperature and Precipitation in Sudan and South Sudan by Using LARS-WG in Future," *Theoretical and Applied Climatology*, Vol. 113, No. 3-4, 2012, pp. 363-375.

[9]   A. Spekat, W. Enke and F. Kreienkamp, "New Development of Regional High Resoluted Circulation Patterns for Germany and Simulation of Regional Climate Scenarios with the Downscaling Model WETTREG on the Basis of Global Climate Simulations of ECHAM5/MPI-OM T63L31 2010 until 2100 for the SRES-Scenarios B1, A1B and A2," Report, Umweltbundesamt, Berlin, 2007.

[10] B. Orlowsky, F.-W. Gerstengarbe and P. C. Werner, "A Resampling Scheme for Regional Climate Simulations and Its Performance Compared to a Dynamical RCM," *Theoretical and Applied Climatology*, Vol. 92, No. 3-4, 2008, pp. 209-223.

[11] M. Safeeq and A. Fares, "Accuracy Evaluation of ClimGen Weather Generator and Daily to Hourly Disaggregation Methods in Tropical Conditions," *Theoretical and Applied Climatology*, Vol. 106, No. 3-4, 2011, pp. 321-341.

[12] M. A. Semenov, R. J. Brook, E. M. Barrow and C. W. Richardson, "Comparison of the WGEN and LARS-WG Stochastic Weather Generators for Diverse Climates," *Climate Research*, Vol. 10, No. 2, 1998, pp. 95-107.

[13] V. Y. Ivanov, R. L. Bras and D. C. Curtis, "A Weather Generator for Hydrological, Ecological, and Agricultural Applications," *Water Resources Research*, Vol. 43, No. 10, 2007, pp. 1-21.

[14] M. Khalili, F. Brissette and R. Leconte, "Effectiveness of Multi-Site Weather Generator for Hydrological Modeling," *Journal of the American water resources Association*, Vol. 47, No. 2, 2011, pp. 303-314.

[15] J. Xia, "A Stochastic Weather Generator Applied to Hydrological Models in Climate Impact Analysis," *Theoretical and Applied Climatology*, Vol. 55, No. 1-4, 1996, pp. 177-183.

[16] C. W. Richardson, "Weather Simulation for Crop Management Models," *Transactions of the American Society of Agricultural Engineers*, Vol. 28, No. 5, 1985, pp. 1602-1606.

[17] C. A. Federer, C. Vörösmarty and B. Fekete, "Sensitivity of Annual Evaporation to Soil and Root Properties in Two Models of Contrasting Complexity," *Journal of Hydrometeorology*, Vol. 4, No. 6, 2003, pp. 1276-1290.

[18] K. Mannsfeld and O. Bastian, "Landscape of Saxony—Catigorization of Manifold," Landesverein Sächsischer Heimatschutz, Dresden, 2005.

[19] F. Haubrich, "The Tharandt Forest as Alegory of the Geology of Saxony," Jahrestagung der Deutschen Bodenkundlichen Gesellschaft, Exkursionführer H5, Dresden, 2007.

[20] German Soil Science Society, "Soils without Borders— Anniversary of the German Soil Science Society, 02. bis 09. September 2007 in Dresden," In: General Excursion Guide/ German Soil Science Society, 2007, Reports of the German Soil Science Society, Göttingen, 2007.

[21] P. Feger, K. Schwärzel, D. Menzer, C. Bernhofer, B. Köstner and F. Katzschner, "Anniversary of the German Soil Science Society—Soils without Borders, Dresden, Saxony, Germany," Reports of the German Soil Science Society, Göttingen, 2007.

[22] H. Leser, "Diercke-Wörterbuch Allgemeine Geographie," Westermann, Braunschweig, 2005.

[23] C. Bernhofer, "Exkursions-und Praktikumsführer Tharandter Wald. Material zum 'Hydrologisch-Meteorologischen Feldpraktikum'," Tharandter Klimaprotokolle 6, TU Dresden, Dresden, 2005.

[24] T. Grünwald and C. Bernhofer, "A Decade of Carbon, Water and Energy Flux Measurements of an Old Spruce Forest at the Anchor Station Tharandt," *Tellus B*, Vol. 59, No. 3, 2007, pp. 387-396.

[25] U. Spank, K. Schwärzel, M. Renner, U. Moderow and C. Bernhofer, "Effects of Measurement Uncertainties of Meteorological Data on Estimates of Site Water Balance Components," *Journal of Hydrology*, Vol. 492, 2013, pp. 176-189.

[26] B. Sevruk, "Methodical Analysis of Systematic Measurement Errors of the Hellmann Rain Gauge for the Summer in Switzerland," Versuchsanstalt für Wasserbau,

Hydrologie u. Glaziologie, 1981.

[27] D. Richter, "Results of the Methodical Analysis of the Correction of Systematic Errors of the Hellmann Rain Gauge," Selbstverlag des Deutschen Wetterdienstes, Offenbach am Main, 1995.

[28] H. Walter and H. Lieth, "Klimadiagramm-Weltatlas," Gustav Fischer Verlag, Jena, 1967.

[29] J. Barron, J. Rockström, F. Gichuki and N. Hatibu, "Dry Spell Analysis and Maize Yields for Two Semi-Arid Locations in East Africa," *Agricultural and Forest Meteorology*, Vol. 117, No. 1, 2003, pp. 23-37.

[30] N. Herrman, "Höhere Mathematik/für Ingenieure, Physiker und Mathematiker," Oldenburg Verlag, München, Wien, 2007.

[31] M. A. Semenov and E. M. Barrow, "Use of a Stochastic Weather Generator in the Development of Climate Change Scenarios," *Climatic Change*, Vol. 35, No. 4, 1997, pp. 397-414.

[32] M. S. Khan, P. Coulibaly and Y. B. Dibike, "Uncertainty Analysis of Statistical Downscaling Methods Using CGCM Predictors," *Hydrological Processes*, Vol. 20, No. 14, 2006, pp. 3085-3104.

[33] S. Dyck and G. Peschke, "Grundlagen der Hydrologie," Verlag für Bauwesen, Berlin, 1995.

[34] K. W. Hipel and A. I. McLeod, "Time Series Modelling of Water Resources and Environmental Systems," El-sevier Scientific Publishing Company, Amesterdam, 1994.

[35] G. E. P. Box, G. M. Jenkins and G. C. Reinsel, "Time Series Analysis: Forecasting and Control," Holden-Day, San Francisco, 1976.

[36] K. E. Taylor, "Summarizing Multiple Aspects of Model Performance in a Single Diagram," *Journal of Geophysical Research*, Vol. 106, 2001, pp. 7183-7192.

[37] R. Kronenberg, K. Barfus, J. Franke and C. Bernhofer, "On the Downscaling of Meteorological Fields Using Recurrent Networks for Modelling the Water Balance in a Meso-Scale Catchment Area of Saxony, Germany," *Atmospheric and Climate Sciences*, Vol. 3, No. 4, 2013, pp. 552-561.

[38] R. Kronenberg, J. Franke and C. Bernhofer, "Comparison of Different Approaches to Fit Log-Normal Mixtures on Radar-Derived Precipitation Data," Meteorological Applications, Wiley Online Library, 2013.

[39] D. Lisniak, J. Franke and C. Bernhofer, "Circulation Pattern Based Parameterization of a Multiplicative Random Cascade for Disaggregation of Daily Rainfall under Non-stationary Climatic Conditions," *Hydrology and Earth System Sciences*, Vol. 17, No. 7, 2013, pp. 2487-2500.

[40] D. Wilks, "Stochastic Weather Generators for Climate-Change Downscaling, Part II: Multivariable and Spatially Coherent Multisite Downscaling," *Climate Change*, Vol. 3, No. 3, 2012, pp. 267-278.

2

Groundwater Chemical Evolution in the Essaouira Aquifer Basin—NW Morocco

Mohammed Bahir[1], Rachid El Moukhayar[1], Najiba Chkir[2], Hamid Chamchati[1], Paula Galego Fernandes[3], Paula Carreira[3]

[1]Geodynamics Laboratory Magmatic Géoressources and Georisks, Université Cadi Ayyad, Marrakech, Morocco; [2]Geography Departement, Faculty of Letters and Humanities, Sfax University, Sfax, Tunisia; [3] Nuclear Technology Institute, Sacavém, Portugal.

ABSTRACT

The sustainability of groundwater resources for drinking water supplies, agriculture, and industry a prime concern in countries dominated by arid and semi-arid climates such as Morocco. The growing demand for groundwater coupled with impacts from land use and climate change make sustainability an even more important water management goal. In order to make sound decisions about water use and protection of water quality, managers and policy makers must have a sound understanding of such factors as the location and amount of groundwater recharge and groundwater ages. Due to the population growth and climate change (causing long periods of drought) in the world, many countries have intensively increased their use of water sources for supplying potable water to population and for their agricultural (irrigation) and industrial developments. Due to the lack of surface waters, people exploit mainly underground water reservoirs. So, it is necessary to study and characterize these water reservoirs to avoid any excess of exploitation. The water resources of the Essaouira basin are characteristic of a semi-arid climate, and are severely impacted by the climate (quantity and quality). Considering the importance of the Essaouira aquifer in the groundwater supply of the region, a study was conducted in order to comprehend this aquifer groundwater evolution. It is an aquifer located on the Atlantic coastline, southern (Morocco), corresponding to a sedimentary basin with an area of near 200 km^2. Covering the Palaeozoic bedrock, the sedimentary series range from the Triassic to the Quaternary. The geological structures delineate a syncline bordered by the Tidzi diapir of Triassic age which outcrops to the East and South. In the Essaouira basin a multi-aquifer was identified constituted by detrital deposits of the Plioquaternary and dolomitic limestones of the Turonian. The Plioquaternary is unconfined below the Senonian marls. However, in some places it can be in direct contact with the other Cretaceous and Triassic units. The Plioquaternary is generally up to 60 m thick. The Turonian is confined by the Senonian marls and in direct contact with the Plioquaternary on the edges of the syncline structure. The main flow direction is from SE to NW towards the Atlantic Ocean, being the recharge area located near the Tidzi diapir. In the Essouaira basin, in spite of the occurrence of calcareous and dolomitic levels, all waters are of Na-Cl-type. The chemical signature of these waters should be the result of the preferential recharge area that is located in the Tidzi diapir. Using a simple mass balance model through the PHREEQC program this scenario was tested. The reaction path was assumed to be such that waters observed at shallow depths evolved to more mineralized waters. It was possible to notice that these waters have an important contribution of water-rock interaction in groundwater mineralization, corroborating the influence of the preferential recharge area located in the Tidzi diaper in the waters signature.

Keywords: Essaouira Basin Aquifer; Hydrogeochemistry; Mass Balance Model

1. Introduction

Morocco is characterized by a semi-arid to arid climate, excluding a humid zone in the north. This climat constraint requires the application of new technologie to supplement conventional hydrological method, in order to improve water resources assessment and management in Morocco.

In coastal regions, the problems related with the increase of salinization and pollution in groundwater systems is generally associated to the effects of seawater (seawater intrusion by overexploitation of the system and by sea-salt-spray) and on the other hand by the anthropogenic activities such as domestic wastes, agriculture and industry. Also, intrinsic properties of aquifers (porous/fractured/karstic media, geological structure, permeability), and external factors such as climate may contribute to mitigate or worsen these problems.

The Essaouira basin is located in a Moroccan semi-arid area with maximum annual rainfall of 300 mm/year and with a high potential evapotranspiration around 920 mm/year [1,2] (**Figure 1**).

This sedimentary basin has an area of about 1200 km² and is filled with Mesozoic and Cenozoic materials, which are overlaid with superficial Plioquaternary terrains (**Figure 2**). One main aquifer was identified, supplying water for drinking and for agricultural activities [3].

2. Geological and Hydrogeological Data

Covering the Palaeozoic bedrock, the sedimentary series range from the Triassic to the Quaternary. The sedimentary sequence begins with Triassic deposits having the same lithology as in Sines basin, outcroping in the E and S of the region. The Carbonate rocks compose the Jurassic and marly sediments of lower Cretaceous to Cenomanian dominate the Cretaceous. The dolomitic limestones of the Turonian are covered by Senonian gypsy marls [4,5], which appear below the Plioquaternary detrital deposits of sands, sandstone and conglomerates. The geological structures delineate a syncline bordered by the Tidzi diapir of Triassic age which outcrops to the E and S.

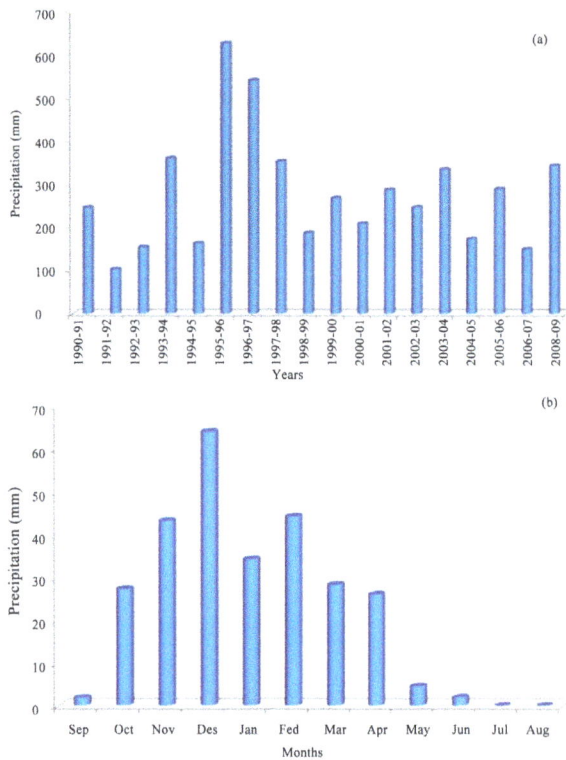

Figure 1. Precipitation (a) annual values and (b) monthly mean values at Essaouira Station (1990-1991 to 2008-2009).

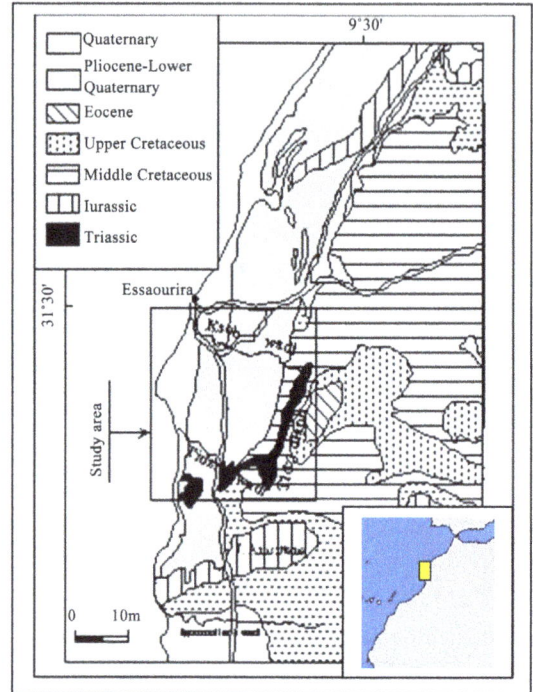

Figure 2. Geological map of essaouira basin.

In the Essaouira basin a multi-aquifer was identified constituted by detrital deposits of the Plioquaternary and dolomitic limestones of the Turonian [6].

The Plioquaternary is unconfined below the Senonian marls. However, in some places it can be in direct contact with the other Cretaceous and Triassic units. The Plioquaternary is generally up to 60 m thick.

The Turonian is confined by the Senonian marls and in direct contact with the Plioquaternary on the edges of the syncline structure.

For a few years, water has been withdrawn through drilling wells to supply the Essaouira City. At present, the Plioquaternary provides 47% of drinking water for Essaouira with about 64,000 inhabitants and rural population. The total rate of extraction in this system is around 97 L/s [7].

The piezometric levels in the Essaouira multi-aquifer present a general standardization through time (1990/2000). However, locally some piezometric variations can be identified. As a consequence of the weak thickness the sensibility to the droughts have an important impact in the water reservoir, as widespread drought periods that are affecting Morocco since 1978 [7].

Morocco has a wide range of climate conditions: the coastal regions generally have a mild climate, the Atlas Mountains can be cold and wet during spring or even in summer, while the desert is hot and dry nearly all year.

The main flow direction is from SE to NW towards the Atlantic Ocean, being the recharge area located near the Tidzi diaper. The piezometric maps, since 1990 to 2000, show, in the N, a line deviation to the NE, indicating a possible contribution of the oued in the aquifer recharge.

In the Essouaira basin, in spite of the occurrence of calcareous and dolomitic levels, all waters are of Na-Cl-type (**Table 1** and **Figure 3**).

The diagram allows not only representation of the hydrochemical data, but also definition of the plausible hydrochemical processes dominating the groundwater chemistry such as mixing, ion exchange and dissolution affecting groundwater composition. Classification of water into "types" according to the dominating cations and anions can then be undertaken.

The chemical signature of these waters should be the result of the preferential recharge area that is located in the Tidzi diapir. A high correlation coefficient was found between electrical conductivity, chlorides and sodium contents, suggesting the large contribution of these elements to the groundwater chemical load. Nevertheless, occulted by chloride the groundwater is highly bicarbonate as a result of the presence of carbonate compounds in the reservoirs matrix.

Analyzing the dispersion of the values of the parameters, it is probable that the difference between the maximum values and the average is a result of a punctual increase. However, the present range is in majority a result of the oued contribution into groundwater recharge, leading to a dilution of the water mineralization. Another hypothesis to explain the range in mineralization could be the available time in the water rock interaction with the diapir that constitutes the basin. It is also important to consider the huge concentrations in chloride and sodium (5019 and 3133 mg/L), not included in the diagram because of the anomaly behaviour of the well number 45 located near the coastline on the NW of the area. These values could be a result of a local intrusion phenomena originated by over exploitation.

On the basis of Durov diagram (**Figure 4**), the water within the aquifer of the Essaouira basin can be considered as mostly homogeneous, among which sodium and chloride are the most dominant facies. These waters types can be described as follows: waters for the two aquifers are plotted in the box 9 indicating increasing salinity by mixing processes and in box 8 highlighting the existence of cation reverse exchange reactions within the Turonian as well as within the Plio-Qaternary aquifers.

3. Groundwater Geochemical Characterization

Hydrogeochemical processes are important in defining groundwater hydrology in complex layered aquifers. Subtle variations of salinity in fresh groundwaters are related to hydrogeochemical processes controlling groundwater composition. Three physical parameters (T, pH,

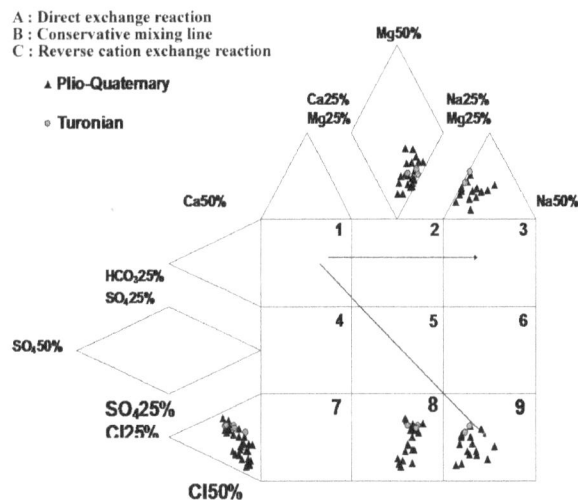

Figure 4. Expanded durov diagram for the essaouira groundwater.

Table 1. Factorial analyses in the essaouira basin.

	Factor 1	Factor 2	Factor 3
HCO_3^-	−0.05209	0.56697	−0.59120
Cl^-	0.91966	−0.24626	0.20629
NO_3^-	0.51705	−0.59032	0.05719
SO_4^{2-}	0.10906	0.82363	0.23708
Na^+	0.83928	0.23035	0.16783
K^+	0.03055	0.24041	0.85555
Ca^{2+}	0.82401	−0.14683	−0.02267
Mg^{2+}	0.83775	0.09662	−0.43083

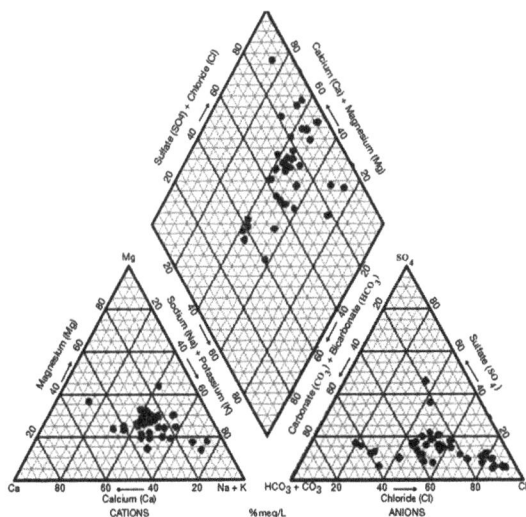

Figure 3. Essaouira groundwater piper diagram.

conductivity) have been measured in situ during sampling with a handling multiparameters equipment. Chemical analysis have been realised according to standard protocols in the Laboratory of 3 Geolab of University Cadi Ayyad of Marrakech.

Unfortunately, levels of physical-chemical parameters such as TDS, EC, Cl^-, SO_4^{2-} are higher than the maximum permissible level prescribed by the World Health Organization standards sets fro drinking water.

3.1. Principal Component Analysis

Considering the numerous species involved in the chemical analyses, the sources of salinization were sought using Principal Component Analysis (PCA). PCA is a Factorial Analysis, in which graphs are generalised, taking into account all the elements involved, in order to achieve optimal data visualization [8]. Using this statistical analysis a reduction of the observation dimensions species obtained in which the given objects are studied, by creating linear combinations of variables that characterize the studied objects [9].

The application of the PCA method to Essaouira groundwater data allows the identification of 3 factors with eigen values greater than 1 (**Tables 1** and **2**):

Factor 1 represents 41% to the total variance of the groundwater system with a positive correlation between chloride, sodium, calcium and magnesium. This pattern materialises the water-rock interaction namely the dissolution of evaporate minerals and/or the contribution of Tidzi diapir in the water recharge.

Factor 2 with a relevance of 19% to the groundwater characterization represents the sulphates content. These values could have natural or anthropogenic origins, resulting from evaporatic dissolution and/or the use of fertilisers.

Factor 3 represents 17% of the total variance and it is probably the result of the influence of the bedrock in the water mineralization; factor 3 shows a high correlation with potassium.

The samples recovered after the exceptional rainfalls in January 1996, show a remarkable increase in Na^+ and Cl^- concentrations when compared with the data of 1995. In fact, the infiltrated water in 1996 remobilized the salts trapped and concentrated in the soil and in the unsaturated zone during the long period of low precipitations.

However, dilution effects can be observed in a particular recharge environment. Indeed, spatial distributions of sodium and chloride for 1996 (**Figure 5**) show that the lowest concentrations are in the NE area immediately southward of *Ksob wadi*. It highlights the aquifer recharge by *Ksob wadi* and the dilution originated by it. This recharge was also confirmed via water flow measurements in this *wadi* [10].

The higher concentrations located in the central part of the area are a result of the influence of the Tidzi diapir in the water recharge, leading to high contents in sodium and chloride, according to the main flow direction. The evolution of these parameters do not indicate the main flow direction probably as related to the *Ksob wadi* recharge and the dilution effect generated by it.

3.2. Anthropogenic Influence on the Aquifers Contamination

In 1995, nitrate concentrations in the groundwater system of Essaouira basin ranged from 1.4 to 187 mg/L (**Table 3**). The mean NO_3^- concentration was 53.5 mg/L and 45 % of the groundwater samples exceeded 45 mg/L [2], which is the World Health Organization maximum amount of NO_3^- content for drinking water.

In 1996 the nitrate content was higher than the one of the previous campaign, the concentration ranged from 1.6 to 295.8 mg/L, with a mean value of 91.5 mg/L being 65% of the groundwater samples over 45 mg/L. The NO_3^- and Cl^- concentration increasing after the rains of 1996's supports the above hypothesis for the external origin of these elements as well as the salt remobilization phenomena triggered by precipitation after a long period of drought.

Table 2. Total variance of factor in the essaouira basin.

	Eigenval	% Total Variance	Cumul. Eigenval	Cumul. %
1	3.27	40.90	3.27	40.90
2	1.49	18.69	4.77	59.59
3	1.40	17.44	6.16	77.03

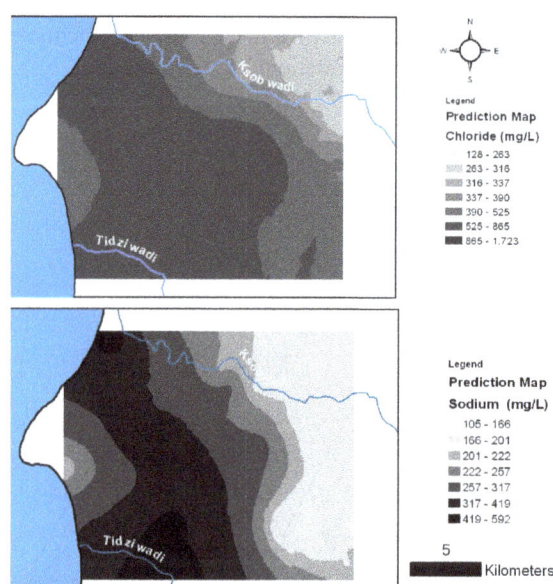

Figure 5. Spatial distribution of sodium and chloride and sodium in the Essaouira basin.

Table 3. Correlation matrix in the essaouira basin.

	HCO_3	Cl^-	NO_3	SO_4^{2-}	Na^+	K^+	Ca^{2+}	Mg^{2+}
HCO_3	1.00	−0.38	−0.18	0.12	0.11	−0.15	−0.19	0.20
Cl^-	−0.38	1.00	0.52	−0.06	0.77	0.08	0.79	0.65
NO_3	−0.18	0.52	1.00	−0.39	0.36	0.04	0.38	0.33
SO_4^{2-}	0.12	−0.06	−0.39	1.00	0.18	0.23	0.04	0.09
Na	0.11	0.77	0.36	0.18	1.00	0.26	0.46	0.59
K	−0.15	0.08	0.04	0.23	0.26	1.00	−0.06	−0.30
Ca	−0.19	0.79	0.38	0.04	0.46	−0.06	1.00	0.66
Mg	0.20	0.65	0.33	0.09	0.59	−0.30	0.66	1.00

The nitrate distribution also indicates the oued contribution in the groundwater mineralization, by dilution the groundwater near this river assume lower concentrations in this pollutant (**Figure 6**).

Based on the different nitrate content at the Essaouira, one might think that important amount of fertilisers were used in intensive agricultural activities in Morocco. The main source of nitrate is associated to wrong wells design, lack of head well protection (e.g. hand-dug wells with no casing and cover), lack of head well protection areas, traditional extraction methods, accumulation of livestock waste nearby the wells. The lack of prevention and environmental programs for the population seriously threatens the groundwater resources and leads to poor quality in the water supplies.

In relation to the sulphates, the origin in the both aquifers could be similar. In Essaouira sulphates are probably the result of evaporites and diapiric dissolution, but they could also be the result of the fertilisers. However, we haven't obtained any correlation between sulphates and other element that supports the different origins (**Table 4**).

3.3. Geochemical Modelation

In the Essouaira basin, in spite of the occurrence of calcareous and dolomitic levels, all waters are of Na-Cl-type. The chemical signature of these waters should be the result of the preferential recharge area that is located in the Tidzi diapir. A high correlation coefficient was found between electrical conductivity, chlorides and sodium contents, suggesting the large contribution of these elements to the groundwater chemical load. Nevertheless, occulted by chloride the groundwater is highly bicarbonate as a result of the presence of carbonate compounds in the reservoirs matrix.

Analysing the dispersion of the values of the parameters is probable that the difference between the maximum values and the average is a result of a punctual increase. However, the present range is in majority a result of the oued contribution into groundwater recharge, leading to a dilution of the water mineralization. Another hypothesis

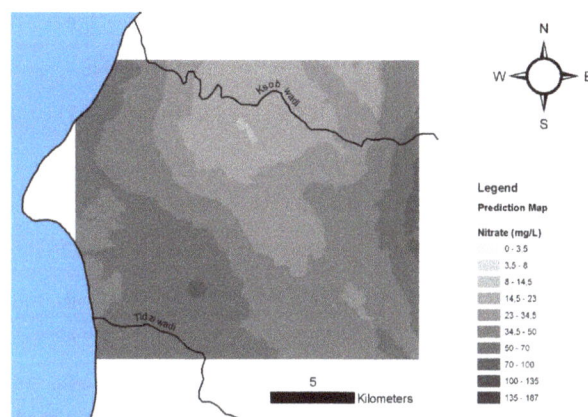

Figure 6. Spatial distribution of nitrate in the Essaouira basin.

to explain the range in mineralization could be the available time in the water rock interaction with the diapir that constitutes the basin.

If we assume that sample 23 is representative of the water, it is expected that waters progressively increase the concentrations of chemical constituents. The higher Na and Cl concentration could be explained by the calcite and halite dissolution. Calculated saturation index (S.I.) indicated that the majority of groundwaters are near equilibrium with respect to calcite and strongly undersaturated relatively with halite. The progressively increasing EC values, Cl and Na concentrations due to increased water rock interaction would ultimately result in higher saturation indexes values with respect to the referred minerals.

Using a simple mass balance model through the PHREEQC program this scenario was tested. The reaction path was assumed to be such that waters observed at shallow depths evolved to more mineralised waters. We considered a path that initiated in sample 23 that was calculated on the basis of the behaviour of groundwater mineralisation with different percentages of halite dissolution (**Table 3**).

Figure 7 shows the results of modelling, with halite dissolution at open system where are projected all the

Table 4. Physical parameters and chemical analyzes of water from the aquifer.

Point d'eau	T (°C)	C.E. 25°C (µs/cm)	HCO_3 (mg/l)	Cl^- (mg/l)	NO_3^- (mg/l)	SO_4^{2-} (mg/l)	Ca^{2+} (mg/l)	Mg^{2+} (mg/l)	Na^+ (mg/l)	K^+ (mg/l)
140/51	19	2700	363.6	738.4	42.5	177.8	179.6	56.6	333.5	3.1
11/51	18	4830	263.5	1654.3	133.4	290.7	172.4	101.0	736.0	35.1
20/51	22	3250	209.8	1029.5	106.3	221.2	159.2	76.3	471.5	7.0
M20	23	3130	383.1	972.7	22.5	163.9	164.0	61.9	414.0	9.8
28/51	23	3360	339.2	1136.0	70.6	137.5	227.2	109.4	333.5	3.5
125/51	24	3330	226.9	1228.3	58.5	106.4	279.6	63.1	368.0	3.1
135/51	22	2500	219.6	788.0	62.0	164.8	155.2	73.7	287.5	2.3
116/51	22	2610	285.5	671.0	165.5	143.5	194.4	69.1	299.0	1.9
104/51	22	2100	268.4	546.7	1.6	168.3	152.0	53.3	287.5	3.5
33/51	21	2340	285.5	646.1	115.2	84.3	172.0	43.2	299.0	2.3
261/51	23	2000	266.0	436.7	79.6	178.4	146.8	48.7	241.5	2.3
327/51	21.5	1870	162.3	422.5	273.2	78.6	92.0	27.8	253.0	5.1
149/51	22	3580	353.8	1075.7		216.3	104.8	101.3	655.5	4.7
272/51	20.5	2280	295.2	600.0		243.3	108.0	98.4	253	2.7
M5	22	1590	205.0	429.6		100.1	112.8	55.2	200.1	12.1
3/51	19	2550	412.4	646.1		163.4	117.2	52.6	391.0	4.7
M10	19	2700	390.4	777.5		131.8	184.4	19.4	402.5	4.7
133/51	22	4710	307.4	1462.6		402.2	295.2	150.0	678.5	3.5
18/51	21	5040	302.6	1693.4		418.3	320.8	76.8	713.0	3.1
132/51	15.5	2580	226.9	830.7		163.4	185.6	75.1	322.0	2.3
137/51	21	3490	175.7	1320.6		129.3	248.0	91.2	471.5	3.9
138/51	21	3960	192.8	1487.5		166.4	298.4	104.6	517.5	3.9
21/51	20	4460	207.4	1728.9		162.0	376.0	61.4	586.5	2.3
124/51	23	1860	241.6	586.8		45.5	140.4	67.9	218.5	2.0
126/51	24.5	3240	222.0	1189.3		132.0	306.0	64.1	356.5	3.1
127/51	19	3570	180.6	1359.7		131.5	325.6	106.3	391.0	2.7
M31	23	3220	102.5	1221.2		147.9	291.2	121.7	368.0	2.7
M33	20.5	1890	346.5	436.7		187.1	107.2	73.0	167.9	1.2
130/51	22	2170	324.5	500.6		205.9	112.0	77.3	299.0	1.2
49/51	20.5	2340	369.7	532.5		244.6	102.0	104.4	287.5	1.6
93/51	21.5	2150	339.2	489.9		202.9	111.2	128.6	230.0	1.6
103/51	21	2100	314.8	532.5		213.3	112.2	85.2	207.0	2.7
260/51	20.5	2090	287.9	539.6		238.9	109.6	71.3	218.5	2.3
227/51	20.5	2280	295.2	600.0		243.3	108.0	98.4	253.0	2.7
27/51	21	2000	358.7	433.1		229.4	108.0	106.1	165.6	2.3
236/51	21	2380	342.8	582.2		241.4	120.0	118.1	264.5	2.0

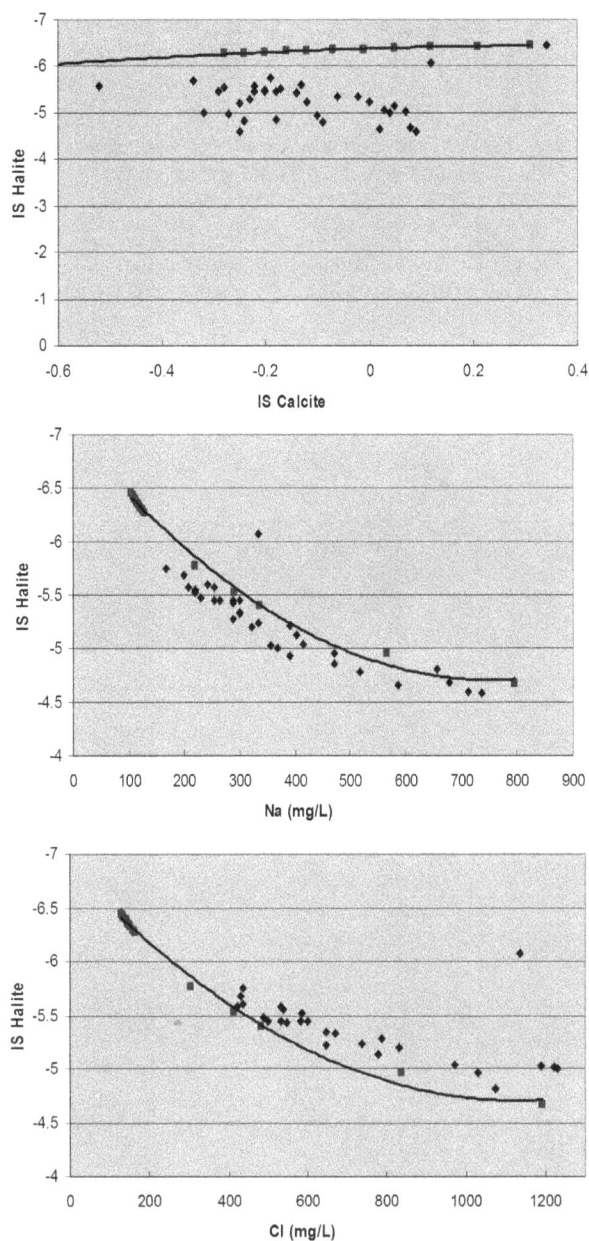

Figure 7. Halite saturation index versus calcite saturation index; sodium and chloride contents.

analysed samples in order to evaluate the behaviour of groundwater from the aquifer. It is possible to notice that the waters are projected along the line that represent the waters evolution in open system, indicated these waters have an important contribution of water-rock interaction in groundwater mineralisation. Corroborating that the chemical signature of these waters should be the result of the preferential recharge area that is located in the Tidzi diaper.

4. Conclusions

The economic and social activities in the Essaouira basin

depend heavily on groundwater. In this basin, the very high evaporation and scarce precipitation of the semi-arid climate activate the processes of salinization and contamination. In this context, the rainfall triggers a remobilization of residual chemical species and contaminants (sodium, chloride, nitrate) trapped and concentrated in the soil or seeped into the unsaturated zone during the long period of drought. This clearly shows the small potential of natural attenuation by precipitation of the semi-arid regions.

According to the climatic and environmental conditions and social and economical factors it is evident that each of the referred basins requires specific management measures to protect and to preserve the groundwater resources and to guarantee the water supplies. [11] suggest three recommendations that might be useful in the Essaouira basin, where less management measures exist:

The use of the water from Ksob wadi for artificial recharge of the aquifers. In fact, the most significant outflows of the wadi are related to occasional floods in winter. A small part feeds the aquifers and the rest is lost to the Atlantic Ocean.

Improvement of the techniques of drilling, constructing and protecting wells.

Public awareness and participation in water management.

REFERENCES

[1] M. Bahir, M. Jalal, A. Mennani and A. Fekri, "Contribution à l'Étude Hydrogéologique et Hydrochimique de la Zone Synclinale d'Essaouira (Contribution to the Study of the Hydrogeology Ande Hydrochemistry of the Essaouira Synclinal Area)," Congrès International sur l'Eau et la Désertification, Le Caire, 1999.

[2] M. Bahir, M. Jalal and A. Mennani, "Pollution Nitratée des Eaux Souterraines du Bassin Synclinal d'Essaouira (Groundwater Pollution by Nitrates of the Essaouira Synclinal Basin)," Journal of Environmental Hydrology, Electronic Journal of the International Association for Environmental Hydrology, Vol. 9, No. 18, 2001.

[3] M. Bahir, A. Mennani, M. Jalal and N. Youbi, "Contribution to the Hydric Resources Study of the Essaouira Synclinal Basin (Morocco)," Estudios Geologicos, Vol. 56, No. 3-4, 2000.

[4] F. Duffaud, L. Brun and B. Planchot, "Bassin du Sud-Ouest Marocain (SW Morocco Basin)," In: Reyre, Ed., Bassin Sédimentaire du Littoral Africain, Vol. 9, 1966, p. 18.

[5] Y. Fakir, "Contribution to the Coastal Aquifers Study—Sahel de Qualidia Case," Ph.D. Thesis, 2001, p. 160.

[6] A. Mennani, V. Blavoux, M. Bahir, Y. Bellion, M. Jalal and M. Daniel, "Chemistry and Isotopes in the Plio-Quaternary and Turoniene Aquifers Funcioning Knowledge in Essaouira Area (Occidental Morocco)," Journal of African Earth Sciences, Vol. 32, No. 4, 2001, pp. 819-835.

[7]   A. Agoumi, "Introduction à la Problématique des Change-ments Climatiques (Introduction to the Climate Changes Problem)," Projet Magrébin sur les Changements Clima-tiques RAB/94/G31, Somigraf, 1999.

[8]   A. J. Melloul, "Use of Principal Components Analysis for Studying Deep Aquifers with Scarce Data—Aplication to the Nubian Sandstone Aquifer, Egypt and Israel," *Hy-drogeologu Journal*, Vol. 3, No. 2, 1995, pp. 19-39.

[9]   A. Mackiewick and W. Ratajcak, "Principal Components Analysis (PCA)," *Computers & Geosciences*, Vol. 19, No. 3, 1993, pp. 303-342.

[10]   A. Fekri, "Contribution to the Hydric Resources Study of the Essaouira Synclinal Basin (Morocco)," 3rd Cycle, 1993, p. 173.

[11]   R. Rajagopal and T. Graham, "Expert Opinion and Ground-water Quality Protection. The Case of Nitrate in Drinking Water," *Groundwater*, Vol. 27, No. 6, 1989, pp. 835-847.

3

Implementing into GIS a Tool to Automate the Calculation of Physiographic Parameters of River Basins

Roberto Franco-Plata[1], Carlos Miranda-Vázquez[1], Héctor Solares-Hernández[1], Luis Ricardo Manzano-Solís[1], Khalidou M. Bâ[2], José L. Expósito-Castillo[2]

[1]Faculty of Geography, Autonomous University of the State of Mexico, Toluca, Mexico; [2]Inter-American Center of Water Resources, Autonomous University of the State of Mexico, Toluca, Mexico.

ABSTRACT

The physiographic characterization of a basin is a fundamental element as it defines the hydrological behavior of that basin. The present work deals with the development and implementation of a tool that allows calculating in an automated manner the physiographic parameters of a basin, as well as those of the surface runoff and main river, besides other graphic elements: hypsometric curve, equivalent rectangle and profile of the main river. Such a tool was developed under Visual Basic 6 programming language and the spatial geographic component ArcObjects by ESRI; they enabled the development of a library as a final product (.dll), which can be loaded and implemented in ArcMap software. In the methodology a Conceptual Model was established, from which it was possible to identify the requirements and methods to obtain the parameters, as well as the conception and implementation of the Logical Model that includes the specific functions and also the input structures, processes and data output. Finally, the tool was tested with actual data from El Caracol river basin, located in central-southern Mexico, which showed the easiness and usefulness of it, besides the effectiveness of the results, not leaving aside the time and resources saved by the user when characterizing a basin, compared with other conventional processes.

Keywords: GIS Programming; ArcObjects; Visual Basic; Physiography of Basin

1. Introduction

Nowadays, obtaining parameters from a hydrographic basin is generally performed manually, using printed cartography and undertaking painstaking processes, and the results are not always accurate, because precision depends on the criterion of the one performing the tasks. Therefore it was deemed necessary to develop a tool that facilitates the obtainment of said parameters in a simple, efficacious manner with good results in some of the most-used Geographic Information System (GIS), this under the premise that the implementation of the tool in GIS will facilitate the capture, preparation of spatial data, and later, the presentation and unfolding of results in an efficacious manner. Moreover, the development of the tool coupled to a GIS offers a powerful computing instrument for hydrologists.

The advancement in the study of hydrological aspects from computing technologies has enabled the development of GIS and hydrologic models to facilitate modeling in management and decision making about water resources. There are diverse advancements as for the development of applications for hydrological issues, this has contributed to the continuous development of this kind of tools together with GIS; but in spite of this, [1] Franco-Plata (2008) and [2] Rodríguez and Santos (2007) make it evident that there are few works oriented to obtain basin parameters.

Among the works referring to the obtaining of physiographic parameters from a basin distinguishable are HecPrepro (Hydrologic Engineering Center-PREPROcessor) and HecPrepro version 2.0 developed by [3] Hellweger (1997). [4] Ehslchlager (1991) developed an application in GRASS software, which generates information similar to that from the application Arc Hec-Pre-pro. [5] Díaz *et al.* (1999) used ArcInfo and ArcView 3.0 software to employ functions of analysis in the calculation of physiographic parameters. Arc-Hydro ([6] Maidment, 2002) is a model of geospatial and temporary data to manage and administrate hydrologic information supported on ArcGIS. [7] Franco-Plata (2006), implemented a geomatic module to extract the physiographic parameters of a basin. [2] Rodríguez and Santos, (2007) retook the above mentioned work making adjustments and incorporating some missing parameters, with the purpose of making the module more efficacious on ArcView 3.2 platform. [1] Franco-Plata (2008) developed a hydro-geomatic module in the raster platform of the GIS Idrisi for the availability of water resources, it is worth underscoring that this application incorporated new hydrological calculations such as evapotranspiration, infiltration and surface runoff, to mention a few, besides the most indispensable physiographic calculations to characterize hydrographic basins, from a DEM and the basin limit.

As it is noticed, the evolution of the development of various platforms for hydrological analysis has experienced and created new concepts and ideas as for the treatment of information, since it is a support to accomplish the unification of some methods and techniques to evaluate hydrological issues.

2. Theoretical-Methodological Support

The reaction of a hydrological basin to precipitations is a phenomenon not still fully grasped; the studies on several components of the hydrological cycle and the relationships between them, in particular the process of rain-runoff, have been the object of many works ([8] Chebani et al., 1992; [9] Llamas, 1993; [10] Ouarda et al., 2008). Hydrologic analysis uses various statistical methods to compare basins in views of evidencing the causes for the variations of hydrologic characteristics. For instance, it is sought to explain why two hydrologic basins under the same climatic conditions may have runoff regimes utterly different; physiographic characteristics have a very important role in hydrologic processes.

Several techniques are nowadays used to relate the design flows to the physiographic characteristics. For instance, [8] Cheibani et al. (1992) established an equation that produces the flow for the return period for ten years with the area of the basin, its mean slope and drainage density. [11] Campos-Aranda (2008) related monthly runoffs from several basins, their respective areas and distances between hydrometric stations in order to transfer the monthly runoffs of these basins to another for which there is no hydrometric information. Separately, [10] Ouarda et al. (2008) related the physiographic parameters of several sub-basins of Balsas, Lerma-Santiago and Panuco Rivers with the quantiles of different return periods.

Additionally, in hydrology there are many empirical equations, either for the calculations of concentration time of a basin or the determination of a design flow, all of these formulas use one or more physiographic parameters. [1] Franco-Plata (2008) mentions that the physic-

ographic characteristics of a basin have a fundamental role in the study and behavior in the components of the hydrological cycle, therefore some parameters are required as input data in most hydrological models. Also, is considered as watershed physiographic parameters (**Table 1**) to that determined by initial data quantification, which involves managing the relief models and hydrological network deriving quantifiable characteristics of a watershed.

It is important to distinguish that numberless calculations can be obtained from a basin, but nonetheless this research focused on the physiographic characterization of a basin from its morphology, drainage and the characteristics of the main river. The morphometric characteristics of a hydrographic basin offer a physical description of their extension and forms, thus enabling comparisons between different hydrographic basins.

As for the technologies applied in the present study, GIS have been defined in various forms ([12] Bernhardsen, 1999; [13] Bosque, 1997; [14] Burrough, 1986; [15] Candeau, 2005; [16] DeMers, 2002), depending on the

Table 1. Main physiographic parameters of a basin, its drainage and main stream ([1] franco-plata, 2008).

Basin	Drainage	Main river
Area	Drainage density	Length of main stream
Real surface	Stream density	Axial length
Perimeter	Bifurcation ratio	Mean slope of the main stream
Centroid	Strahler orders	Maximum height
Factor of shape	Number of segments by Strahler order	Minimum height
Compactness coefficient	Stream of maximum order in the basin	Difference between maximum and minimum heights
Circularity ratio	Longitudinal addition of all streams	Main river profile
Elongation radius	-	-
Hypsometric relation	-	-
Mean height	-	-
Mean slope	-	-
Confluence mean relation	-	-
Hypsometric curve	-	-
Equivalent rectangle	-	-
Time of Concentration	-	-

point of view of the author assumed in this field, when comparing the definitions however, elements in common appear: spatial information, spatial data, spatially referenced data; all pinpoints at the spatial data as that which differences GIS from other specialized databases, representing the center around which all the possible applications of GIS orbit; hence, spatial data contains, in its most elemental concept, characteristics of localization (X, Y) and sort of thematic characteristic (Z), on which rests the base of all the operations possible to perform in a GIS.

Usually, the use of GIS applied to hydrological modeling offers benefits in the representation and simulation of problems that require interpretation and analysis of spatial information ([17] Farías de Reyes and Reyes, 2001). [1] Franco-Plata (2008) mentions that GIS can actuate as a platform to experiment new ideas and concepts as for the processing of hydrological information, as they become a valuable support to attain the systemic integration of methods and techniques to carry out the evaluation of water resources as for water availability, from the natural water balance of a basin. Thus, the use of GIS has had a great boom in the hydrological sphere for the development of interfaces or applications of spatial and temporary simulation.

3. Material and Method

In the present work, ArcGIS Desktop was used because of the broad range of applications that can be generated by means of resorting to the component of ArcObjects with Visual Basic 6 (VB6), which allowed creating a DLL (Dynamic Link Library) that can be integrated into ArcGIS Desktop to carry out the calculation of the physiographic parameters of a basin.

VB6 programming language is oriented to create programs for Windows, being able to incorporate all the elements of this computing platform: windows, buttons, dialog and text boxes, option or selection boxes, displacement bars, graphics and menus, among other. Said objects were utilized with the intention of providing the tool with a better graphical appearance, since the properties and methods of these objects do not allow the developer the manipulation of geographical data or spatial objects for they are solely oriented to develop pure computing applications, thereby it was necessary to install the spatial component ArcObjects, which enabled the use of ESRI specific libraries, to manipulate the spatial objects and so being able to calculate the physiographic parameters of a basin. Then VB6 together with ArcObjects allowed automating some of the most important tasks, such as the selection of libraries, references, compilation and registration, generator of line number, generator of errors in code and the implementation of the tools that were employed, and the most important was the generation of

a code that enabled the automation of all the processes to produce the physiographic parameters, among other things.

It is important to mention that ArcObjects technology uses and meets the regulations of the component object model (COM) and using it allows developing new tools, functions or creating work flows on ArcGIS. To fulfill the stated objective, the project underwent the following methodological stages proposed by [18] Franco et al. (2012): requirement analysis, conceptual model, geomatic model and implementation.

3.1. Requirements Analysis

As previously indicated, the obtaining of physiographic parameters of a basin provides the adequate bases to accomplish an assessment of the hydrologic resources in a basin; such is the case referring to water availability. Therefore, the development of the application was glimpsed as one that will supply effectiveness and repercussion in saving time and resources oriented to obtain physiographic parameters, which would help hydrologists and all those specialists in water sciences, as for decision making and planning, as well as for the integrated management of these resources.

Moreover, noticeable was the necessity of having sufficient knowledge on hydrology, physiographic parameters, programming in Visual Basic 6 and ArcObjects in views of materializing the GIS tool for the automated and efficacious calculation of the physiographic parameters of a basin; we dealt with these issues resorting to experts in each of the indicated topics.

3.2. Conceptual Model

The conceptual model as a theoretical and strategic backbone of the project development allowed identifying the different interactions between each of the processes and parameters when obtaining the tool, with the aim of choosing the methods to develop for each element on the basis of the requirement analysis and information availability. In **Figure 1**, there is an instance of the conceptual model, in which the set of processes necessary to generate the physiographic parameters considered in the research is schematically represented; the inputs that will feed the processes are identified. By the end of this stage, the conceptual models to obtain the physiographic parameters, hypsometric curve, equivalent rectangle and profile of the main river were produced.

3.3. Geomatic Model

The geomatic model is the representation of a conceptual model, which indeed helps schematize and understand the stages of solving a problem, yet from a geomatic environment and perspective, in which the spatial aspect is

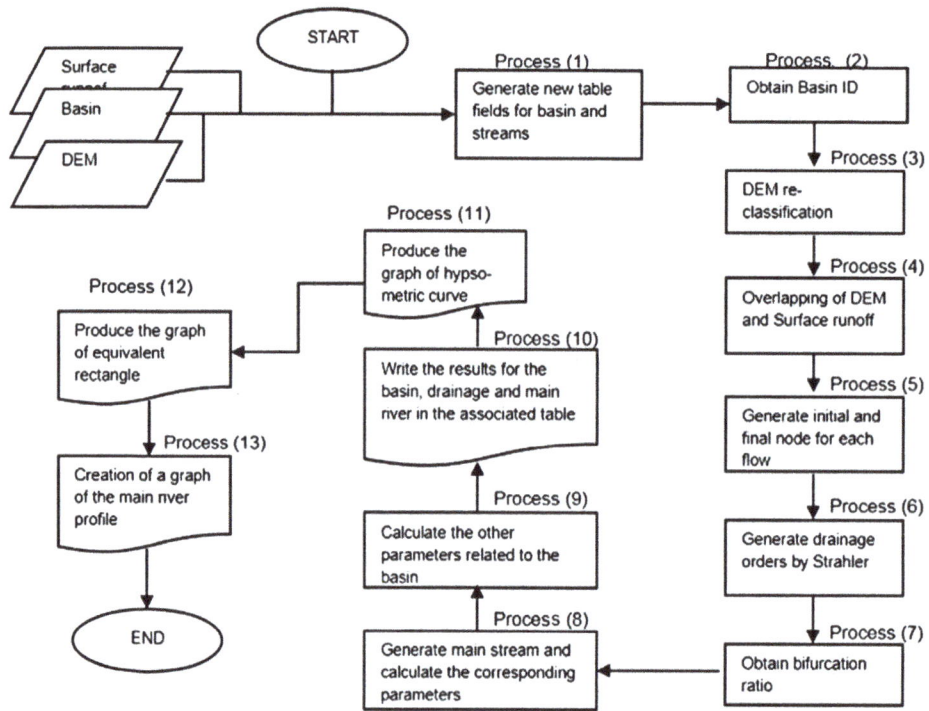

Figure 1. Example of a conceptual model.

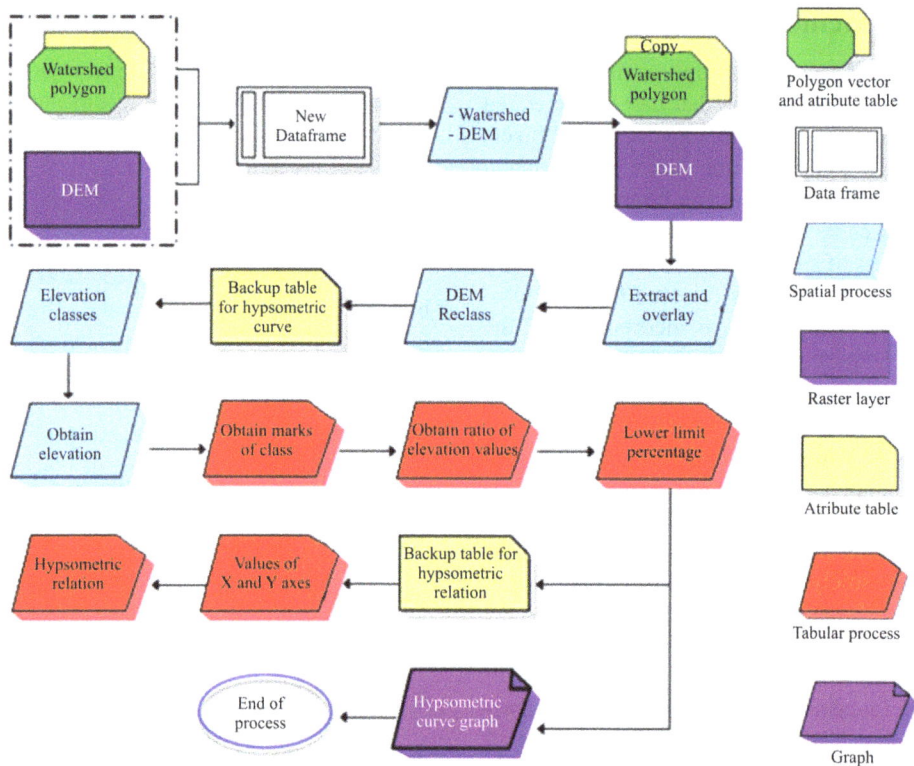

Figure 2. Example of a geomatic model.

distinguished. For the general schematization flow diagrams were employed to represent the logical sequence followed to solve each of the conceptual models stated in the previous stage. In **Figure 2**, an instance of the geomatic models generated for the project is shown.

3.4. Implementation

From the geomatic model obtained in the previous section were selected the tools that allowed translating the model into the computing language to perform the automated processes required by the application. It was considered that the user interface was of the utmost importance to familiarize the user with the application and its elements, and thus make its adequate use and handling easy.

The hydrogeomatic interface called "HidroCuenca" was implemented in Visual Basic 6 with the ArcObjects spatial geographic component, in such manner that it is composed of 4 class modules, 22 modules with 65 scripts; as for Help, FlashMx 2004 was used. Once the implementation was installed, the tool was incorporated in ArcMap as a menu to amicably show the application tools (**Figure 3**).

4. Results

The main window of the GIS tool (**Figure 4** and **Table 2**)

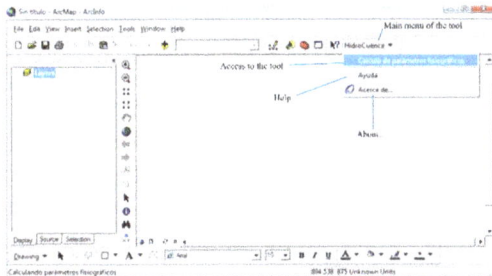

Figure 3. Access through a menu to the implemented tool.

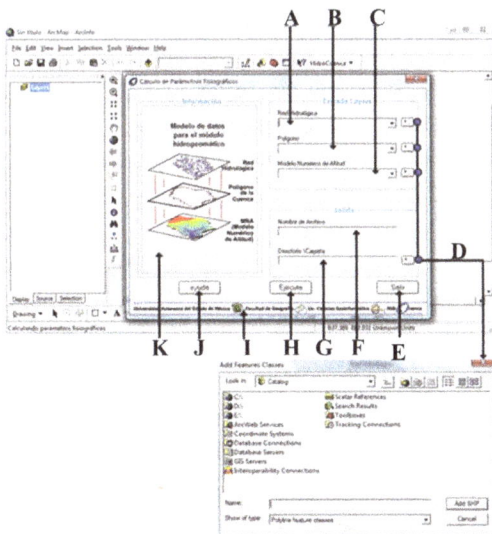

Figure 4. User interface of the GIS tool, the explanation for its components can be seen in Table 2.

shows the interface for the calculation of physiographic parameters; likewise, the necessary objects that allow processing input and output data are also displayed.

4.1. Application to a Case Study

In order to prove the functionality of the new generated GIS tool, an application exercise to El Caracol sub-basin in Mexico was made; this basin is found within that of Balsas River, which on its own belongs to the 18th Hydrological Region established by the program of Priority Hydrological Regions of the National Commission for the Knowledge and Use of Biodiversity (ComisiónNacional para el Conocimiento y Uso de la Biodiversidad, CONABIO) of the Federal Government. The study zone is located between the geographic coordinates 19°42'04.39"N, 17°04'04.84"N, and 99°38'11.51"W, 97°38'11.51"W (**Figure 5**).

To evaluate the physiographic parameters of the river basin three files were utilized; two in vector format (hydrological network and basin polygon) and one in raster format (DEM) (**Figure 6**). In this point is important to underline the exposed by [19] Pineda *et al.* (2012) when they said that it is very important to use data of good quality because results depend of these data. To carry out the automated calculation, the user must start in the "HidroCuenca" menu and click on "Cálculo de parámetros fisiográficos" to open the window shown in **Figure 7**, which is where the aforementioned files are entered.

Once the layers have been entered, is necessary to capture in the tool the name that will be assigned to the results and the folder in which these data will be stored; this in the "Salidas" (Outputs) section (shown in the "interface window" in **Figure 7**). Once the input and output have been indicated, by clicking on "Procesar" (To process) button so that the tool starts calculating the physiographic parameters of the basin under study and the results are unfolded: hypsometric curve (**Figure 8**); equivalent rectangle (**Figure 9**); Strahler drainage order (**Figure 10**); main river (**Figure 11**); and the longitudinal profile of the main river (**Figure 12**).

Likewise, the tool generates attribute tables associated to the results; these are the attributes with the physicographic parameters of the basin, those of the main river, and the attributes of the drainage network (**Figure 13**).

In views of having a means to compare and validate the obtained results, the same parameters were produced in the ordinary manner, in an exercise carried out manually from printed maps from the National Institute of Statistics and Geography (Instituto Nacional de Estadística y Geografía, INEGI). The results obtained from both procedures are shown in **Table 3**.

A comparative analysis of the values shown in the previous table allows verifying that processes carried out

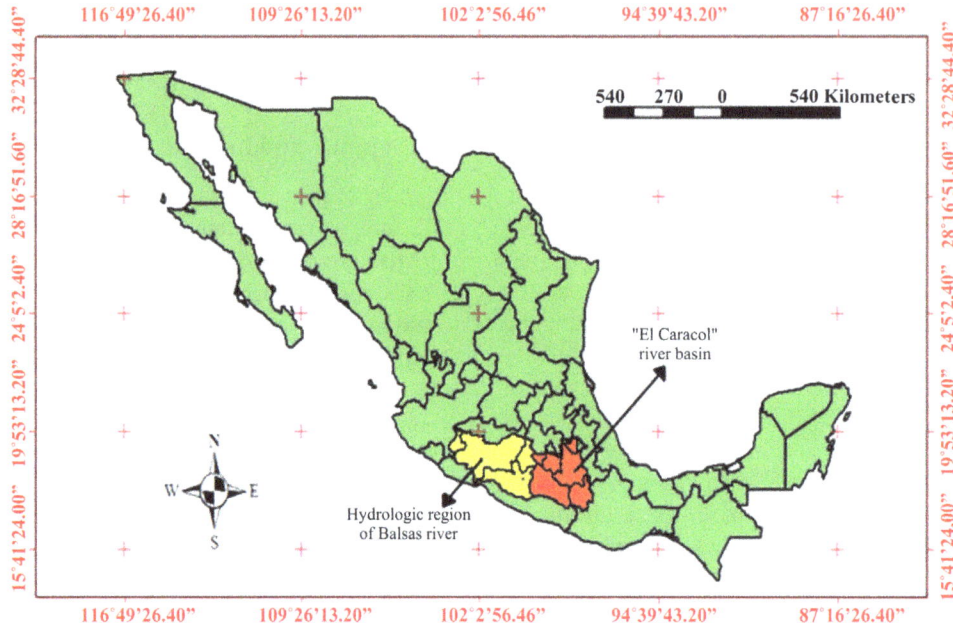

Figure 5. Localization of El Caracol river basin.

Table 2. Explanation of the user interface components of the GIS tool to calculate the physiographic parameters of a basin (Figure 4).

Letter	Object	Function
A	ComboBox	List of shapefile and layer files (polylines).
B	ComboBox	List of shapefile and layer files (polygons).
C	ComboBox	List of files grid, raster dataset (*DEM*).
D	Button (CommandButton)	Shows a textbox to enter data and establish an exit path.
E	Button (CommandButton)	Terminates the application.
F	Text box (TextBox)	Space established for the user to assign a word (letters or numbers) so that they identify their information, after being processed.
G	Text box (TextBox)	Specifies the storing path of the final files.
H	Button (CommandButton)	Executes the program to obtain the physiographic parameters of the basin.
I	Movie (ShockwaveFlash)	Area that shows the hydrographic model in a flash file.
J	Button (CommandButton)	Calls the help system of the application.
K	Movie (ShockwaveFlash)	Displays the name of the institutions that developed the project and the name of the authors.

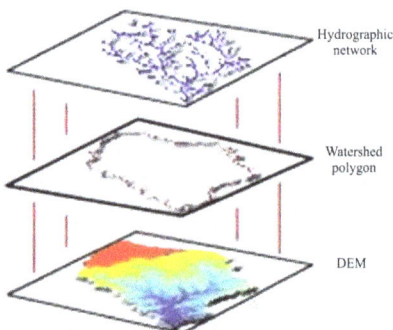

Figure 6. Necessary input layers for the GIS tool.

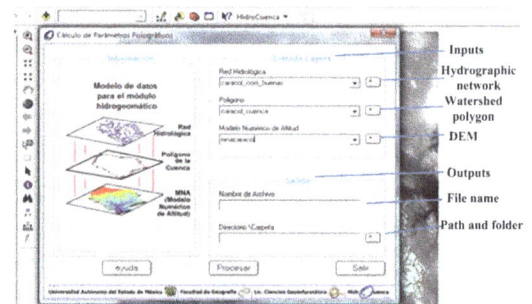

Figure 7. Window to enter the necessary data to calculate the physiographic parameters of a basin.

Table 3. Comparison of the physiographic parameters of El Caracol river basin, obtained from two ways of working out the data: ordinary and using the GIS tool.

Parameter	Ordinary fashion	Automated version
Surface	49671.17 km^2	49671.174 km^2
Perimeter	1399.33 km	1399.327 km
Mean slope	0.018	9.3
Compactness coefficient	1.758	1.759
Elongation ratio	0.92	0.428
Hypsometric relation	2.25	2.5
Drainage density	0.354 km/km^2	0.35 km/km^2
Hydrographic density	0.035 km^{-2}	0.05 km^2
Maximum length of all streams	17606.06 km	17605.901 km
Mean slope of the main river	0.004	0.004
Maximum height	5500 masl	5500 masl
Minimum height	400 masl	400 masl

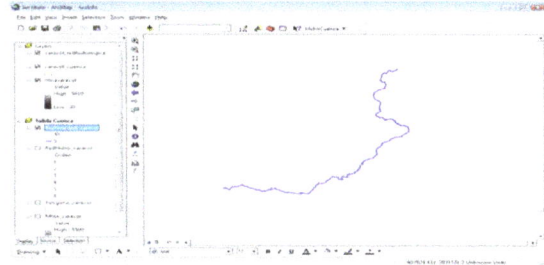

Figure 8. Graph of the hypsometric curve.

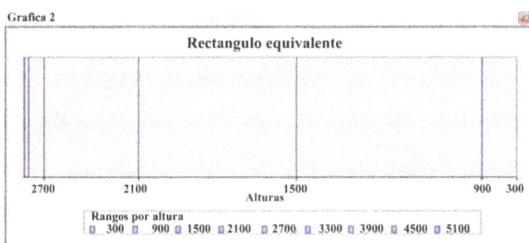

Figure 9. Equivalent rectangle of El Caracol river basin.

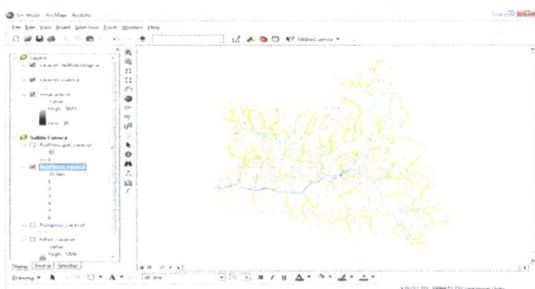

Figure 10. Determination of Strahler drainage order for El Caracol river basin, the maximum order was six.

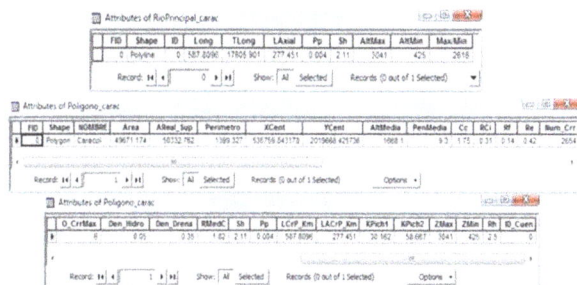

Figure 11. Determination of main river from the input drainage network.

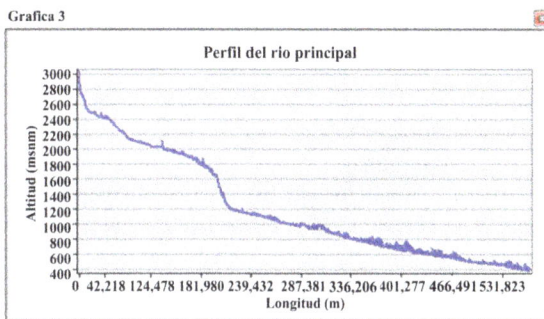

Figure 12. Longitudinal profile of the main river.

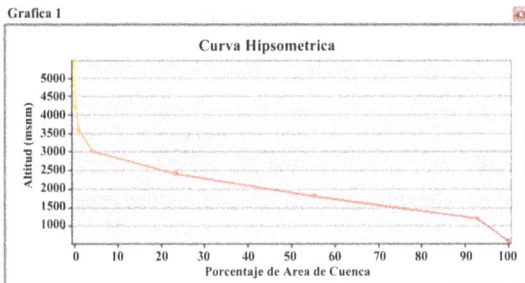

Figure 13. Attribute tables with the main parameters.

in the GIS tool were correct, whose processing time lasted for about two hours.

5. Conclusions

The method employed in the research allowed designing and implementing a GIS tool for the calculation of the physiographic parameters of a basin; by means of applying said tool, it was possible to corroborate and mainly verify the saving of time and resources compared with other systems devoted to the obtaining of parameters. Even if it is true that the results may vary in a negligible way with other commercial systems utilized to obtain physiographic parameters, this can be due to the difference between the applied formulas or method. When comparing the obtained values with others generated by means of different frameworks, they reflect the usefulness of HidroCuenca tool and its application in a GIS platform and in projects in which the hydrological analysis of basins is necessary.

Although the development of applications in the hydrological sphere has been developing in a significant manner, it is worth distinguishing that the numbers of commissioned applications to calculate physiographic parameters are very few; therefore, the present work represents an important and innovative effort to calculate said parameters for a basin. The implementation of the developed GIS tool automated and simplified diverse tasks as for the calculation of physiographic parameters (to name a few: Strahler orders, direction of the flow of rivers, obtaining of main river, hypsometric curve, equivalent rectangle, profile of main river, etc.), thus preventing the users from using time resorting to other spatial analysis modules to obtain said parameters, instead a single interface is offered for this task in a simple, efficacious and automated manner.

In the international, national and regional spheres the topic of basin management becomes more important by the day; not only is it the interest and concern of the actors and those directly involved: communities, local organizations, municipalities, national institutions, etc., but also of donor and cooperating organizations. It is intended, due to this reason, to use information and geographic technologies to develop automated products that facilitate decision making for better water management.

Finally, it is important to underscore that in the presence of poor-quality input data, the interface does not guarantee the correct acquisition of physiographic parameters.

REFERENCES

[1] R. Franco-Plata, "Concepción e Implementación de un Módulo Hidrogeomático Para la Evaluación de Disponibilidad de Recursos Hídricos," Ph.D. Dissertation, Autonomous University of State of Mexico, Mexico City, 2008.

[2] G. Rodríguez and A. Santos, "Diseño e Implementación de un Módulo Hidrogeomático Para la Estimación de Parámetros Fisiográficos de Cuencas Hidrográficas," Undergraduate Dissertation, Autonomous University of State of Mexico, Mexico City, 2007.

[3] F. L. Hellweger and D. R. Maidment, "HEC-PREPRO: A GIS Preprocessor for Lumped Parameter Hydrologic Modeling Programs," Center for Research in Water Resources, The University of Texas, Austin, 1997. http://www.crwr.utexas.edu/gis/gishydro03/Library/hellweg/ferdi.pdf

[4] C. Ehlschlaeger, "The GRASS/Mathematical link: Developing Hydrologic Models in Geographic Information Systems Interfaced with Computer Algebra Systems," US Army Construction Engineering Research Lab, Champaign, 1991.

[5] C. Díaz, K. M Bâ, A. Iturbe, M. V. Esteller and F. Reyna, "Estimación de las Características de una Cuenca con la Ayuda de SIG y MEDT: Caso del Curso Alto del río Lerma, Estado de México," Ciencia Ergo Sum, Vol. 6, No. 2, 1999, pp. 124-134.

[6] D. Maidment, "ArcHydro-GIS for Water Resources," ESRI Press, Redlands, 2002.

[7] R. Franco-Plata, "Concepción de un Módulo Hidrogeomático Para el Análisis de Cuenca," Toluca, Mexico, Unpublished, 2006.

[8] A. Chebani, J. Llamas and C. Díaz-Delgado, "Estimation de la Cruedécenale par les Caractéristiques Physiographiques des Bassins," Le Clima, Vol. 10, No. 2, 1992, pp. 24-37.

[9] J. Llamas, "Hidrología General: Principios y Aplicaciones," Servicio Editorial de la Universidad del País Vasco, Bilbao, 1993.

[10] T. B. J. M. Ouarda, K. M. Bâ, C. Díaz-Delgado, A. Cârsteanu, K. Chokmani, H. Gingras, E. Quentin and E. Trujillo, "Intercomparasion of Regional Flood Frequency Estimation Methods at Ungauged Sites for Mexican Case Study," Journal of Hydrology, Vol. 348, No. 1-2, 2008, pp. 40-58.

[11] D. F. Campos-Aranda, "Contraste de un Método Simple de Transferencia de Información Para Estimación de Volúmenes Escurridos Mensuales," Ingeniería Hidráulica en México, Juitepec, 2008.

[12] T. Bernhardsen, "Geographic Information System, an Introduction", Asplan Viak, Toronto, 1999.

[13] J. Bosque, "Sistemas de Información Geográfica," Ediciones Rialp S. A., Alcala, 1997.

[14] P. A. Burrough, "Principles of Geographical Information Systems for Land Resources Assessment," Clarendom, Oxford, 1986.

[15] R. Candeau, "Regionalización Socioeconómica del Parque Nacional Nevado de Toluca y su Relación con el Deterioro Ambiental," Ms. C. Dissertation, Autonomous University of State of Mexico, Mexico City, 2005.

[16] M. N. DeMers, "Fundamentals of Geographic Information Systems," John Wiley, Hoboken, 2002.

[17] C. M. Farías de Reyes and J. Reyes, "Modelación de Lluvia Escorrentía Usando Sistemas de Información Geográfica (GIS) en Situaciones de Información Escasa," Proceedings of the 7th National Congress of Civil Engineering—PUNO, Lima, 2001.

[18] R. Franco-Plata, L. R. Manzano-Solís, M. A. Gómez-Albores, J. I. Juan-Pérez, N. B. Pineda-Jaimes and A. Martínez-Carrillo, "Using a GIS Tool to Map the Spatial Distribution of Population for 2010 in the State of Mexico, Mexico," Journal of Geographic Information Systems, Vol. 4, No. 1, pp. 1-11.

[19] N. B. Pineda, J. Bosque, M. Gómez, R. Franco, X. Antonio and L. R. Manzano, "Determination of Optimal Zones for Forest Plantations in the State of Mexico Using Multi-Criteria Spatial Analysis and GIS," Journal of Geographic Information System, Vol. 4, No. 3, pp. 204-218.

4

Calibration of Hydrognomon Model for Simulating the Hydrology of Urban Catchment

Haruna Garba[1], Abubakar Ismail[2], Folagbade Olusoga Peter Oriola[1]

[1]Department of Civil Engineering, Nigerian Defence Academy, Kaduna, Nigeria; [2]Department of Water Resources and Environmental Engineering, Ahmadu Bello University, Zaria, Nigeria.

ABSTRACT

Due to chaotic nature of flow in natural open channels and the physical processes and the unknown in a river basin variable, the parameters to be used in studying the behavior of river basin to a given rainfall data cannot be measured directly for this reason, a hydrological model was calibrated and applied to simulate the hydrology of Kaduna River (7112 sq miles) North West Nigeria. Prior to the model calibration, a sensitivity analysis of the hydrognomon model to the parameters was carried out to gain a better understanding of the correspondence between the data and the physical processes being modeled.

Keywords: Calibration; Model; Optimization; Flood; Stream Flow; Parameter

1. Introduction

Across the globe, human populations are becoming increasingly urban with approximately fifty percent of the worlds population residing in urban areas as observed by [1]. Continued land development and land use changes within cities present considerable challenges for environmental management. [2] observed that hydrologic changes including increase impervious area, soil compaction and increased drainage efficiency generally lead to increase direct runoff, decreased ground water recharge and increase flooding cited by [3]. Increase stream stage and discharge variability are the common responses to urbanization. [4] observed that urban streams are susceptible to the occurrence of extreme flow event than their rural counterpart. Furthermore, according to [4] changes in peak flow due to urbanization vary with the degree of urban development, recurrence interval of the peak flow and location within the water shed.

Hydrological models especially the simple rainfall-runoff types are widely used in understanding and quantifying the impacts of land use changes and provide information that can be used in land use decision making. Furthermore [5], observed that the frequency of occurrences of extreme flood due to climate changes have increased the need for better understanding of floods.

Understanding the behavior of a river basin to a given

rainfall volume depends on the analysis of rainfall data which helps to provide information about data with regards to the variable under study. The probability distribution is a hydrological tool most widely used in flood estimation and prediction Probability distribution functions have been used to model phenomenon characterized by significant variability such as rainfall in contrast to deterministic approach determined by physical principles. Parameters of distribution functions are characteristics of rainfall data which helps to provide information about the data with regards to the variable under study. Due to parametization of the physical processes and the unknown in the basin characteristics parameters to be used cannot be measured directly; they are therefore obtained by a process of model calibration.

Model calibration according to [6] is the procedure of demonstrating that the model can produce field measured quantities such as hydraulic heads and flows. It is carried out by finding set of parameters and boundary conditions that produce simulation results that is in agreement with field measured data. The objective of model calibration is maximize the coefficient of efficiency by reducing the percentage difference in mean, standard deviation and coefficient of variation between the predicted and observed stream flow. Calibration of a rainfall-runoff model with respect to local observational data will help to improve model predictability as observed by [3]. When model results match observed values from stream flow measurement, there is greater confidence in the reliability of the model.

The model was calibrated and applied to simulate the hydrology of Kaduna River under various climatic conditions as part of a PhD study by [5].

2. Materials and Methods

2.1. The Model

Hydrognomon is an open sources software tool used for the processing of hydrological data (**Figure 1**). Data are usually imported through standard text files, spread sheets or by typing. The available processing techniques for the tool includes time step aggregation and regularization, interpolation, regression analysis and infilling of missing values, consistency test, data filtering, graphical and tabular visualization of time series. Hydrognomon support several time step from the finest minutes scales up to decades. The programme also include common hydrological application such as evapotranspiration modeling, stage discharge analysis, homogeneity test, areal integration of point data series, processing of hydrometric data as well as lumped hydrological modeling with automatic calibration facilities as contained in the hydrognomon user manual (http:www.hydrognomon.org) [7].

2.2. The Study Area

Kaduna Township located between latitudes 10°27'15"N - 10°13'5"N and Longitudes 7°21'48"E - 7°29'36"E in the high plains of North western Nigeria region (**Figure 2**). The general relief of the area is an undulating plain land at a height of between 582 m to 640 m [8]. Kaduna state experiences a typical continental climate with distinct seasonal regimes. The spatial and temporal distribution of rainfall varies over Kaduna. The soils and vegetation

area typical of red brown to red yellow tropical soils and savanna grass land comprising of scattered trees and woody scrubs.

Kaduna river is the main tributary of Niger river in central Nigeria. It rises on the Jos plateau south west of Jos town in a North West direction to the northeast of Kaduna town (**Figure 3**). It then adopts a south westerly and southerly course before completing its flow to the Niger River at Mureji. Most of its course passes through open savanna woodlands but its lower section cut several gorges including the granite ravine at Shiroro above its entrance into the extensive Niger flood plains.

2.3. Data Used

The historical data used for the calibration were recorded rainfall and gauge height levels from 1975-2000 (26 years) of record at a gauging point (Datum at 582.96 m) located in the study area at Kaduna south water works. The data are totals on monthly basis spanning the calibration period. The steps of the data collection process

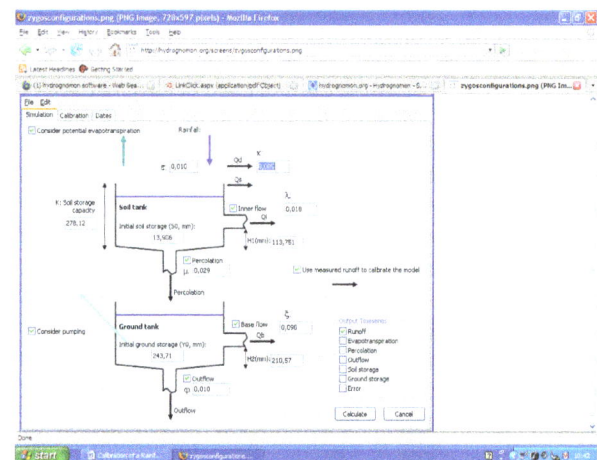

Figure 2. Location of Kaduna state on map of Nigeria.

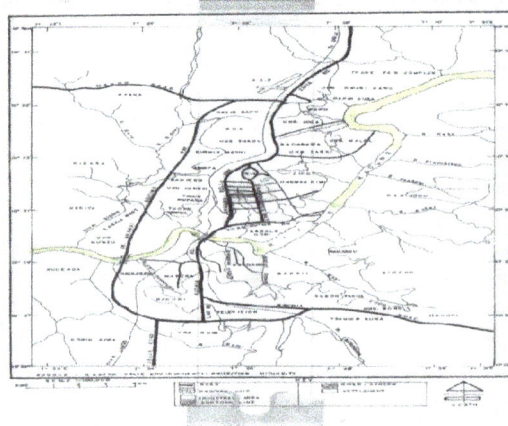

Figure 1. Structure of the simulation/calibration module (http: www.hydrognomon. org).

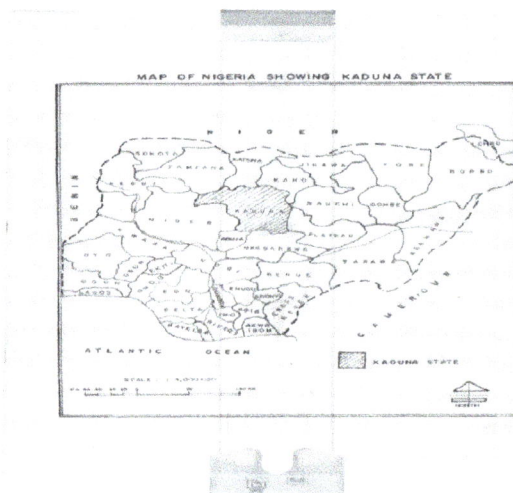

Figure 3. Location map of the study area (Kaduna River).

involve the following:

- The daily stream flow was read as gauge height while the daily rainfalls were read for each of the stations.
- The monthly maximum stream flow values and rainfall values were extracted from the daily values.
- The gauged levels measured were used to scale the flow to runoff volume of the watershed by using the expression below [9] in calibrating the model.

$$Q = ICA$$

where

Q = calculated runoff;

I = gauged water levels;

C = a factor (distribution coefficient) the ratio maximum gauge level at a point tothe mean gauge levels of Kaduna river.

A = drainage area of Kaduna river.

3. Results and Discussions

Before calibrating, the sensitivity analysis of the hydrognomon model to the flow parameters was tested. The hydrognomon model is most sensitive to the flow recession parameters in both the initial soil storage and the initial ground water storage. The recession parameters control the ability of the soil to retain water in the two storages, the model is insensitive to the capillary flow and the coefficient of base flow is insignificant in simulating the flow.

The physical processes on the basin scale and the unknowns in the basin characteristics consist of too many parameters which cannot be measured directly, for this reason the model was calibrated. **Figure 4** shows the observed and simulated runoff. The actual runoff was calibrated with the calculated runoff at an objective function of 0.993 [10]. Prior to the model calibration, a sensitivity analysis of the hydrognomon model to the soil parameters have been tested. For the test of each parameter, all the others were fixed and the tested parameter was change from the lower to the upper boundary on a discrete value and applied to the model (**Figure 4**). The table below shows parameters calibrated in the study together with their upper and lower boundaries. The model was calibrated at an objective function of 0.390 with 8001 iterations. The parameter calibrated in this study with lower and upper boundaries is in the **Table 1** below.

4. Conclusion

Due to the fact that the physical processes at the river basin scale and the unknown in the basin characteristics consist of too many parameters which cannot be measured directly the model was thus calibrated. The model is most sensitive to the flow recession parameters of both the lower and the upper tanks. The calibrated results showed the Nash and Sutcliffe efficiency of >9. The pa-

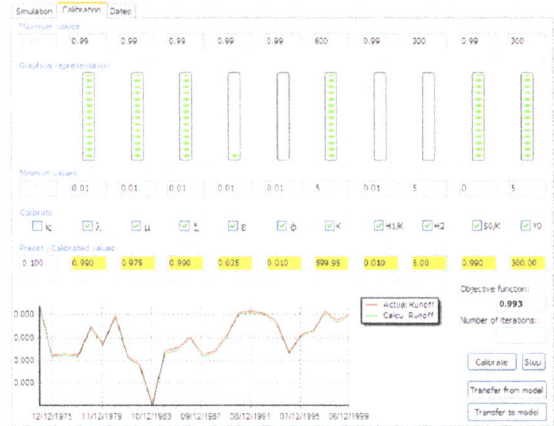

Figure 4. Predicted and simulated monthly flow for Kaduna River.

Table 1. Parameters calibrated in the study with their minimum and maximum values.

Parameter	Lower boundary	Upper boundary	Sensitivity level
Soil storage capacity (κ)			Insensitive
Coefficient for inner flow (λ)	0.01	0.990	Highly sensitive
Percolation coefficient (μ)	0.01	0.975	Highly sensitive
Evaporation (ε)	0.01	0.035	Insensitive
Coefficient for out flow (υ)	0.00	0.010	Insensitive
Coefficient for surface runoff flow (κ)	5	599.95	Highly sensitive
H1/κ	0.01	0.01	Insensitive
Threshold for creating base flow (H2)	5.0	5.0	Insensitive
S0/κ	0.00	0.990	Highly sensitive
Initial ground storage (γ_o)	5	300.00	Highly sensitive
Coefficient for base flow (ξ)	0.01	0.990	Highly sensitive

rameters of λ (0.99), μ (0.975), κ (599.95), S0/κ (0.990), γ_o (300) and ξ (0.990) can be transferred to the model and applied to simulate the hydrology of Kaduna River for hydrological investigation.

REFERENCES

[1] J. Cohen, "Human Populations: The Next Centaury," *Science*, Vol. 302, No. 5648, 2003, pp. 1172-1175.

[2] D. Booth, "Urbanization and Natural Drainage System—Impact, Solutions and Prognoses," *Northwest Environ-*

mental Juornal, Vol. 7, pp. 93-118.

[3]   S. Muthukrisman, J. Harbour, K. J. Lim and B. A. Engel, "Calibration of a Simple Rainfall-Runoff Model for Long-Term Hydrological Impact Evaluation," *URISA Journal*, Vol. 18, No. 2, 2006, pp. 35-42.

[4]   O. M. Driscol, S. Clinton and A. Jefferson, "Urbanization Effect on Watershed Hydrology and In-Stream Processes in the Southern United States," *Water*, Vol. 2, No. 3, 2010, pp. 605-645.

[5]   H. Garba, "Predicting the Potential Effect of Climate Change on the Hydrological Response of Kaduna River North West Nigeria Using Hydrological Model," Ph.D. Thesis, Department of Civil Engineering, Nigerian Defence Academy, Kaduna, 2013.

[6]   D. Potcrajac and K. Howard, "Advanced Simulation and Modeling for Ground Water Management," The United Nations Education, Scientific and Cultural Organization (UNESCO), Paris, 2010.

[7]   EGU, "Hydrognomon," 2010, Vienna. http://www.hydrognomon.org

[8]   N. D. Jeb and S. P. Aggarrwal, "Flood Inundation Hazard Modeling of the River Kaduna Using Remote Sensing and Geographical Information Systems," *Journal of Applied Science Research*, Vol. 4, No. 12, 2008, pp. 1822-1833.

[9]   K. D. W. Nandala, "CE 205 Engineering Hydrology Lecture Notes," Department of Civil Engineering, University of Peradeniya, Peradeniya.

[10] J. E. Nash and J. V. Sutcliffe, "River Flow Forecasting through Conceptual Models, a Discussion of Principles," *Journal of Hydrology*, Vol. 10, No. 3, 1970, pp. 282-291.

5

Streamflow Decomposition Based Integrated ANN Model

Nikhil Bhatia, Laksha Sharma, Shreya Srivastava, Nidhish Katyal, Roshan Srivastav

School of Mechanical and Building Sciences, VIT University, Vellore, India.

ABSTRACT

The prediction of riverflows requires the understanding of rainfall-runoff process which is highly nonlinear, dynamic and complex in nature. In this research streamflow decomposition based integrated ANN (SD-ANN) model is developed to improve the efficacy rather than using a single ANN model for the flow hydrograph. The streamflows are decomposed into two states namely 1) the rise state and 2) the fall state. The rainfall-runoff data obtained from the Kolar River basin is used to test the efficacy of the proposed model when compared to feed-forward ANN model (FF-ANN). The results obtained in this study indicate that the proposed SD-ANN model outperforms the single ANN model in terms of both the statistical indices and the prediction of high flows.

Keywords: Artificial Neural Network; Rainfall-Runoff Modeling; Streamflow Decomposing; Black Box Modelling

1. Introduction

A wide variety of rainfall-runoff models have been developed and applied for water resources planning which is vital in terms of flood control and management. Traditionally, the hydrologists and water resources researchers have used conventional modeling techniques either deterministic models that includes physics of the underlying process or systems theoretic (black box) models. However these models require a large quantity of data and a complex methodology for its calibration. Most of the hydrological models either show unsuccessful results or become cumbersome. Many researchers report that these models fail to capture the high flows in a hydrograph [1,2] due to limited data sets available in the high flow domain (5% of total calibrating patterns) for capturing the nonlinear dynamics.

Recently the researchers have focused to decompose the data corresponding to flow hydrograph to enhance the performance of the hydrologic models. Mostly the studies have concentrated on using either the statistical techniques or soft decomposing techniques for data decomposition [3]. Studies include automated base flow separation and recession analysis [4], spectral analysis [5], wavelet transforms and runoff time series analysis [6-9], modular neural network (MNN) [10], self-organizing map (SOM) classifier [11,12] and self organizing linear output map (SOLO) [13]. Most of these studies conclude that the decomposition and partitioning of data resulted in better model performance.

Artificial neural network (ANN) has been proposed by researchers which is a system theoretic model that has gained momentum in the last few decades as it has been successfully applied to a wide range of problems in hydrology [2,14-16]. It is used to develop relationship between input and output variables using the existing data. Jain and Srinivasulu [3] proposed an integrated approach to model decomposed flow hydrograph using ANN and conceptual techniques. The streamflow decomposition was carried out based on physical processes which divide the input-output and fit the models for each of the segments [3]. However, the models developed using the distributed approach would have made the solution procedures complex significantly [3]. In this study, efforts are made to develop a simplified ANN based decomposed streamflow model without requiring any prior knowledge or understanding of physical processes. In this study the data is divided into two states namely rise and fall, based on the current state. The proposed model is compared with the feed forward ANN model, on a real case example of Kolar basin, India.

This paper is organized as follows. Section 2 provides a brief introduction on ANN. Section 3 describes proposed methodology. Section 4 illustrates the case study on Kolar basin, India. Section 5 includes Results and Discussion and the paper is concluded with summary and conclusions presented in Section 6.

2. Artificial Neural Network

The ANNs are highly interconnected mathematical models with its structure analogous to that of the human brain.

It attempts to develop the massively parallel local processing and the distributed storage properties which are believed to exist in the human brain [17]. Simple processing units of an ANN are called "neurons". Neurons having similar characteristics are grouped in one single layer (neurons in an input layer receive an input from the external source, and transmit the same to a neuron in an adjacent layer, which could either be a hidden layer or an output layer). Structure of the ANN Model is shown in **Figure 1**.

The general mathematical form of an ANN Model is given as:

$$\hat{y} = f\left[V_i u\left(\sum_{i=1}^{n} x_i W_i + \alpha \right) + \beta \right] \qquad (1)$$

where, x_i is the input of the ANN Model, W_i is the weight connecting input nodes to hidden nodes, V_i is the weight connecting hidden nodes to output nodes, α, β are the bias at hidden and output layer respectively and $u(\ ), f(\ )$ are the activation functions at hidden and output layer respectively.

The weights W_i and V_i are usually determined by minimizing the quadratic error function,

$$E = \frac{1}{2n} \sum (\hat{y} - y)^2 \qquad (2)$$

Once the ANN Model is executed then the error at the output layer from an ANN can be computed if output is known.

$$\varepsilon = y - \hat{y} \qquad (3)$$

where, ε is the error at the output layer, y is the observed stream flow and y is the estimated stream flow.

Using the process of the feed-forward calculations and back-propagation of the errors the connection strengths are updated and an acceptable level of output is predicted. This is called as training of an ANN. Once the network has been trained, it can be tested using the testing data.

3. Model Development

Determination of significant input variables is a very essential step in ANN Modeling [18,19]. Cross correlation is used to find the relationship between the variables [2,19-21] and is used to represent the most popular analytical techniques for selecting appropriate inputs [18]. Observed relationships between the training samples and the connection weights enhance generalization ability of an ANN model [22].

The inputs to the SD-ANN model were selected on the basis of cross- and auto-correlation method as proposed by Sudheer *et al.* [2]. The significant input variables were found to be the effective rainfalls at lag time steps of $t - 9$, $t - 8$, and $t - 7$ (P_{t-9}, P_{t-8} and P_{t-7}) using the cross-correlation and the river flow values at lag time steps of t

$- 1$ and $t - 2$ (Q_{t-1} and Q_{t-2}) using the autocorrelation function [23]. The output of the model is the riverflow at time t (Q_t). Thus Q_t is represented as

$$Q_t = f\left(P_{t-9}, P_{t-8}, P_{t-7}, Q_{t-1}, Q_{t-2} \right) \qquad (4)$$

In this study, the hourly input data are divided into two cases based on the previous data sets,

1) Rise: In the rise pattern the value of runoff at time t is greater than that of time step $t - 1$, *i.e.*, $Q_t > Q_{t-1}$.

2) Fall: In the fall pattern the value of runoff at time $t - 1$ is greater than that of time step t, *i.e.*, $Q_{t-1} > Q_t$.

Figure 2 shows the proposed methodology in which Model 1 decomposes the data into classes (*i.e.*, rise and fall) based on the input variables, Model 2 is the calibrated ANN, model for the rise and Model 3 is the calibrated ANN model for the fall.

Statistical indices like the coefficient of correlation (C_c), root-mean-square error ($RMSE$) and Nash-Sutcliffe efficiency (NSE) [24] are used to evaluate the performance of the model. The equations of these statistical indices are,

$$C_c = \left[\frac{\sum_{i=1}^{n}(y_i - \bar{y})(\hat{y}_i - \bar{\hat{y}})}{\sqrt{\sum_{i=1}^{n}(y_i - \bar{y})^2}\sqrt{\sum_{i=1}^{n}(\hat{y}_i - \bar{\hat{y}})^2}} \right] \qquad (5)$$

$$RMSE = \sqrt{\frac{\sum_{i=1}^{n}(y_i - \hat{y}_i)^2}{n}} \qquad (6)$$

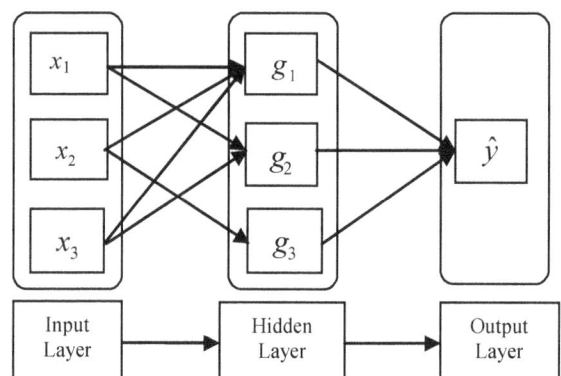

Figure 1. Typical model structure of the FF-ANN model.

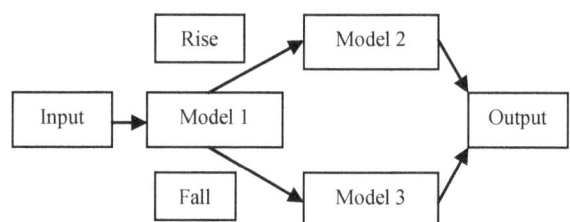

Figure 2. Methodology of the streamflow decomposed based ANN model.

$$NSE = 1 - \frac{\sum_{i=1}^{n}(y_i - \hat{y}_i)^2}{\sum_{i=1}^{n}(y_i - \overline{y})^2} \qquad (7)$$

where y_i is the Observed Runoff Value, y_i is the Predicted Runoff Value, y is the mean of the observed runoff values and $\hat{y}$ is the mean of the predicted runoff values.

4. Case Study

A case study on the Kolar River basin is chosen to demonstrate the proposed SD-ANN method. FF-ANN and SD-ANN models for forecasting the runoff values at 1-hour lead time have been developed. Data relating to monsoon season (i.e., July, August, and September) for 3 years period (from 1987 to 1989). Note that areal average values of rainfall data for three rain gauge stations were used in the study.

The Kolar River is a tributary of the river Narmada that drains an area about 1350 km^2 before its confluence with Narmada near Neelkant (**Figure 3**). In this study the catchment area up to the Satrana gauging site is considered, which constitutes an area of 903.87 km^2. The 75.3-km-long river course lies between north latitude $21°09' - 23°17'$ and east longitude $77°01' - 77°29'$ Further more details on the basin are given by Nayak et al. [25].

From the total available data for 3 years, 6525 patterns (input-output pairs) were identified for the study and were split into calibration (5500 sets, 1987-1988 data sets) and validation (1025 sets, 1989 data sets). Note that the 1025 sets considered for validation were corresponding to a continuous hydrograph.

The activation function used at the hidden layer and at the output layer is sigmoid function as it is easily differentiable.

$$y = \frac{1}{1 + e^{-\alpha x}} \qquad (8)$$

To calculate the network parameters back propagation algorithm [26] has been used. Adaptive learning and momentum rates have been employed for the model training [25].

5. Results and Discussions

As discussed earlier, the SD-ANN model developed is used for forecasting the river flow for Kolar Basin at a lead time of 1 hour. The performance of the proposed SD-ANN model and FF-ANN model have been evaluated by means of a variety of statistical criteria such as coefficient of correlation (C_C), coefficient of efficiency (NSE) and the root-mean-square error (RMSE) between the actual and estimated flow values. The various statistics stated below in **Table 1** indicates that the predicted value of runoff by the SD-ANN is more accurate than that of FF-ANN Model. Performance of both the models in terms of statistical indices is very similar and satisfactory as the correlation coefficient of both the models are very close to the unity. Further it is observed from the **Table 1** that the efficiency of both the models is greater than 90% which is highly satisfactory according to Shamseldin [27]. In addition, it is worth noting that the RMSE of the proposed model is less when compared to the FF-ANN model. Also the prediction of high flows is well modeled by the proposed SD-ANN model when compared to the FF-ANN model (**Figure 4**). It is evident from the results that the decomposition of the streamflow has considerable impact on the performance of models.

Table 1. Statistical indices—comparison between SD-ANN and FF-ANN model.

S.No.	Statistical Indices	FF-ANN Model	SD-ANN Model
1.	Coefficient of Correlation (C_c)	0.990569	0.991947
2.	Root-Mean-Square Error (RMSE)	26.26719	23.81039
3.	Nash-Sutcliffe Efficiency (NSE)	0.980432	0.983921

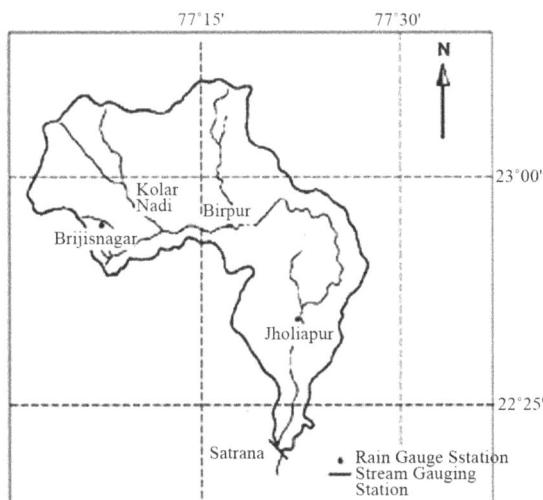

Figure 3. Map of Kolar river basin [25].

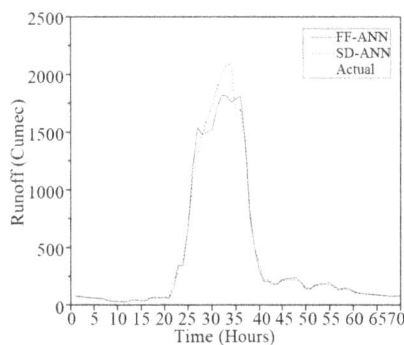

Figure 4. Computed streamflows for a typical event during validation.

6. Summary and Conclusion

In this study, a simplified ANN based decomposed streamflow model is developed. The proposed data decomposition does not require any prior knowledge or understanding of physical processes. In this study the data is divided into two states namely rise and fall, based on the current state. The performance of the proposed SD-ANN model is compared to that of the feed-forward ANN model in terms of statistical indices such as coefficient of correlation, coefficient of efficiency and root means square error. The exercise was carried out for the hourly data in Kolar river basin, India. It is observed that the proposed SD-ANN model and the FF-ANN model show similar results in terms of statistical indices except the case of RMSE where the former outperforms the latter. Further, the SD-ANN model outperforms the FF-ANN model in prediction of high flows. The results show the significance of the streamflow decomposition when compared to single hydrograph. The performance of the SD-ANN models has to be tested on various time scales. Further extensions of this model can be examined to improve the forecasting accuracy [28].

7. Acknowledgements

The authors thank the Vellore Institute of Technology, Vellore, India, for providing the necessary facilities to carry out this research work.

REFERENCES

[1] C. E. Imrie, S. Durucan and A. Korre, "River Flow Prediction Using Artificial Neural Networks: Generalization beyond the Calibration Range," *Journal of Hydrology*, Vol. 233, No. 1-4, 2000, pp. 138-153.

[2] K. P. Sudheer, A. K. Gosain and K. S. Ramasastri, "A Data-Driven Algorithm for Constructing Artificial Neural Network Rainfall-Runoff Models," *Hydrological Processes*, Vol. 16, No. 6, 2002, pp. 1325-1330.

[3] A. Jain and S. Srinivasulu, "Integrated Approach to Model Decomposed Flow Hydrograph Using Artificial Neural Network and Conceptual Techniques," *Journal of Hydrology*, Vol. 317, No. 3-4, 2005, pp. 291-306.

[4] J. G. Arnold, P. M. Allen, R. Muttiah and G. Bernhardt, "Automated Base Flow Separation and Recession Analysis Techniques," *Ground Water*, Vol. 33, No. 6, 1995, pp. 1010-1018.

[5] M. E. Spongberg, "Spectral Analysis of Base Flow Separation with Digital Filters," *Water Resources Research*, Vol. 36, No. 3, 2000, pp. 745-752.

[6] L. C. Smith, D. L. Turcotte and B. L. Isacks, "Streamflow Characterization and Feature Detection Using a Discrete Wavelet Transform," *Hydrological Processes*, Vol. 12, No.

2, 1998, pp. 233-249.

[7] D. Labat, R. Ababou and A. Mangin, "Rainfall-Runoff Relations for Karstic Springs. Part II: Continuous Wavelet and Discrete Orthogonal Multiresolution," *Journal of Hydrology*, Vol. 238, No. 3-4, 2000, pp. 149-178.

[8] S. Y. Liu, X. Z. Quan and Y. C. Zhang, "Application of Wavelet Transform in Runoff Sequence Analysis," *Progress in Nature Science*, Vol. 13, No. 7, 2003, pp. 546-549.

[9] F. Anctil and D. G. Tape, "An Exploration of Artificial Neural Network Rainfall-Runoff Forecasting Combined with Wavelet Decomposition," *Journal of Environmental Engineering and Science*, Vol. 3, No. 1, 2004, pp. S121-S128.

[10] B. Zhang and S. Govindaraju, "Prediction of Watershed Runoff Using Bayesian Concepts and Modular Neural Networks," *Water Resources Research*, Vol. 36, No. 3, 2000, pp. 753-762.

[11] D. Furundzic, "Application Example of Neural Networks for Time Series Analysis: Rainfall Runoff Modeling," *Signal Processing*, Vol. 64, No. 3, 1998, pp. 383-396.

[12] R. J. Abrahart and L. See, "Comparing Neural Network and Autoregressive Moving Average Techniques for the Provision of Continuous River Flow Forecasts in Two Contrasting Catchments," *Hydrological Processes*, Vol. 14, No. 11-12, 2000, pp. 2157-2172.

[13] K. L. Hsu, H. V. Gupta, X. Gao, S. Sorooshian and B. Imam, "Self-Organizing Linear Output Map (SOLO): An Artificial Neural Network Suitable for Hydrologic Modeling and Analysis," *Water Resources Research*, Vol. 38, No. 12, 2002, pp. 1-17.

[14] K. Hsu, V. H. Gupta and S. Sorooshian, "Artificial Neural Network Modeling of the Rainfall-Runoff Process," *Water Resources Research*, Vol. 31, No. 10, 1995, pp. 2517-2530.

[15] N. Sajikumar and B. S. Thandaveswara, "A Nonlinear Rainfall-Runoff Model Using an Artificial Neural Network," *Journal of Hydrology*, Vol. 216, No. 1-2, 1999, pp. 32-55.

[16] K. P. Sudheer, P. C. Nayak and K. S. Ramasastri, "Improving Peak Flow Estimates in Artificial Neural Network River Flow Models," *Hydrological Processes*, Vol. 17, No. 3, 2003, pp. 677-686.

[17] M. J. Zurada, "An Introduction to Artificial Neural Systems," West Publishing Company, St Paul, 1997.

[18] G. J. Bowden, G. C. Dandy and H. R. Maier, "Input Determination for Neural Network Models in Water Resources Applications: 1. Background and Methodology," *Journal of Hydrology*, Vol. 301, No. 1-4, 2004, pp. 75-92.

[19] G. J. Bowden, G. C. Dandy and H. R. Maier, "Input Determination for Neural Network Models in Water Resources Applications: 2. Background and Methodology,"

Journal of Hydrology, Vol. 301, No. 1-4, 2004, pp. 93-107.

[20] K. C. Luk, J. E. Ball and A. Sharma, "A Study of Optimal Model Lag and Spatial Inputs to Artificial Neural Network for Rainfall Forecasting," *Journal of Hydrology*, Vol. 227, No. 1-4, 2000, pp. 56-65.

[21] D. Silverman and J. A. Dracup, "Artificial Neural Networks and Long-Range Precipitation Prediction in California," *Journal of Climate and Applied Meteorology*, Vol. 39, No. 1, 2000, pp. 57-66.

[22] H. R. Maier and G. C. Dandy, "Neural Networks for the Prediction and Forecasting of Water Resources Variables: A Review of Modeling Issues and Applications," *Environmental Modelling & Software*, Vol. 15, No. 1, 2000, pp. 101-124.

[23] R. K. Srivastav, K. P. Sudheer and I. Chaubey, "A Simplified Approach to Quantifying Predictive and Parametric Uncertainty in Artificial Neural Network Hydrologic Models," *Water Resources Research*, Vol. 43, No. 10, 2007, Article ID: W10407.

[24] J. E. Nash and J. V. Sutcliffe, "River Flow Forecasting through Conceptual Models: 1. A Discussion of Principles," *Journal of Hydrology*, Vol. 10, No. 3, 1970, pp. 282-290.

[25] P. C. Nayak, K. P. Sudheer, D. M. Rangan and K. S. Ramasastri, "Short-Term Flood Forecasting with a Neurofuzzy Model," *Water Resources Research*, Vol. 41, No. 4, 2005, Article ID: W04004.

[26] D. E. Rumelhart, G. E. Hinton and R. J. Williams, "Learning Representations by Back-Propagating Errors," *Nature*, Vol. 323, No. 6088, 1986, pp. 533-536.

[27] A. Y. Shamseldin, "Application of a Neural Network Technique to Rainfall-Runoff Modelling," *Journal of Hydrology*, Vol. 199, No. 3-4, 1997, pp. 272-294.

[28] P. Mittal, S. Chowdhury, S. Roy, N. Bhatia and R. Srivastav, "Dual Artificial Neural Network for Rainfall-Runoff Forecasting," *Journal of Water Resource and Protection*, Vol. 4, No. 12, 2012, pp. 1024-1028.

6

Simulation of a Daily Precipitation Time Series Using a Stochastic Model with Filtering

Chieko Gomi[1,2], Yasuhisa Kuzuha[1]

[1]Graduate School of Bioresources, Mie University, Tsu, Japan; [2]Aichi Prefectural Government, Nagoya, Japan.

ABSTRACT

After we modified raw data for anomalies, we conducted spectral analysis using the data. In the frequency, the spectrum is best described by a decaying exponential function. For this reason, stochastic models characterized by a spectrum attenuated according to a power law cannot be used to model precipitation anomaly. We introduced a new model, the e-model, which properly reproduces the spectrum of the precipitation anomaly. After using the data to infer the parameter values of the e-model, we used the e-model to generate synthetic daily precipitation time series. Comparison with recorded data shows a good agreement. This e-model resembles fractional Brown motion (fBm)/fractional Lévy motion (fLm), especially the spectral method. That is, we transform white noise X_t to the precipitation daily time series. Our analyses show that the frequency of extreme precipitation events is best described by a Lévy law and cannot be accounted with a Gaussian distribution.

Keywords: e-Model; Daily Precipitation Time Series; Filtering; Fractional Brownian Motion; Fractional Lévy Motion; Stochastic Model

1. Introduction

Just as turbulence and clouds have been described using (random) fractals, geoscientific fields such as topographical fields, temporal or spatial rainfall fields, and earthquake-slip fields are often modeled using fractals (Gagnon, et al. [1]; Lavallée and Archuleta [2]; Lavallée [3]; Lovejoy and Schertzer [4]; Schertzer and Lovejoy [5]; Tchiguirinskaia, et al. [6]). If we specifically examine simulations of temporal or spatial rainfall field as one example, then two approaches might be used (Over and Gupta [7]): stochastic approaches, or physical or dynamical approaches. Regarding the former, many stochastic models of temporal and spatial rainfall fields have been developed. According to Over and Gupta [7], a pioneering research effort for modeling of this type was that of LeCam [8]. Many stochastic models have been proposed, examples of which are the AR model (Brockwell and Davis [9]), ARMA model (Box and Jenkins [10]), NSRP model (Cowperhwait [11]), and the WGR model (Waymire, et al. [12]). Regarding precipitation models before 1990, which are not based on the fractal theory, see Valdes [13]. In recent twenty years, several mono-fractal or multifractal models have been used to model the scaling property of rainfall fields (Over and Gupta [7]; Lovejoy and Schertzer [4]). One of these multifractal models, the discrete cascade model is used to model rain data in Over and Gupta [7]. An alternative multifractal model, the continuous cascade model (e.g. Lovejoy and Schertzer [14]) is used to model rain fields in Wilson et al. [15].

As indicated above, geoscientific fields are often modeled by more physical or dynamical methods. Regarding precipitation, rainfall fields are often generated using meso-scale meteorological models or a global climate model. An important shortcoming of physical or dynamic approaches is their nature: they are time-consuming and resource-consuming. Simulation using stochastic approaches is based on random number generation. A simulation result by the latter is not deterministic but stochastic. Once one can generate random numbers, however, the random numbers must be transformed to the required data.

As described herein, we specifically examine the simulation of daily precipitation time series because simulation using daily precipitation is extremely important for flood design. For generating daily precipitation time series at specific observation station, we believe that the stochastic approach is superior to physical and dynamical approaches from the viewpoint of the cost of resources including time.

Although stochastic models of various kinds exist, models based on fractal theory or scaling theory present great advantages because of their capability of simply describing phenomena. Fractals were originally conceptualized by Mandelbrot [16]. Multifractals were proposed later (Frisch and Parisi [17]). Regarding the general theory of fractals and multifractals, see Feder [18] and Falconer [19]. For instance, Gagnon *et al.* [2] described the evolution from the (mono-) fractal model to the multifractal model, and explained the latter's superiority in terms of the topographical field. Lovejoy, Schertzer and others (e.g. [14]) have clarified that some kind of (temporal or spatial or both) rainfall data can be modeled by multifractal model. Their multifractal model is characterized by the terms "continuous cascades", "universal model", and "the fractional integrated flux model (FIF)", and so on. Hereinafter, we simply refer to their simulation model as "FIF". Because the FIF model is generally defined by the three parameters of α, C1, and H, it is necessary to estimate these three parameters before conducting simulations using the multifractal FIF. Usually, two steps exist for obtaining fractal parameters. The first is Fourier analyses. From results of these analyses, one can confirm whether or not the field is fractal field. To parameterize some multifractal models, it is possible to use the Double Trace Moment (DTM; Lavallée [20]) and Fourier analysis (it should be noted that the DTM has only be validated with discrete cascade model (personal communication, Daniel Lavallée, 2013)). Regarding mono-fractals, we will interpret a spectral method for fBm/fLm, which is one example for obtaining parameters or the nature of a mono-fractal, in Section 3. Using these steps, we can confirm the fractality of the field and obtain relevant parameters.

We have outlined the flow of our research here.
1) Construction of data: Raw data is strongly affected by seasonal and periodic changes. We modified the data and made our data for analyses using raw data.
2) Confirmation of fractality: Fourier analyses were conducted. The power spectra were obtained. If ω (angular frequency) versus $E(\omega)$ is log-log linear, then the data field is mono-fractal or multifractal.
3) If the data field is fractal, we can obtain three parameters of the universal model (multifractal), or try to apply fBm or fLm (a mono-fractal model). Subsequently, we must ascertain which model is appropriate.
4) If the data field is not fractal, then we must first seek the appropriate filter. Then we must construct a new model.

2. Precipitation Data

2.1. Raw Data

For precipitation observations, more than 150 manned observation stations (offices of the Japan Meteorological Agency) exist throughout Japan. Although their beginning times of operation mutually differ, only 51 manned stations have operated for more than 100 years. For this study, we used daily precipitation time series data (hereinafter, often designated simply as "daily precipitation") at these 51 manned stations during 1901-2011. At 10 stations, however, the data are not complete because of natural disasters and wars. In **Figure 1**, we portray these 10 stations using red (Naha), orange (Kure), and yellow (Miyako, Fukui, Yokohama, Tsuruga, Kofu, Hamamatsu, Kobe and Sakai) circles. Missing periods are, respectively 10, 2, and 1 year. Green circles show stations that have complete (111 yr) data.

2.2. Anomaly ΔR

We analyze the time series (which contains 365 data = 1 yr) of daily precipitation R. The total time series includes data of 111 years at most observation stations. First, we carry out the Fourier transform and investigate the relation between ω versus the power spectrum $E(\omega)$. As we described earlier, the relation is log-log linear: $E(\omega) = \omega^{-\beta}$ if a field of R is a fractal field. As one

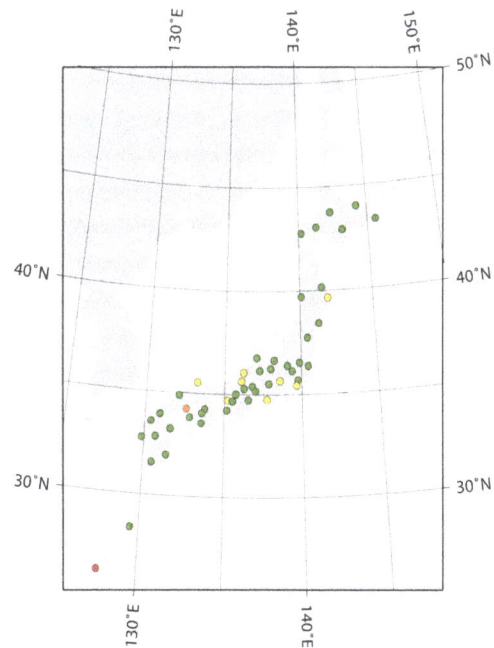

Figure 1. Locations of observation stations: observation stations having complete data obtained during 1901-2011, stations with 1 year of data were lost, stations with 2 years of data lost, and stations with 10 years of data lost are denoted respectively with green circles, yellow circles, orange circles, and red circles.

might imagine, however, the relation does not show log-log linearity because of various factors. As described later, we obtained a relation (an exponential function) other than the log-log linearity using anomaly. However, in reality, we were unable to obtain a smooth function if we used R itself.

Therefore, we conducted modification of the data as described to remove these factors in the following manner:

1) Assuming that the Tsu Observation Station has $m \times n$ daily precipitation data, where m is the year and n is the Julian day number, then we can use a variable $R(m,n)$. First, we calculate the average value of each day over a period of 111 years.

$$R(n) = \sum_{m=1901}^{2011} R(m,n) \Big/ 111 \quad (1 \leq n \leq 365) \quad (1)$$

2) Secondly, we conducted smoothing (low-pass filtering) the $R(n)$. We obtained five-day moving averages $\bar{R}(n)$ $(3 \leq n \leq 363)$. We obtained 361 values for Tsu Observation Station.

3) For each year with index m, we calculated anomaly $\Delta R(m,n) = R(m,n) - \bar{R}(n)$. So we have 111 time series available for analysis.

In **Figure 2**, black lines and red lines respectively represent $R(n)$ and $\bar{R}(n)$ (upper panel). The lower panel shows the anomaly time series of $\Delta R(m, n)$ in 2011 at Tsu Observation Station. We will analyze this $\Delta R(m, n)$ in the following sections.

3. Spectral Method for fBm/fLm

As we described in the previous chapter, first we carry out Fourier analyses to confirm the fractality of ΔR. If the field of ΔR has a fractal nature, then the relational Equation (2) holds between ω (angular frequency) and $E(\omega)$ (power spectrum).

$$E(\omega) = \left[F_S(\Delta R) \right]^2 \propto \omega^{-\beta} \quad (2)$$

In that equation, β is the scale exponent and $\omega = 2\pi(s-1)/N$, s is the discrete variable, N represents the number of data ΔR (=361 in this study), and F_s is the discrete Fourier transform.

The spectral method, discussed in Lavallée [3] can be used to generate Bm/fBm/fLm. See Saupe [21] for details and other methods. Lavallée's method is the following:

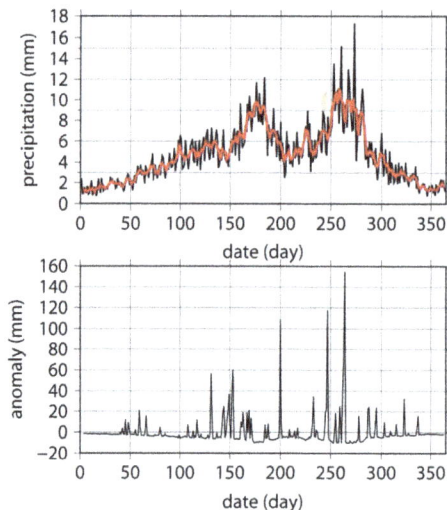

Figure 2. (Upper) $R(n)$ **(black line) and** $\bar{R}(n)$ **(red line); (Bottom) anomaly time series,** $\Delta R(m,n)$ **in 2011 at Tsu observation station.**

$$Y_t \propto \sum_{s=2-N/2}^{1+N/2} \left| \frac{\omega}{2\pi} \right|^{-\beta/2} F_s(X_t) \exp\left(\frac{-2\pi i (x-1)(s-1)}{N} \right) \quad (3)$$

In this equation, X_t represents independent and identically distributed random variables (white noise). Y_t shows the generated fBm/fLm data. If X_t is generated using the Gauss law, then Y_t is fBm. If X_t is generated using the Lévy law, then Y_t is fLm. Here, one can imagine an inverse procedure: for certain data Y_t, one might want to ascertain whether Y_t is fBm or fLm.

The following Equation (4) is an inverse equation of Equation (3) in which F_t^{-1} is the Fourier inverse.

$$X_t \propto F_t^{-1} \left[F_s(Y_t) \cdot \omega^{\beta/2} \right] \quad (4)$$

Particularly, we collect Y_t from a certain geoscientific field. After estimating β using Equation (2), we can obtain random variables X_t using Equation (4). Subsequently, we can confirm which probability density function (PDF) is appropriate for X_t. We assume that the candidate of the PDF is a Gaussian or Lévy distribution in this study.

The PDF of the Lévy distribution cannot be expressed explicitly. The characteristic function is the following (Robust Analysis Inc. [22]; Nolan [23]);

$$E \exp(iuX)$$

$$= \begin{cases} \exp\left(-\gamma^\alpha |u|^\alpha \left[1 + i\beta \left(\tan \frac{\pi\alpha}{2} \right)(\text{sign } u)\left(|\gamma u|^{1-\alpha} - 1 \right) \right] + i\delta u \right) & \alpha \neq 1 \\ \exp\left(-\gamma |u| \left[1 + i\beta \frac{2}{\pi}(\text{sign } u) \ln(\gamma |u|) \right] + i\delta u \right) & \alpha = 1 \end{cases} \quad (5)$$

Therein, X is a random variable, α is the tail index $(0 < \alpha \leq 2)$, β is a skewness parameter $(-1 \leq \beta \leq 1)$, γ is a scale parameter (positive) and δ is a location parameter (an arbitrary real number). The Gauss distribution is a special case of the Lévy distribution and $\alpha = 2$ for the Gauss distribution. Detailed explanations have been made by Nolan [23], Zolotarev [24] and Uchiaikin and Zolotarev [25].

4. Results

4.1. Spectral Analysis

First, we conducted spectral analysis using ΔR. In practice, ΔR is $\Delta R(m, n)$, and $3 \leq m$ 363 and $1901 \leq n \leq 2011$. Furthermore, N in Equations (3) is 361, and 111 $E(\omega)$ were averaged. **Figure 3** (lower) shows the relation between ω and $E(\omega)$ on a double logarithm chart. The relation does not show log-log linearity. **Figure 3** shows the relation at the Tsu Observation Station. However, a similar behavior has been observed for all the other observation stations considered in this study. These results suggest that stochastic process characterized by a spectrum attenuation given by a power law—for instance, fBm, fLm and other multifractal models (e.g. FIF)—cannot properly model the average spectrum $E(\omega)$ observed for the anomaly ΔR considered in this study.

Although the fractal model is inapplicable to daily precipitation, the procedure explained by Lavallée [3] is applicable to the data. **Figure 3** (upper) shows ω versus

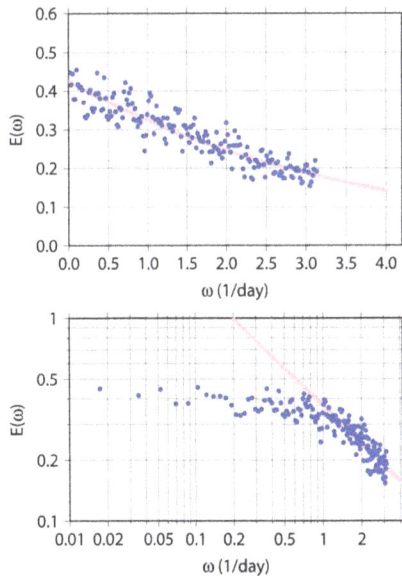

Figure 3. (Upper) Power spectra of ΔR in Tsu on the normal chart. The blue dots are power spectra of ΔR in Tsu. The red curve is a regression curve; (Bottom) The same figure as the upper panel, but on the log-log chart. Blue dots represent power spectra of ΔR. The red line is a linear regression line.

$E(\omega)$ on a normal chart. Not log-log linearity, but a certain relation is apparent. The relation is investigated in the following section.

4.2. Exponential Filter

We cannot generate ΔR using the fBm/fLm model. Therefore, we tried to derive a filter other than the filter for fractal model to simulate ΔR. We strove to apply the following Equation (6) for regression analysis. First, "a" and "b" are parameters which we must estimate.

$$E(\omega) = a \cdot \exp(-b \cdot \omega) \qquad (6)$$

We conducted regression analyses for all 51 observation stations. The range of correlation coefficient of ω versus $\log E(\omega)$ was -0.74 - -0.97 (-0.92 ± 0.04; mean $\pm$ standard deviation). Therefore, we conclude that Equation (6) is applicable as a regression curve of ω versus $E(\omega)$ of precipitation anomaly. Liu *et al.* [26] discussed a similar stochastic process but with a filter given by a von Karman function.

4.3. Probability Law

In the previous section, we derived a new filter instead of a log-log filter (Equation (6)). Therefore, Equation (4), which was discussed in the preceding section, is modified as shown below.

$$X_t \propto F_t^{-1}\left[F_s(\Delta R) \cdot [a \cdot \exp(-b \cdot \omega)]^{-1/2} \right] \qquad (7)$$

In that expression, ΔR is used instead of Y_t in Equation (4). Equation (7) shows that X_t is computed using ΔR. We tried to clarify X_t's nature, specifically, which of the Gauss law or Lévy law is appropriate for generating X_t. We compared histograms of X_t and PDF of the Gaussian and Lévy law. Parameters of the Gaussian law are estimated using the L-moment method (Hosking and Walllis [27]). Parameters of the Lévy law were estimated using three methods: maximum likelihood, quantile, and the empirical characteristic function. In practice, we used Nolan's software [22,23]. **Figure 4** (upper) presents a histogram of X_t (green circles) and PDFs of the Gaussian law (blue solid line) and Lévy (dashed line) in Tsu. Regarding **Figure 4**, parameters of the Lévy law were calculated using the maximum likelihood function. According to **Figure 4** (upper), the Lévy law is more appropriate than the Gaussian law. Furthermore, we focused on the PDF tail, as earlier studies have done: Lavallée [3]. There is an additional justification to focus on the tail of the distribution. In Geophysics, extreme but less frequent events are the ones mainly responsible for significantly perturbing the system under consideration and the potential cause of much damage to nature and society. Properly quantifying the frequency of these extreme events is

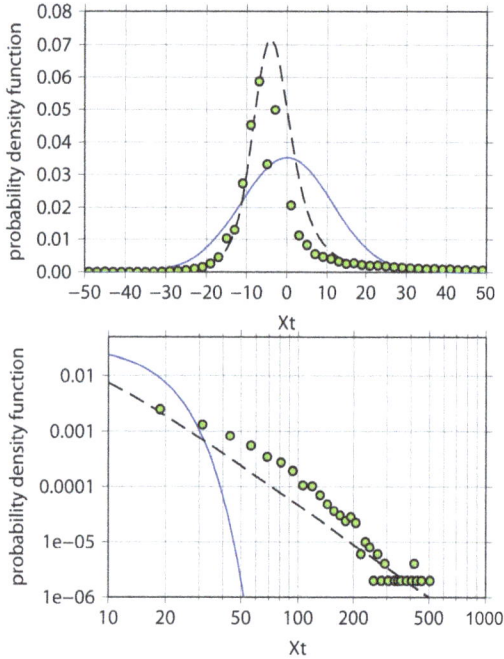

Figure 4. (Upper) A histogram of X_t (green dots), a PDF of the Gaussian law (blue curve) and the Lévy PDF (dashed curve). The histogram size has been reduced for comparison to PDF. Therefore, to be precise, the histogram is PDF; (Bottom) The same figure as shown in the upper panel, but on the double logarithm chart. The curve of the Lévy PDF (dashed curve) provides a better fit to the tail of the histogram (PDF curves) of X_t.

a potential outcome of the method discussed in the paper (personal communication, Daniel Lavallée, 2013).

Figure 4 (bottom) is the same graph as that shown in the upper panel, but on double logarithm graph to emphasize the tail's nature. The lower panel indicates the superiority of Lévy's law more clearly than the upper panel: the lower panel clearly shows that the Lévy law provide a better description of the extreme events located in the tail of the PDF of X_t. X_t is clearly related to daily precipitation. We intend to emphasize the reproducibility of extreme daily precipitation. Therefore, it is concluded that the Lévy's law is more appropriate.

We can quantitatively estimate goodness-of-fit of the PDFs of Lévy's law. We calculated the correlation coefficient between the histogram and PDFs of Lévy's law for 51 observation stations. Ranges of coefficients of correlation of normal graph were 0.86 ± 0.08 (1.00 ± 0.02), 0.80 ± 0.07 (0.92 ± 0.04) and 0.86 ± 0.06 (1.01 ± 0.04) by the three methods described above (numbers in the parentheses are slopes of the regression lines). Sufficiently high correlation and slopes that are almost unity

were shown for all 51 stations for three methods. Regarding tail characteristics (double logarithm graph), similar results were obtained. Furthermore, these analyses demonstrate that Lévy's law is quantitatively superior to the Gaussian law.

Finally, we show four parameters for each observation station (**Table 1**). The ranges of the most important parameter α are 1.22 ± 0.24 (mean $\pm$ standard deviation), 1.03 ± 0.04 and 1.17 ± 0.06, respectively, as obtained using the maximum likelihood, the quantile and the empirical characteristic function ("Method 1", "Method 2" and "Method 3", respectively, in **Table 1**).

5. Simulation

In the previous section, we show that the Lévy law properly describes the distribution of X_t—especially extreme values. In this section we discussed how to use the e-model and Lévy law to simulate daily precipitation. This e-model resembles fBm/fLm, especially the spectral method, but it is not characterized by spectrum attenuation given by a power law but the spectrum is best described by a decaying exponential function.

5.1. Lévy Random Number

An expression for synthetic precipitation anomaly can be obtained by modifying Equation (3) in the following way (Equation (8)): the power law filter is replaced by the exponential filter given in Equation (6).

Therein, the random number X_t is generated using the Lévy law. Four parameters of the distribution for each observation station were obtained as explained in the preceding section. $\Delta R'$ is a generated precipitation anomaly. We intend to reproduce ΔR (observed anomaly) using Equation (8). We use the truncated Lévy law (Lavallée [3]) in practice: overly large or small values of X_t, which are outside of a certain allowable range, are removed. The allowable range is found using actual precipitation data.

5.2. Proportionality Factor of Generated ΔR

We conducted spectral analysis of the generated $\Delta R'$. Results show that the spectral $E(\omega)$ of $\Delta R'$ is much smaller than that of ΔR. Here, it is noteworthy that an equality is "$\propto$" in Equation (3) but "$=$" in Equation (8), and that ΔR is an observed real value and that $\Delta R'$ is a generated value which is expected to reproduce ΔR. Theoretically, $\Delta R'$ is expected to be proportional to ΔR (in the statistical sense). In fact, the cause of disagreement is apparently differences of two "a" values. How-

$$\Delta R' = \sum_{s=2-N/2}^{1+N/2} \left| a \cdot \exp(-b \cdot \omega) \right|^{1/2} F_s(X_t) \exp\left(\frac{-2\pi i (x-1)(s-1)}{N} \right) \tag{8}$$

Table 1. Estimated parameters of the Lévy law. Regarding the number of methods, see the text. Each parameter value is shown as the mean ± standard deviation.

Method	α	β	γ	δ
1	1.22 ± 0.24	0.18 ± 0.43	3.45 ± 0.79	-3.25 ± 1.14
2	1.03 ± 0.04	0.39 ± 0.08	3.15 ± 0.32	-2.61 ± 0.52
3	1.17 ± 0.06	-0.27 ± 0.17	3.36 ± 0.41	3.13 ± 0.54

ever, as for "b" in regression curves, both values of b well coincide.

Then we assume the following equation.

$$\Delta R \approx k \cdot \Delta R' \qquad (9)$$

In that equation, k is a proportional factor that stands as a correction coefficient of "a" to reproduce ΔR. We obtained correction factor k for $E(\omega)$ of $\Delta R'$ to coincide with $E(\omega)$ of ΔR. The distribution of k obtained from 51 observations was 2.44 ± 0.43. After we estimated the correction factors "k", we conducted spectral analysis of $k \cdot \Delta R'$. The $\omega - E(\omega)$ relation of $k \cdot \Delta R'$ well coincides with that of observed ΔR. We conclude that we are able to simulate ΔR using our e-model including the k-correction procedure.

5.3. Correction of Negative Values

For this study, we generated 300 sets of $\Delta R'$. Then, the simulated daily precipitation (R') is calculable as $R' = \bar{R} + k \cdot \Delta R'$. **Figure 5** depicts a chart of the interannual change of the simulated $R' = \bar{R} + k \cdot \Delta R'$ (blue line) and observed R in 2011 (red line) at Tsu Observation Station located in Mie. According to **Figure 5**, their characteristics are comparable. However, R' shows a negative value on some days. Furthermore, we defined the new variable R'' as shown below.

$$R'' = \begin{cases} R' & \text{if } R' \geq 0 \\ 0 & \text{if } R' \text{ is negative} \end{cases} \qquad (10)$$

In this case, the simulated anomaly ($\Delta R''$) is estimated as $\Delta R'' = R'' - \bar{R}$. We conducted spectral analysis of $\Delta R''$. Then we compared the spectra of $\Delta R''$ and ΔR from the viewpoint of the $\omega - E(\omega)$ relation.

Results (**Figure 6**) show that $E(\omega)$ of $\Delta R''$ is almost identical to that of ΔR for 51 observation stations: removing negative values from simulated time series does not affect the result.

6. Conclusions

We strove to apply a stochastic model to daily precipitation time series recorded at 51 observation stations in Japan. Fourier analysis of precipitation anomaly suggests that the spectrum of precipitation anomaly is attenuated as an exponential function. Thus stochastic model of the fractal variety with a spectrum characterized by a spec-

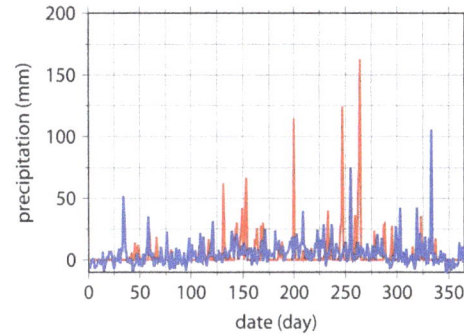

Figure 5. Interannual change of precipitation R' (blue line) generated using our method and parameters obtained in the preceding section, and actual precipitation R (red line) in 2011 at the Tsu Observation Station. Both lines show similar characteristics. However, R' (blue line) has some negative values.

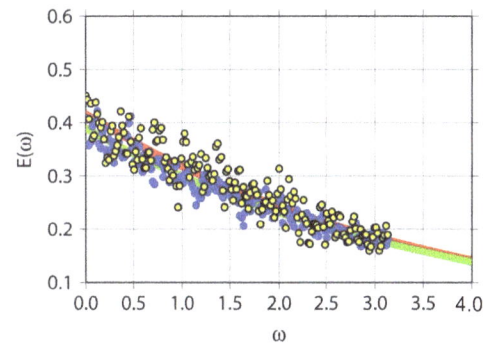

Figure 6. The same figure as Figure 3 (upper). Yellow dots represent power spectra of ΔR in Tsu. The red line is a linear regression line. The Blue dots and green line are for $\Delta R''$.

trum following a power law must be disregarded. To model precipitation anomaly, we propose a new stochastic model: the e-model. Our findings are the following.

1) The e-model closely resembles fBm/fLm, but an exponential type filter is applicable to our data: anomaly time series data.

2) The e-model use an exponential type of filter: $E(\omega) = a \cdot \exp(-b \cdot \omega)$.

3) We generated daily precipitation time series using the e-model. The model generates some negative values. Therefore, we modify the negative values to zero values. This modification did not affect the reproducibility of the relation between ω and $E(\omega)$.

We emphasize the reproducibility of extreme daily precipitation. The e-model with Lévy's law can reproduce extreme daily precipitation. We intend to search for other phenomena to which the e-model and Lévy's law can be applied and which are predicted by the model.

7. Acknowledgements

We are grateful for Dr. Daniel Lavallée's many kind and

valuable comments. This research was supported by scientific research funds (No. 21560539 and No. 23360244) from the Japanese Ministry of Education, Culture, Sports, Science and Technology (MEXT) and the "Precise Impact Assessments on Climate Change" of the Program for Risk Information on Climate Change (SOUSEI Program), supported by MEXT.

REFERENCES

[1] J. S. Gagnon, S. Lovejoy and D. Schertzer, "Multifractal Earth Topography," *Nonlinear Processes in Geophysics*, Vol. 13, No. 5, 2006, pp. 541-570.

[2] D. Lavallée and R. J. Archuleta, "Stochastic Modeling of Slip Spatial Complexities for the 1979 Imperial Valley, California, Earthquake," *Geophysical Research Letters*, Vol. 30, No. 5, 2003, p. 1245.

[3] D. Lavallée, "On the Random Nature of Earthquake Sources and Ground Motions: A United Theory," *Advances in Geophysics*, Vol. 50, 2008, pp. 427-461.

[4] S. Lovejoy and D. Schertzer, "Multifractals and Rain. New Uncertainty Concepts," In: A. W. Kundzewicz, Ed., *Hydrology and Hydrological Modelling*, Cambridge Press, Cambridge, 1995, pp. 62-103.

[5] D. Schertzer and S. Lovejoy, "Physical Modeling and Analysis of Rain and Clouds by Anisotropic Scaling Multiplicative Processes," *Journal of Geophysics Research*, Vol. 92, No. D8, 1987, pp. 9692-9714.

[6] I. Tchiguirinskaia, S. Lu, F. J. Molz, T. M. Williams and D. Lavallée, "Multifractal versus Monofractal Analysis of Wetland Topography," *Stochastic Environmental Research and Risk Assessment*, Vol. 14, No. 1, 2000, pp. 8-32.

[7] T. M. Over and V. K. Gupta, "Statistical Analysis of Mesoscale Rainfall: Dependence of a Random Cascade Generator on Large-Scale Forcing," *Journal of Applied Meteorology*, Vol. 33, No. 12, 1994, pp. 1526-1542.

[8] L. LeCam, "A Stochastic Description of Precipitation," *4th Berkeley Symposium on Mathematical Statistics and Probability*, Statistical Laboratory of the University of California, Berkeley, 20 June-30 July 1960, pp. 165-186.

[9] P. J. Brockwell and R. A. Davis, "Time Series: Theory and Methods," 2nd Edition, Springer-Verlag, New York, 1991.

[10] G. Box and G. M. Jenkins, "Time Series Analysis: Forecasting and Control," 2nd Edition, Holden-Day, San Francisco, 1976.

[11] P. S. P. Cowperthwait, "Further Developments of the Neyman-Scott Clustered Point Process for Modelling Rainfall," *Water Resources Research*, Vol. 27, No. 7, 1991, pp. 1431-1438.

[12] E. C. Waymire, V. K. Gupta and I. Rodriguez-Iturbe, "A Spectral Theory of Rainfall Intensity at the Meso-β Scale," *Water Resources Research*, Vol. 20, No. 10, 1984, pp. 1453-1465.

[13] J. B. Valdes, "Issues in the Modelling of Precipitation," *Stochastic Hydrology and Its Use in Water Resources Systems Simulation and Optimization*, NATO ASI Series Vol. 237, 1993, pp. 217-220.

[14] S. Lovejoy and D. Schertzer, "Scale, Scaling and Multifractals in Geophysics: Twenty Years on," In: A. A. Tsonis and J. Elsner, Eds., *Nonlinear Dynamics in Geosciences*, Springer, New York, 2007, pp. 311-337.

[15] J. Wilson, D. Shertzer and S. Lovejoy, "Continuous Multiplicative Cascade Models of Rain and Clouds," In: D. Schertzer and S. Lovejoy, Eds., *Non-Linear Variability in Geophysics*, Kluwer, Dordrecht, 1991, pp. 185-207.

[16] B. B. Mandelbrot, "Fractal Geometry in Nature," W. H. Freeman and Company, San Francisco, 1982.

[17] U. Frisch and G. Parisi, "Turbulence and Predictability of Geophysical Flows and Climate Dynamics," Varenna Summer School LXXXVIII, 1983.

[18] J. Feder, "Fractals (Physics of Solids and Liquids)," Springer, New York, 1988.

[19] K. Falconer, "Fractal Geometry: Mathematical Foundations and Applications," Wiley, Chichester, 2003.

[20] D. Lavallée, "Multifractal Analysis and Simulation Techniques and Turbulent Fields," Ph.D. Dissertation, McGill University, Montreal, 1991.

[21] D. Saupe, "Algorithms for Random Fractals," In: H.-O. Peitgen and D. Saupe, Eds., *The Science of Fractal Images*, Springer, New York, 1998, pp. 71-136.

[22] Robust Analysis Inc., "User Manual for STABLE 5.3 C Library Version," Robust Analysis Inc., Takoma Park, 2012.

[23] J. P. Nolan, "Stable Distributions—Models for Heavy Tailed Data, Chapter 1 (Online Version)," Birkhauser, Boston, 2013.

[24] V. M. Zolotarev, "One-Dimensional Stable Distributions," American Mathematical Society, Providence, 1986.

[25] V. V. Uchiaikin and V. M. Zolotarev, "Chance and Stability," VSP International Science Publishers, Utrecht, 1999.

[26] P. Liu, R. Archuleta and S. Hartzell, "Prediction of Broadband Ground-Motion Time Histories: Hybrid Low/High-Frequency Method with Correlated Random Source Parameters," *Bulletin of the Seismological Society of America*, Vol. 96, No. 6, 2006, pp. 2118-2130.

[27] J. R. M. Hosking and J. R. Wallis, "Regional Frequency Analysis: An Approach Based on L-Moments," Cambridge University Press, Cambridge, 1997.

7

HEC-RAS Model for Mannnig's Roughness: A Case Study

Prabeer Kumar Parhi

Center for Water Engineering and Management, Central University of Jharkhand, Ranchi, India.

ABSTRACT

Channel roughness is considered as the most sensitive parameter in development of hydraulic models for flood fore-casting and flood inundation mapping. Hence, it is essential to calibrate the channel roughness coefficient (Mannnig's "n" value) for various river reaches through simulation of floods. In the present study it is attempted to calibrate and validate Mannnig's "n" value using HEC-RAS for Mahanadi River in Odisha (India). For calibration of Mannnig's "n" value, the floods for the years 2001 and 2003 have been considered. The calibrated model, in terms of channel rough-ness, has been used to simulate the flood for year 2006 in the same river reach. The performance of the calibrated and validated HEC-RAS based model has been tested using Nash and Sutcliffe efficiency. It is concluded from the simula-tion study that optimum Mannnig's "n" value that can be used effectively for Khairmal to Barmul reach of Mahanadi River is 0.029. It is also verified that the peak flood discharge and time to reach peak value computed using Mannnig's "n" of 0.029 showed only an error of 5.42% as compared with the observed flood data of year 2006.

Keywords: Hydrodynamic Model; Flood Simulation; Flood Forecasting; HEC-RAS; River Mahanadi

1. Introduction

For flood forecasting, flood plane mapping and flood volume estimation, various hydrodynamic models, based on hydraulic routing, have been developed and applied to different rivers in the past using computer technology and numerical techniques. For flood warning, the dis-charge and river stage were chosen as the variables, which along with other hydraulic properties are interre-lated to each other [1]. Among various hydraulic pa-rameters, the channel roughness plays very important role in the study of open channel flow particularly in hydraulic modeling. Channel roughness is highly vari-able which depends upon number of factors like surface roughness, vegetation cover, channel irregularities, chan-nel alignment etc. [2]. It also depends on such factors as: bed material, vegetation, channel irregularity and align-ment, scour and deposition, obstructions, channel size and shape, stage and discharge, seasonal changes, sus-pended material and bed load [3].

Earlier, good numbers of researchers including Patro et al. [4], Usul and Turan [5], Vijay et al. [6], Parhi et al. [7] and Wasantha Lal [8] have calibrated channel rough-ness for different rivers for the development of hydraulic model for flood forecasting and flood plane mapping. Ramesh et al. [2] estimated single channel roughness

value for open channel flow using optimization method, taking the boundary condition as constraints. Timbadiya et al. [9] calibrated channel roughness for Lower Tapi River, India using HEC-RAS model. Ross Doherty [10] calibrated the channel roughness for large number of semiarid rivers of Western Australia having variable channel characteristics for development of rating curves.

In the above context, there is a need to calibrate the channel roughness coefficient (Mannnig's "n" value) for the River Mahanadi, Odisha through simulation of floods, using HEC-RAS. It will be pertinent to mention that the river Mahanadi experiences severe floods frequently causing huge loss to life and property. Hence the present study attempts to accurately estimate the channel rough-ness of the upstream reach of river Mahanadi beyond Hirakud reservoir from Khairmal to Barmul gauging sta-tions.

2. Model Description

In the present study, unsteady, gradually varied flow simulation model, which is dependent on finite differ-ence solutions of the Saint-Venant equations (Equations (1) and (2), has been used to simulate the flood in the Mahanadi River. Here HEC-RAS has been used to per-form one dimensional hydraulic calculation for full network of natural and constructed channels [3].

$$\frac{\partial A}{\partial t} + \frac{\partial Q}{\partial x} = 0 \qquad (1)$$

$$\frac{\partial Q}{\partial t} + \frac{\partial \left(Q^2/A\right)}{\partial x} + gA\frac{\partial H}{\partial x} gA\left(S_0 - S_f\right) = 0 \qquad (2)$$

where A = cross-sectional area normal to the flow; Q = discharge; g = acceleration due to gravity; H = elevation of the water surface above a specified datum, also called stage; S_0 = bed slope; S_f = energy slope; t = temporal coordinate and x = longitudinal coordinate. Equations (1) and (2) are solved using the well known four-point implicit box finite difference scheme [11].

3. Study Reach

In the context of flood scenario, the Mahanadi system can be broadly divided into two distinct reaches: 1) Upper Mahanadi (area upstream of Mundili barrage, intercepting a catchment of 132,100 sq·km), which does not have any significant flood problem 2) Lower Mahanadi (area downstream of Mundili barrage, intercepting a catchment of 9304 sq·km). The key area downstream of Hirakud up to Munduli intercepting a catchment of 48,700 sq·km is mainly responsible for flood havoc in the deltaic area of Mahanadi [12]. **Figure 1** shows the details of catchments of Mahanadi Basin inside and outside of Odisha. In the present study, river reach in the Mahanadi system extending over a length of 106 km from Khairmal to Barmul is considered for analysis.

4. Geometric and Hydrologic Data

The channel geometry, upstream and downstream bound-

ary conditions and channel resistance are required for conducting flow simulation through HEC-RAS. The cross-section data at 8 to 10 Kilometer intervals from Khairmal to Barmul extending over a length of 106 km were collected from the Department of Water Resources Odisha. The cross section data of the down stream catchment of Hirakud reservoir used for the present analysis was collected from the Department of Water Resources, Odisha, which was surveyed during 1997-1998 by Department of Water Resources, Odisha, for dam break analysis of Hirakud Dam and preparation of emergency action plan The flood hydrograph at Khairmal and the friction slope of the reach have been considered as up-stream and downstream boundary conditions respectively. The flood hydrograph at Barmul has been used for validation of the model.

5. Calibration of HEC-RAS Model for Manning's Roughness Coefficient "n"

The data pertaining to the floods for years 2001 and 2003 have been used for calibration of Manning's roughness coefficient "n". In the present study, effort has been made to calibrate Manning's roughness coefficient for single value using aforesaid data and subsequently, different values of "n" (from 0.04 to 0.025) have been used to justify their adequacy for simulation of flood in the study reach along the channel. Nash and Sutcliffe efficiency test [13] has been used for comparison of simulated flow hydrograph (computed using different Manning's roughness coefficient "n") with the observed flow hydrograph at Barmul gauging station where gauge discharge data is available. **Table 1** shows the flood year, flow duration, name of gauging station and various single values of "n"

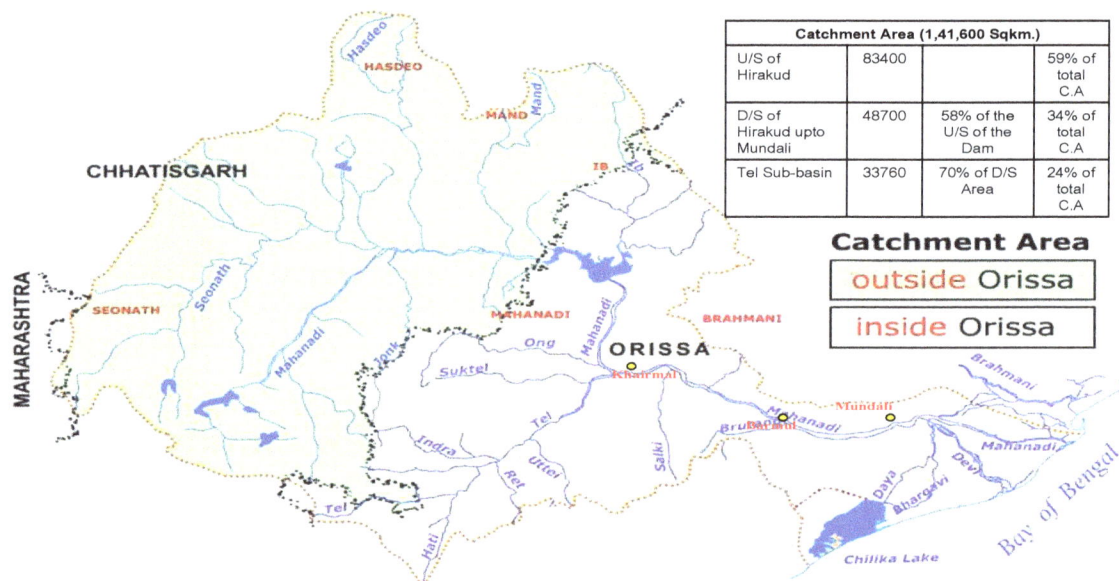

Catchment Area (1,41,600 Sqkm.)		
U/S of Hirakud	83400		59% of total C.A
D/S of Hirakud upto Mundali	48700	58% of the U/S of the Dam	34% of total C.A
Tel Sub-basin	33760	70% of D/S Area	24% of total C.A

Figure 1. Details of catchments of Mahanadi system inside and outside of Odisha.

(from 0.04 to 0.025) used for model calibration. The comparison of observed and simulated flow hydrograph (calibration) at Barmul gauging station for Manning's "*n*" value of 0.028 is also shown in **Figure 2**. From **Table 1** it is clearly visible that for the flood of the year 2001 Manning's "*n*" value of 0.03 yields maximum efficiency of 88.61 and that of 0.028 yields maximum efficiency of 89.21 for the flood year 2003.

6. Performance of Calibrated Model in Simulation of Flood for Year 2006

Taking the mean of the optimum Manning's "*n*" values estimated for the flood years of 2001 and 2003, as 0.029 for the focus reach, HEC-RAS based model has been used to simulate the flood for year 2006. It is found from the simulation that Manning's "*n*" value of 0.029 yields the maximum Nash and Sutcliffe efficiency of 92.39. **Table 2** shows the simulated flood hydrograph at Barmul gauging station for various Manning's "*n*" values. The

comparison of observed and simulated flow hydrograph at Barmul gauging station for Manning's "*n*" value of 0.029 is also shown in **Figure 3**.

Further, considering Manning's "*n*" value as 0.029, the flood peak and time to peak for the flood year 2006 is computed and it is observed that there is a close agreement between the observed and computed values. **Table 3** shows the comparison between the observed and computed values of the flood peak and time to peak for the flood year 2006 for different values of Manning's "*n*". It is clearly visible from **Table 3** that the flood peak and time to peak estimated using Manning's "*n*" value as 0.029 shows minimum percentage error.

In the above context, it shall be pertinent to mention that the Manning's "*n*" value as detailed by Chow [14] lies between 0.025 to 0.035 for flood planes having short grasses and also for straight clean having no deep pools, which shows close resemblance with the channel characteristics of focus reach (Khairmal to Barmul) of Mahanadi River.

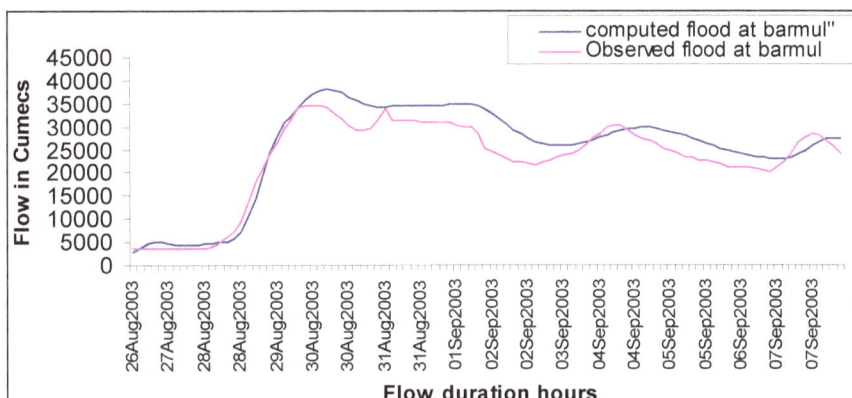

Figure 2. Observed and simulated flow hydrograph at Barmul (calibration) for flood year 2003 using Manning's "*n*" value of 0.028.

Table 1. Flow year, simulation duration, Manning's "*n*" and gauge station used for calibration.

Flow year	Simulation duration	Roughness coefficient Manning's "*n*"	Nash and Sutcliffe efficiency	Guage station used for calibration
2001	July 14, 00:00 to July 26, 09:00	0.04	84.23	Barmul (calibration)
		0.035	85.68	
		0.03	88.61	
		0.028	88.53	
		0.025	88.01	
2003	Aug. 27, 00:00 to Sep. 8, 09:00	0.04	87.15	Barmul (calibration)
		0.035	87.88	
		0.03	88.53	
		0.028	89.21	
		0.025	89	

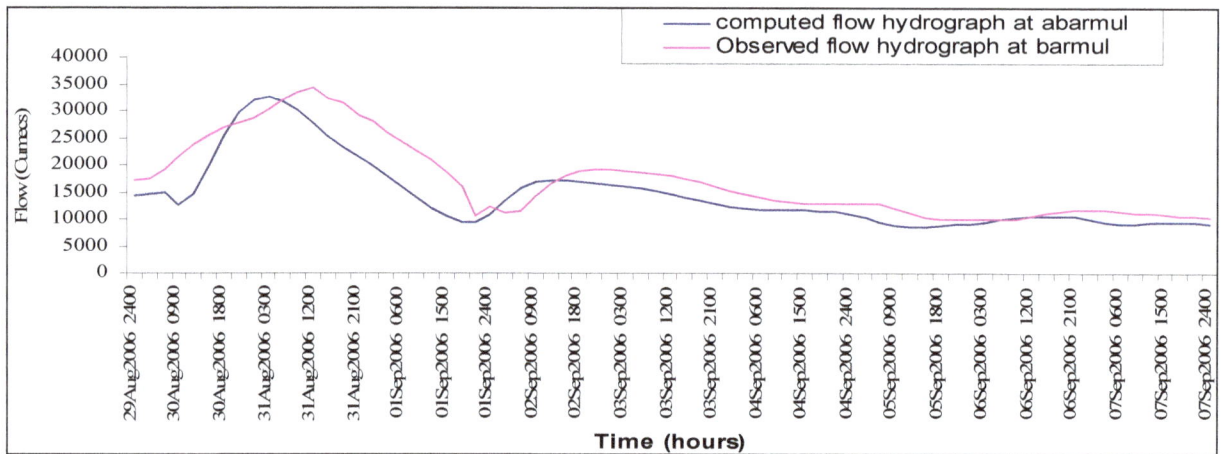

Figure 3. Observed and simulated flow hydrograph at Barmul (validation) for flood year 2006 for Manning's "*n*" value of 0.029.

Table 2. Simulation period, Manning's "*n*" and gauge station used for validation at Barmul.

Flow year	Simulation period	Roughness coefficient Manning's "*n*"	Nash and sutcliffe efficiency
2006	Aug. 30, 00:00 to Sep. 4, 09:00	0.029	92.39
		0.035	92.02
		0.025	92.32
		0.02	92.12

Table 3. Observed and computed values of flood peak and time to peak for different values of Manning's "*n*".

Flow year	Simulation duration	Manning's "*n*"	Observed		Computed		% Error
			Flood peak (cumecs)	Time to peak	Flood peak (cumecs)	Time to peak	
2006	Aug. 30, 00:00 to Sep. 4, 09:00	0.029	34,368	Aug. 31, 12:00	32,505	Aug. 31, 06:00	5.42
		0.035	34,368	Aug. 31, 12:00	32,194	Aug. 31, 03:00	6.33
		0.025	34,368	Aug. 31, 12:00	31,969	Aug. 31, 06:00	6.98
		0.02	34,368	Aug. 31, 12:00	31,193	Aug. 31, 03:00	6.31

7. Conclusions

Based on the simulation study carried out for the down stream catchment of Hirakud Reservoir of Mahanadi River (Khairmal to Barmul reach) following conclusions can be summarized:

1) The most effective single Manning's roughness coefficient calibrated (on flood data of the years 2001 and 2003) and validated (on flood data of the year 2006) for the Khairmal to Barmul reach of the Mahanadi River comes out to be 0.029.

2) The performance of calibrated and validated model has been verified using Nash and Sutcliffe (1970). A close agreement (92.39% efficiency) is seen between the simulated and observed flows at Barmul gauging station.

3) Furthermore, the calibrated Manning's roughness co-

efficient of 0.029 also works best for the estimation of flood discharge peak and time to peak at Barmul reach of the Mahanadi River, as these values can be computed only with an error of 5.42% (compared with the observed flood data of the year 2006).

4) The calibrated Manning's roughness coefficient value of 0.029 for the focus reach between Khairmal to Barmul of Mahanadi River having short grasses, straight and no deep pools can be further supported by the "*n*" value detailed by Chow [14] for flood planes having similar channel characteristics as above.

REFERENCES

[1]   W.-M. Bao, X.-Q. Zhang and S.-M. Qu, "Dynamic Correction of Roughness in the Hydrodynamic Model," *Journal of Hydrodynamics*, Vol. 21, No. 2, 2009, pp. 255-263.

[2] R. Ramesh, B. Datta, S. Bhallamudi and A. Narayana, "Optimal Estimation of Roughness in Open-Channel Flows," *Journal of Hydraulic Engineering*, Vol. 126, No. 4, 1997, pp. 299-303.

[3] HEC-RAS, "User Manual," US Army Corps of Engineers, Hydrologic Engineering Center, Davis Version 4.0, 2008.

[4] S. Patro, C. Chatterjee, S. Mohanty, R. Singh and N. S. Raghuwanshi, "Flood Inundation Modeling Using Mike Flood and Remote Sensing Data," *Journal of the Indian Society of Remote Sensing*, Vol. 37, No. 1, 2009, pp. 107-118.

[5] N. Usul and T. Burak, "Flood Forecasting and Analysis within the Ulus Basin, Turkey, Using Geographic Information Systems," *Natural Hazards*, Vol. 39, No. 2, 2006, pp. 213-229.

[6] R. Vijay, A. Sargoankar and A. Gupta, "Hydrodynamic Simulation of River Yamuna for Riverbed Assessment: A Case Study of Delhi Region," *Environmental Monitoring Assessment*, Vol. 130, No. 1-3, 2007, pp. 381-387.

[7] P. K. Parhi, R. N. Sankhua and G. P. Roy, "Calibration of Channel Roughness of Mahanadi River (India) Using HEC-RAS Model," *Journal of Water Resources and Protection*, Vol. 4, No. 10, 2012, pp. 847-850.

[8] A. M. Wasantha Lal, "Calibration of Riverbed Roughness," *Journal of Hydraulic Engineering*, Vol. 121, No. 9, 1995, pp. 664-671.

[9] P. V. Timbadiya, P. L. Patel and P. D. Porey, "Calibration of HEC-RAS Model on Prediction of Flood for Lower Tapi River, India," *Journal of Water Resources and Protection*, Vol. 3, No. 11, 2011, pp. 805-811.

[10] R. Doherty, "Calibration of HEC-RAS Models for Rating Curve Development in Semi Arid Regions of Western Australia," *AHA* 2010 *Conference*, Perth, 2010.

[11] HEC-RAS, "Hydraulic Reference Manual," US Army Corps of Engineers, Hydrologic Engineering Center, Davis Version 4.0, 2008.

[12] P. K. Mishra and S. Behera, "Flood Management Planning in the Mahanadi River Basin Odisha," 7th *International R&D Conference*, Bhubaneswar, 4-6 February 2009, pp. 149-150.

[13] J. E. Nash and J. V. Sutcliffe, "River Flow Forecasting through Conceptual Models, Part I—A Discussion of Principles," *Journal of Hydrology*, Vol. 10, No. 3, 1970, pp. 282-290.

[14] V. T. Chow, "Open Channel Hydraulics," McGraw Hill Book Company, New York, 1959.

8

Effect of Terrestrial LiDAR Point Sampling Density in Ephemeral Gully Characterization

Henrique G. Momm[1]*, Ronald L. Bingner[2], Robert R. Wells[2], Seth M. Dabney[2], Lyle D. Frees[3]

[1]Department of Geosciences, Middle Tennessee State University, Murfreesboro, USA; [2]National Sedimentation Laboratory, United States Department of Agriculture—Agricultural Research Service, Oxford, USA; [3]United States Department of Agriculture—Natural Resources Conservation Service, Salina, USA.

ABSTRACT

Gully erosion can account for significant volumes of sediment exiting agricultural landscapes, but is difficult to monitor and quantify its evolution with traditional surveying technology. Scientific investigations of gullies depend on accurate and detailed topographic information to understand and evaluate the complex interactions between field topography and gully evolution. Detailed terrain representations can be produced by new technologies such as terrestrial LiDAR systems. These systems are capable of collecting information with a wide range of ground point sampling densities as a result of operator controlled factors. Increasing point density results in richer datasets at a cost of increased time needed to complete field surveys. In large research watersheds, with hundreds of sites being monitored, data collection can become costly and time consuming. In this study, the effect of point sampling density on the capability to collect topographic information was investigated at individual gully scale. This was performed through the utilization of semi-variograms to produce overall guiding principles for multi-temporal gully surveys based on various levels of laser sampling points and relief variation (low, moderate, and high). Results indicated the existence of a point sampling density threshold that produces little or no additional topographic information when exceeded. A reduced dataset was created using the density thresholds and compared to the original dataset with no major discrepancy. Although variations in relief and soil roughness can lead to different point sampling density requirements, the outcome of this study serves as practical guidance for future field surveys of gully evolution and erosion.

Keywords: Ephemeral Gully; Ground-Based LiDAR; Soil Erosion; Point Sampling Density; Remote Sensing

1. Introduction

In agricultural fields, gully erosion is significant and often similar to or exceeding sheet and rill erosion volume. A large number of modeling tools have been developed over the years to estimate sediment transport from agricultural fields to streams and lakes [1-3]. These tools play an important role in assessing existing and planned conservation practices and are accepted by various regulatory and management agencies such as the United States Environmental Protection Agency (EPA) and the United States Natural Resource Conservation Service (NRCS). However, at the current stage of development, sediment loss estimation from gullies is either limited or neglected. Unlike sheet and rill erosion, which occurs as a result of the impact of raindrops and water flowing on the soil surface, gully erosion in agricultural fields occurs as a result of concentrated flow of surface runoff along a defined channel, and also by subsurface flow by seepage and flow through preferential pathways [4]. Gullies in agriculture fields are often classified into either ephemeral or classical gullies.

Ephemeral gullies are defined as small channels located in agricultural fields eroded primarily from concentrated overland flow that can be easily filled by normal tillage, only to reform again in the same location by additional runoff events [5]. Due to their small dimensions, producers often reshape the channel's topography, by refilling the channel through tillage, to maintain regular farming operations [6]. Because the field topography is often unchanged between seasons, ephemeral gullies have a tendency to re-form at the same or nearby location [7]. An ephemeral gully becomes permanent, referred to as classic gully, in situations where a headcut migrates upstream faster than the time interval between farmers tilling operations, followed by widening of the channel and which forces producers to operate around the gully channel.

This dynamic behavior poses a challenge in the understanding and estimation of the soil erosion of ephemeral

*Corresponding author.

gullies in agricultural watersheds [7]. Studies designed to understand ephemeral gully formation and development typically use Digital Elevation Models (DEM) as the basis of formulations explaining the relationship between field topography and ephemeral gully occurrence [8-10]. Despite the availability of DEMs at regional and local scales (spatial resolution ranging from 1 to 30 meters), these datasets often do not offer the necessary spatial resolution to quantify gullies at field scales. In order to capture the micro-topography impacting ephemeral gully formation, DEMs with spatial resolution ranging between 5 mm to 5 cm have found to be necessary [11]. Digital representations at such a detailed scale provide the necessary information to accurately quantify ephemeral gully soil loss and channel morphology.

Recent developments in laser scanner technology, provides new opportunities for scientific investigation of ephemeral and classical gullies. Although, laser scanners have been previously used in similar investigations such as large-scale classical gullies in different locations such as mountain-side sites [12], forest sites [13], desert sites [14], and landslides [15,16], its application to ephemeral gully investigations is in the early stages of development. This can be partially attributed to the dynamic nature and small scale of ephemeral gullies that limit the identification and description of such features using airborne-based Light Detection And Ranging (LiDAR) systems. Due to reduced cost and increased portability of laser scanner technology, terrestrial LiDAR surveys are now practical.

Ground-based LiDAR systems provide the tools for detailed multi-temporal analysis of micro-topography of ephemeral gullies. These systems are capable of generating terrain representations with sub-centimeter vertical accuracy (**Table 1**). This is especially important in research sites where the same location needs to be surveyed multiple times over lengthy intervals as conditions change due to precipitation, runoff events, field management changes, and/or implementation of conservation practices. However, as the number of data points increase (point sampling density) with increased resolution-rich surveys, so does the time required for field data collection, post processing, and the size of datasets. This problem is compounded when monitoring multiple sites with different topographic and farming practices. The objecttive of this study is to identify the minimum point sampling density that provides the necessary topographic information to efficiently produce reproducible field surveys. The results of this study may be used as a guideline in using laser scanner technology to characterize topographic conditions associated with the evolution of ephemeral and classical gullies and, more importantly, accurate estimates of quantities of sediment eroded or conserved by erosion control practices.

Table 1. General specifications of the TOPOCON GLS-1000 laser scanner.

Parameter	Value	Unit	Condition
Single point acc. distance	4.0	(σ) mm	1 to 150 m
Single point acc.ver. angle	6.0	s	1 to 150 m
Single point acc. hor. angle	6.0	s	1 to 150 m
Maximum scan rate	3000	Hz	
Scan density spot size	6.0	mm	1 to 40 m
Scan max.sample density	1.0	mm	up to 100 m
Wavelength	1535	Nm	
Laser pulse duration	3.6	nano s	

2. Background

2.1. Point Sample Density

LiDAR technology measures the laser pulse travel time from the transmitter to the target and back to the receiver [17]. Because the speed of light is a known constant, the distance can then be calculated. In this process, a very accurate timing system is needed to guarantee precise estimates since the laser pulses are generally sent at the rate of thousands of times per second. Additionally, the transmitter and the receiver must be located at the same physical location, so in effect containing a single-ended system [18]. During the field data collection process, ground-based laser scanners record vertical and horizontal angles, intensity of the returned signal, and the traveled distance for each laser pulse.

These systems are capable of collecting information with a wide range of ground point sampling densities as a result of operator controlled factors such as the scan angles (area covered by individual scans), average point density of individual scans, and degree of overlap between scans (**Figure 1**). Higher point density can be achieved by higher sensor resolution, smaller vertical field of view angles, and multiple scans of the same ground location. The instrument resolution is often controlled by an imaginary plane located at the middle range of the vertical field of view and is orthogonal to the sensor's normal sight. The vertical scan angle also influences sampling density. Surveying with large vertical angles covers more area and often yields shorter survey times, but at the expense of point sampling density (see **Figure 1**). Multiple scans can be used to collect data over the same geographical location resulting in increased sampling density and often overcomes problems such as shadowing and limited coverage due to vegetation.

2.2. Spatial Variability

Modeling tools designed to define and to understand the pathways of surface water movement across fields and even across watersheds rely on raster-based digital elevation models to derive the required topographic attributes [19]. The conversion of irregularly spaced and uneven distributed laser points into a regularly spaced grid requires the use of interpolation and re-sampling techniques. The choice of a specific interpolation method and its corresponding parameters introduce uncertainties that are propagated to the hydrologic model [19] and studies have been conducted to study such phenomena [20].

Among all the available interpolation techniques, kriging is often used because of the ability to provide unbiased estimates with minimum and known variance [21]. The core element of kriging interpolation techniques is the variogram. The variogram explores spatial independence [22] and quantitatively relates variance to space separation [21]. Variogram analysis aids our understanding of the effects of scale in spatial variability on the data being interpolated.

As an illustration of the variogram computation concept, consider a one-dimensional set of laser points (Figure 2). Variograms are computed using a set of pair of points, which are selected, based on a pre-defined separation distance h, referred to as lag distance. A practical equation to compute the variance of differences is the use of semivariogram equation, which is given by:

$$\gamma\left(h\right) = \frac{1}{2N_h}\sum_{i=1}^{N_h}\left[\left(Z_i - Z_{(i+h)}\right)^2\right] \qquad (1)$$

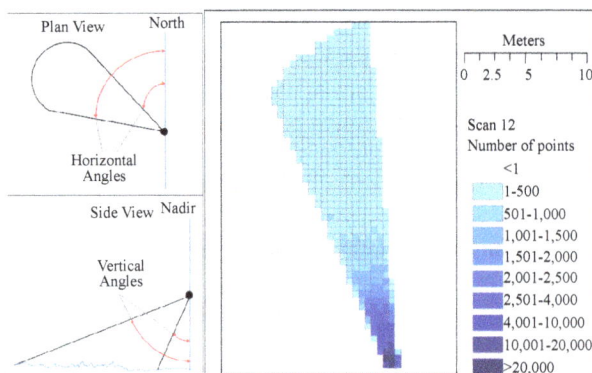

Figure 1. Illustration of spatial variation of laser point sampling density, represented as number of points per grid unit. Laser point sampling density decreases with vertical scanning angle of ground-based laser scanners.

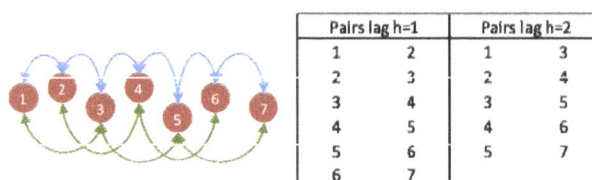

Figure 2. Conceptual illustration of the pairing of points with varying separation distance h. The list of points is used in the computation of experimental variogram.

The semivariogram equation represents the average semi-variance for a lag distance h between points and a total number of pairs N_h. In this equation Z_i and $Z_{(i+h)}$ represent the elevation value of points at location i and i plus the separation distance h $(i + h)$, respectively. The smaller the γ value the more related the points are. In other words, the semivariogram represents the average squared difference of any pair of points located h distance from each other [21].

Plotting semi-variance using an increasing range of lag distances generates a semivariogram graphic (Figure 3). The range of lag distances is plotted on the x-axis and the semi-variance on the y-axis. In the ideal case, points close to each other should have small difference values (small semi-variance) and as the separation distance increases the differences between points should also increase [23]. The plotted curve resulting from the semivariogram is referred to as experimental semivariogram (red points in Figure 3). When a mathematical function is used to model the experimental semivariogram it is then referred to as theoretical semivariogram. Also, in the ideal variogram two properties are regarded as the most important curve characteristics: sill and range or correlation length (Figure 3). The sill should be equal to the sample variance [24] and should match the values where the semivariogram curve "levels off". The range, in the semivariogram plot, should correspond to the value in the x-axis matching the sill and indicates the distance where the samples become independent of each other. Experimental variograms can be considered isotropic or anisotropic, where isotropic variograms depend only on the separation distance, while anisotropic variograms depend on the separation distance and orientation. A detailed description of variograms and its role in kriging interpolations is beyond the scope of this manuscript and additional information on this topic can be found in many textbooks [23,25,26].

3. Methodology

3.1. Study Site

The study site selected for this investigation is located within the Cheney Lake Reservoir watershed near the town of Hutchinson in South Central Kansas. The predominant land use is agriculture (>73%) in the form of cropland and rangeland. The gully within the study site was 96 meters long oriented North-South, approximately 1.3 meters wide and from 10 to 50 cm deep. The channel is free of vegetation and crop residues, while the surroundings are covered by crop residues resulting from no-till management used in winter wheat followed by sorghum (milo) in the 2010 crop rotation. Historical cultivation practices indicates that initially this ephemeral gully did not disrupt farming operations; however, as no-tillage practices were adopted in 2005, the channel grew wider and deeper to the point that the farming equipment could not be used to travel across the gully and the ensuing cropping activity was performed around the main channel (**Figure 4**).

Two locations with known geographic coordinates within the study site were used to provide reference geographical coordinates. This is an important step to translate the equipment local coordinates into geographic coordinates, thus providing a means to compare surveys performed at different times. The equipment used was a TOPCON GSL-1000 series and its general specifications are listed on **Table 1**. Initially, the operator scans the pre-defined targets installed at the reference points. Based on the known geometry of the targets, the instrument is then capable of calculating its location in relation to the reference points (geographical coordinates). Four standard targets were installed in the far outmost corners of the gully being investigated. These four static targets are surveyed and their coordinates computed and recorded. Each subsequent scan starts with surveying the four targets to locate the laser scanner in the local coordinate system. A total of eleven scans in eleven set ups (one scan for each equipment set up) were used to describe the gully. In the post processing steps, each scan with local coordinates are translated into geographical coordinates using the relation between the four targets and the reference points.

3.2. Accuracy Assessment

During data collection in the field it is possible to survey the same geographical location from different scans. This practice is often used to increase sampling density and to avoid problems such as shadowing and/or limited coverage due to vegetation. The overlap in LiDAR point sampling can also be used to evaluate survey accuracy. Given that the overlapping scans were collected with enough point density, it is possible to identify neighboring points, collected from different scans, and compute elevation differences between these points (**Figure 5**). A similar approach is used to evaluate the accuracy of airborne LiDAR sensors by comparing elevation values of nearby points collected by different flight paths in overlapping areas.

Cross-validation was performed considering two distance threshold values of 1 and 5 millimeters. Histograms of the elevation difference of the selected point pairs are shown in **Figure 6** and the summary statistics in **Table 2**. The mean difference for both cases is less than 2 cm; however, some points showed absolute differences above 5 cm. These isolated differences could be attributed to sharp elevation differences caused by features such as crop residues and standing vegetation.

3.3. Sampling Density Investigation

Studies have been performed to identify the optimum

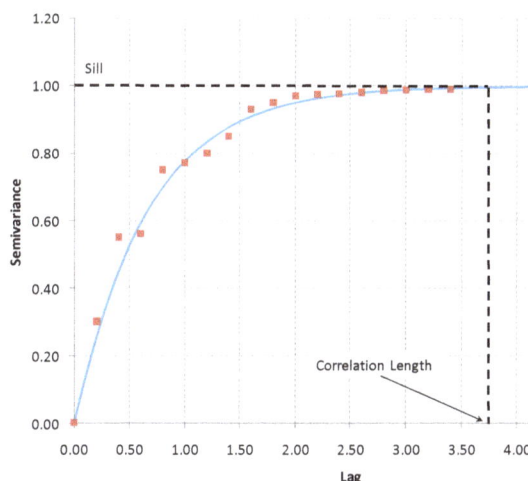

Figure 3. Ideal shape of semivariogram plot. Red squares represent the experimental semivariogram plotted from the sample data. Blue line represents the theoretical semivariogram curve obtained from fitting a mathematical model to the experimental variogram.

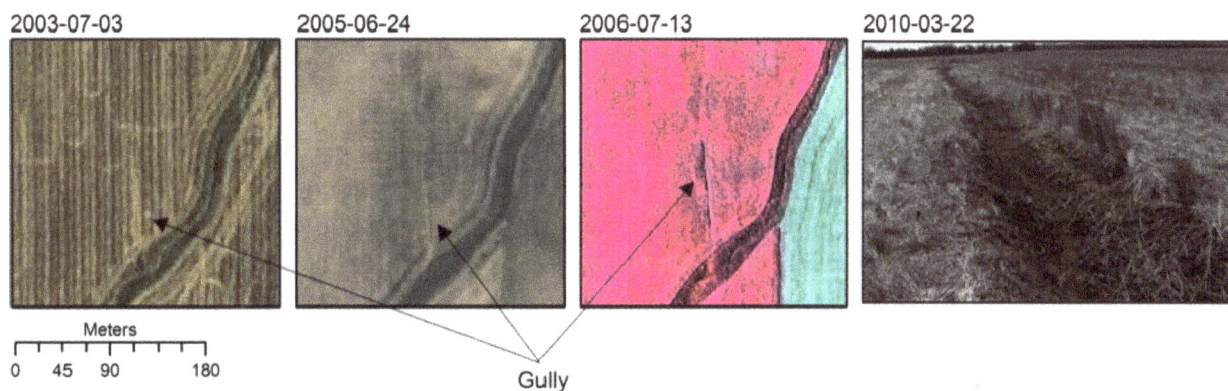

Figure 4. Ephemeral gully evolution into classical gully and its consequent disruption to producer's operations. Imagery data for years 2003, 2005, and 2006 obtained from the National Agriculture Imagery Program (NAIP) and 2010 from field visit.

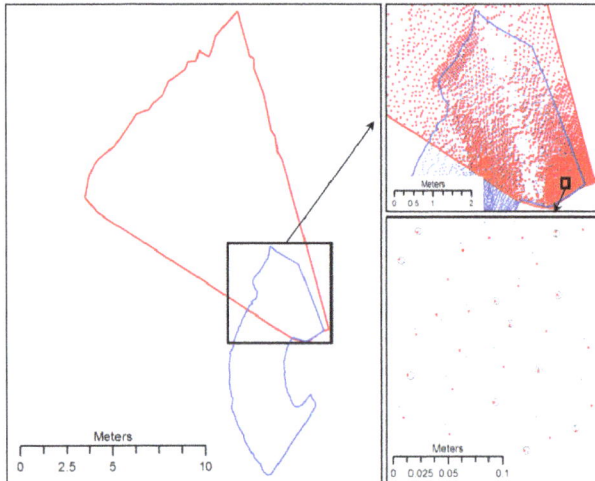

Figure 5. Overlap between different scans of ground-based LiDAR. The pair of points highlighted by a circle in the lower right map indicates the points with distances smaller or equal than 5 mm. These points were used in the survey accuracy assessment.

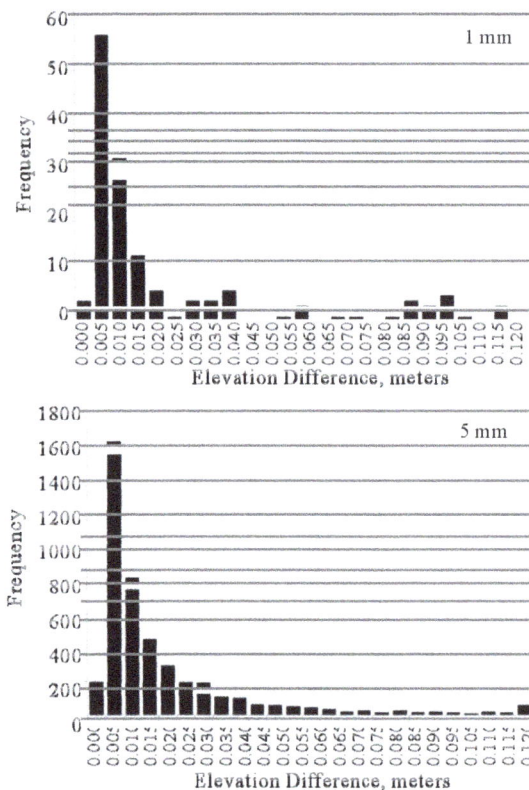

Figure 6. Histograms of elevation differences between ground-based LiDAR points collected in different scans and with distanced less 1 mm (left-hand side plot) and 5 mm (right-hand side plot) apart.

balance between point density at small gully scales and volume of data with the goal of optimizing data collection and cost. Guo [20] provides a detailed investigation of the relationship associated with airborne LiDAR point

density reduction, interpolation methods, and resolution of digital elevation models. There have not been studies that investigate the relationship of laser point sampling density collected using ground-based laser scanners on topographic information tailored to gully investigations in agricultural fields. Ground-based systems differ from airborne systems as a result of encountering a wider range of incidence angles [27]. Also, the finer resolution in investigations of gullies formed in agricultural fields requires the determination of micro-topography, combined with the presence of crop residues, produces various levels of terrain roughness, posing a challenge to interpolation techniques.

Point sampling density was investigated by tiling the entire LiDAR point cloud into one-meter square grids. Sampling density was computed by counting the number of points in each tile. This information can be utilized when verifying spatial coverage of sampling points to identify gaps or under-sampled regions (**Figure 7**). Areas with specific features, such as gully headcuts, should be scanned with higher point density, whereas featureless areas can be scanned at lower point density. An area with high point density designed to detail the gully active headcut as accurately as possible was obtained through multiple overlapping scans of the same area (inset in **Figure 7**). In contrast, there is an under-sampled region in the mid-section of the gully as result of large vertical scan angles and the selection of the location for the laser scanner (**Figure 7**).

A total of 5032 tiles were generated (many of them containing no points) (**Figure 7**). Tiles were ranked by standard deviation of elevation values and divided in three groups based on the data quartile values. In each group, the tiles with the highest number of points were selected, 2175 (155,373 points with $\sigma_{elev} = 0.13$), 2177 (40,923 points with $\sigma_{elev} = 0.06$), and 2304 (17,144 points

Table 2. Descriptive statistics of absolute value of the elevation differences between LIDAR points collected from different scans.

	1 mm	5mm
Mean	0.01899	0.01834
Standard Error	0.00249	0.00042
Median	0.00600	0.00800
Mode	0.00100	0.00100
Standard Deviation	0.02885	0.02761
Sample Variance	0.00083	0.00076
Skewness	2.03700	3.29071
Minimum	0.00000	0.00000
Maximum	0.11500	0.28200
Count	134	4425

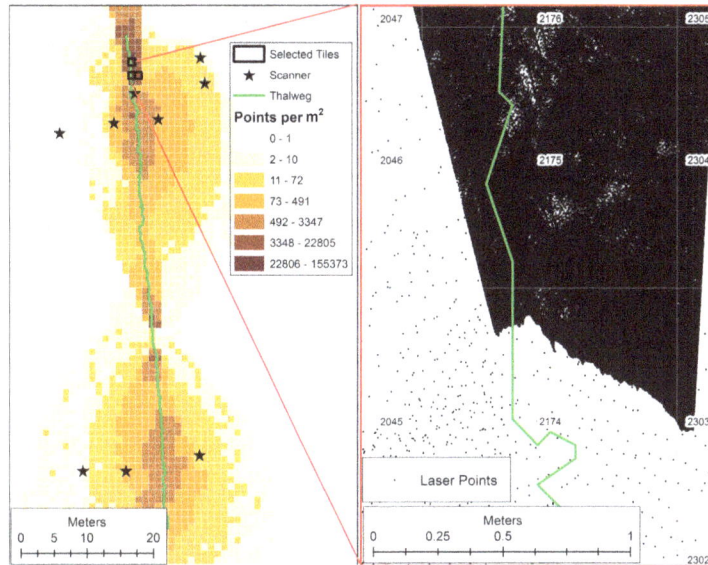

Figure 7. Ground-based LiDAR point density investigation. Left hand-side map shows the spatial variation of the sampling density in square meter cell size grid. The right hand-side map illustrates the differences in point density between tiles. Tile 2174 has 61,912 laser points per square meter while tile 2175 has 155,373 laser points per square meters (both represented by black dots).

with $\sigma_{elev} = 0.01$). The same variation in elevation represented by standard deviation values can be observed on histograms (**Figure 8**). Tile 2175 has the largest elevation range ($\cong$ 50 cm) as the gully active headcut is located in this tile. Histogram plots of the distance values to the nearest neighbor depict point density of each plot. The vast majority of the points are within 5 mm of other points.

Experimental semivariograms for each of the three tiles selected were computed using the algorithm *gamv* available in the Geostatistical Software LIBrary (GSLIB) due to the irregularly spaced nature of the laser points [28]. The lag separation distance (distance between two points used to create the point pair database) and lag tolerance were selected to be 2 cm and 1 cm respectively (2 cm ± 1 cm). The omnidirectional variogram was considered throughout our investigation.

The experimental semivariograms were computed using all the available laser points in each tile with the *gamv* algorithm (black dots in **Figure 9**). The theoretical semivariograms were generated using the Levenberg-Marquardt optimization algorithm [29] for determining the set of parameters that provides the best fit to the experimental variogram through minimization of the sum of squares of the residuals. Different mathematical models were selected to represent the theoretical semivariogram curves. For tile 2175 a composite Gaussian model was used (Equation (2)).

$$\gamma(h) = -0.13\left[1 - e^{\left[-(0.31*h)^{3.44}\right]}\right] + 1.10\left[1 - e^{\left[-(0.12*h)^{1.47}\right]}\right] \quad (2)$$

Variations of the standard Gaussian model were used to generate theoretical semivariogram curves for the remaining two tiles 2304 (Equations (3)) and 2177 (Equation (4)). The three curves of the theoretical semivariograms are plotted in **Figure 10**. These theoretical semivariograms were considered as reference in the subsequent spatial continuity experiment.

$$\gamma(h) = 0.0001\left[1 - e^{\left[-\frac{h^2}{0.76^2}\right]}\right] \quad (3)$$

$$\gamma(h) = 0.0046\left[1 - e^{\left[-\frac{3h^2}{0.274}\right]}\right] \quad (4)$$

Randomly selecting a subset of points for each tile and then evaluating their variogram was utilized to quantify of the influence of the sampling density on the topographic information. Large number of repetitions for the random creation of the subsets was adopted to minimize the odds, and possible influence, of one "bad" selection of points. A Monte Carlo type investigation was performed by creating a series of independent simulations of reduced datasets containing a smaller number of laser points than the original number in each tile. The reduced dataset was generated by randomly selecting laser points based on a pre-defined percentage. A percentage of 100% represents all the laser points available in the tile while a reduced set using a percentage of 50% would yield half of the available points in the tile. For each pre-defined percentage, a total of 100 independent realiza-

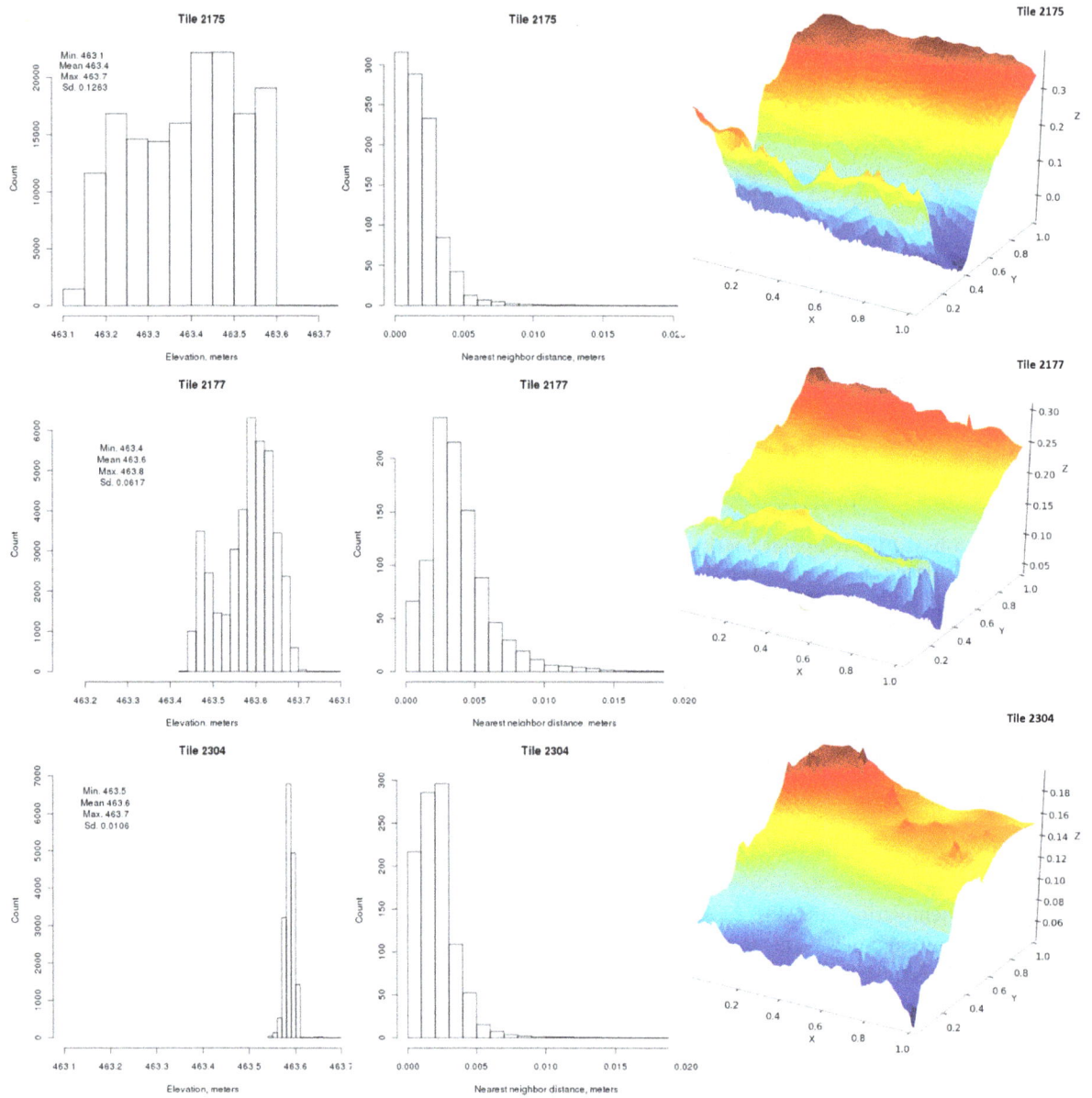

Figure 8. Three tiles selected for the point sampling density investigation. Left hand-side column presents the histograms of elevation values, center column the distance to the nearest neighbor in each of the square meter tile considered, and the right hand-side column three-dimensional grids of each tile.

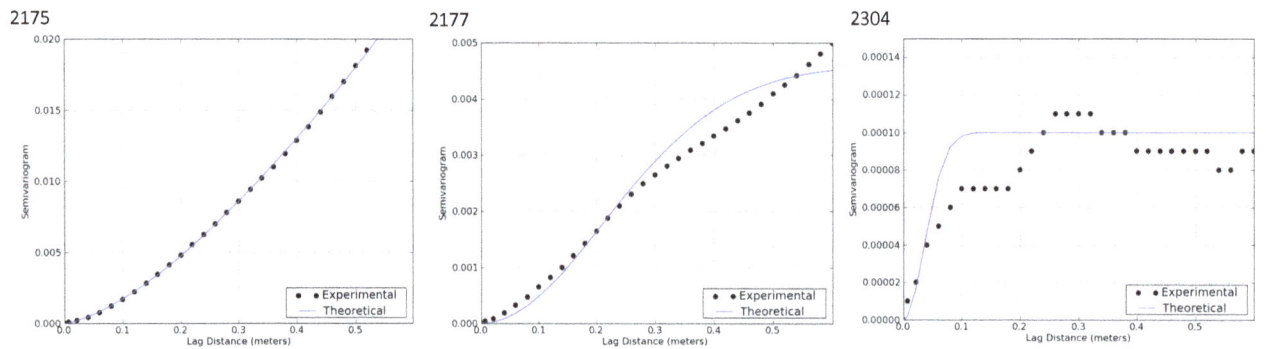

Figure 9. Plots contrasting experimental and theoretical semivariogram for tiles 2175, 2177, and 2304 of irregularly spaced elevation points collected using ground-based laser scanner.

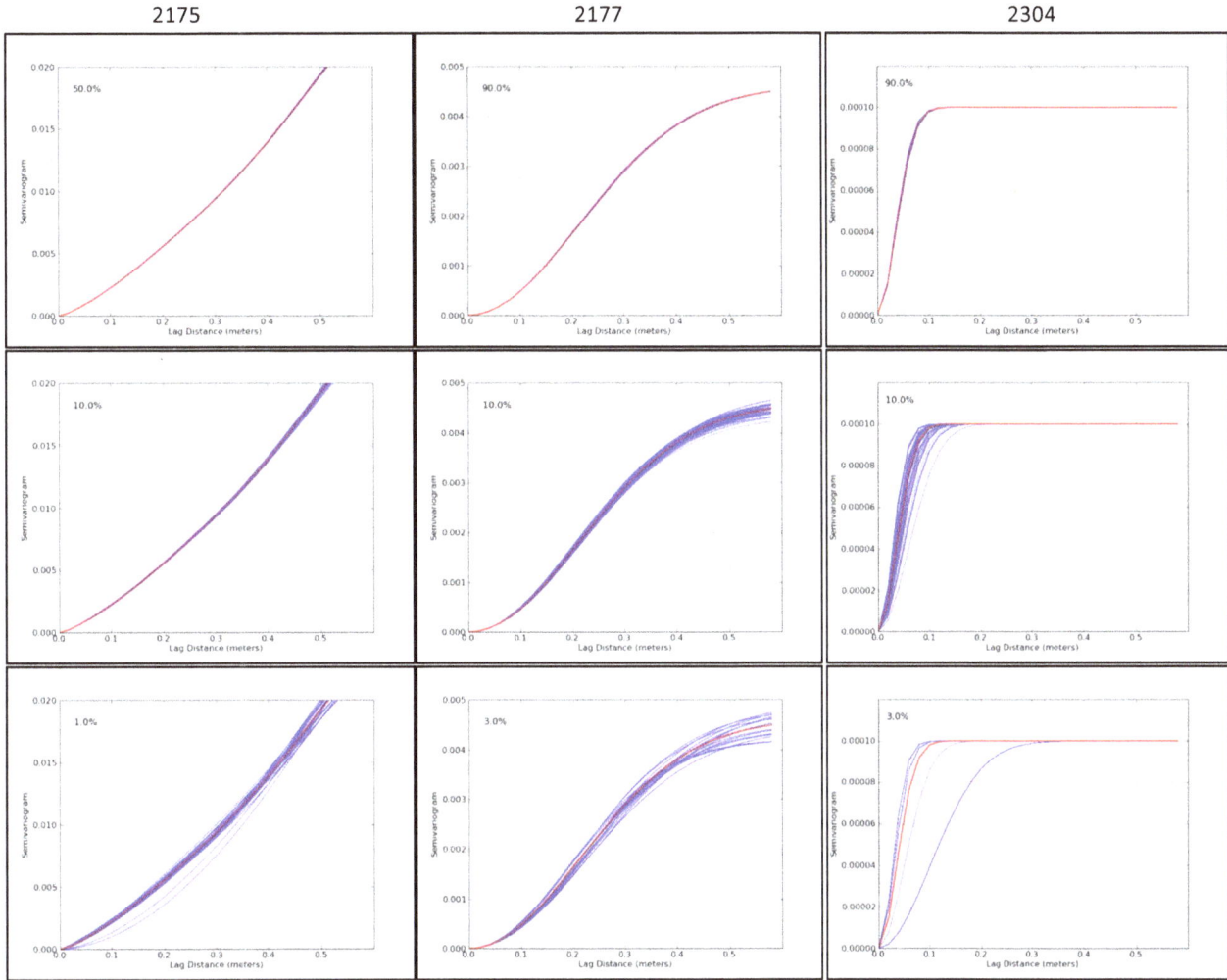

Figure 10. Comparison of theoretical variograms generated using all available data points (red) and reduced dataset (blue). Percentage values indicate the amount of points randomly selected for variance analysis. A total of 100 realizations were performed for each percentage of points considered.

tions were performed (100 independent randomly selected reduced sets). Each reduced set was used in the computation of experimental and theoretical semivariogram curves (**Figure 10**). Multiple percentage threshold values were used (90%, 80%, 70%, 60%, 50%, 40%, 30%, 20%, 10%, 9%, 8%, 7%, 6%, 5%, 4%, 3%, 2%, and 1% for tile 2177; 90%, 80%, 70%, 60%, 50%, 40%, 30%, 20%, 10%, 9%, 8%, and 7% for tile 2304; and 50%, 40%, 30%, 20%, 10%, 9%, 8%, 7%, 6%, 5%, 4%, 3%, 2%, and 1% for tile 2175). Smaller percentage threshold values introduce higher levels of uncertainty represented by the increased variability of the curves.

4. Discussion of Results

The theoretical semivariogram curves generated with reduced data points were quantitatively evaluated by individual comparison to the theoretical semivariogram-curve, obtained using all collected laser points, through

the calculation of root mean squared deviation (RMSD) as shown in Equation (5).

$$\text{RMSD}_{(V_{100}, V_P)} = \sqrt{\frac{\sum_{i=1}^{n}\left(X_{100,i} - X_{p,i}\right)^2}{n}} \qquad (5)$$

In this equation, V_{100} represents the theoretical semivariogram curve developed using all available laser points, n is the total number of points in the curve (total number of lag interval considered), V_P represents theoretical semivariogram curves generated using a reduced dataset with percentage P. A total of 100 RMSD values for each percentage threshold were calculated and averaged. The resulting set of averaged RMSD values are graphically displayed in **Figure 11** for each tile.

The three curves display similar shape with the largest discontinuities found in the plot for tile 2304. Points representing the percentage of 10% and 8% yielded higher averaged RMSD values than the point with the lowest number of points (7%). This can be partially explained

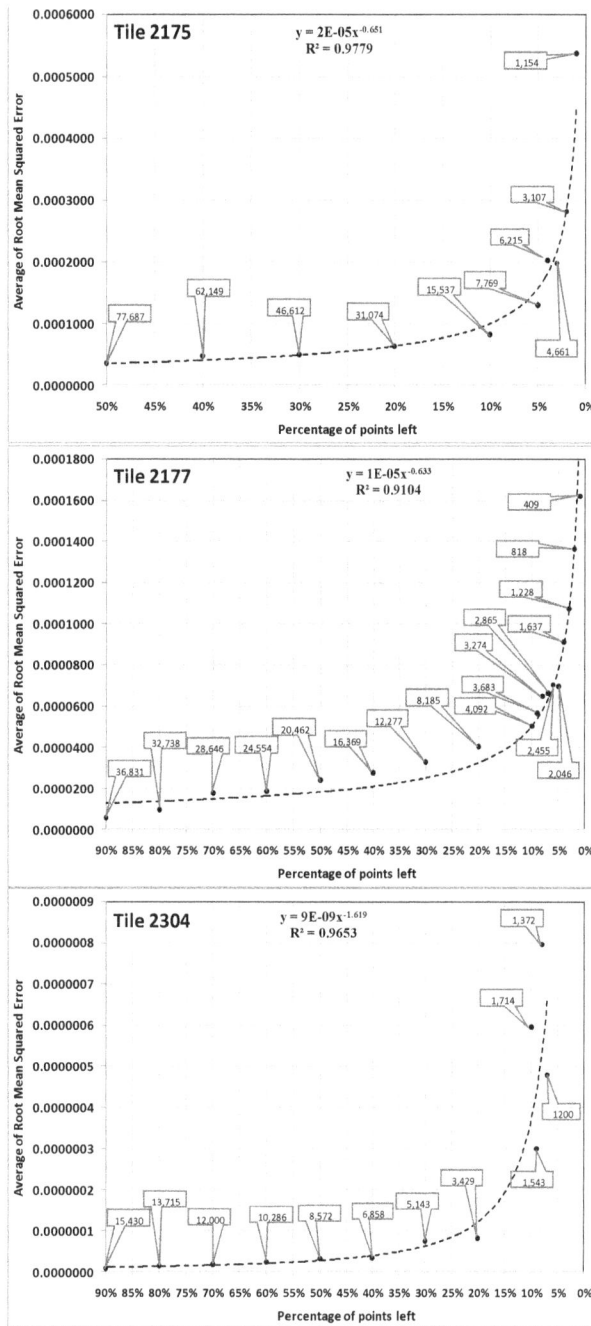

Figure 11. Comparison of the goodness of fit between theoretical variogram using 100% of the available laser points in each tile and 100 realizations of reduced sets of points. Numbers in the callout boxes represent the number of points per square meter left in the tile for each percentage considered.

by the procedure from which a reduced set was created. A standard random sampling technique was used, therefore, it is possible that selected points were not uniformly distributed throughout the tile (forming clusters) and as result, the theoretical variogram curve differs from the reference, yielding large RMSD values. Just a few reali-

zations of clustered points could significantly increase the average value. Nonetheless, despite these two discontinuities it is possible to identify a general trend. The curves start with a gentle slope and as the number of points becomes smaller, curves tend towards to increase rapidly. In other words, results indicate that, in the scale considered, there is an upper threshold of point density where topographic information provided by the LiDAR point cloud does not increase (or increases very little) despite the increased point sampling density. Additionally, it can be observed a positive relationship between this minimum number of points and the tile standard deviation of elevation, as higher sampling densities are needed to topographically describe locations with higher relief, as expected.

To further evaluate the effect of point sampling on topographic information, these curves were used to select three threshold values to reduce the remaining tiles in the survey, 7500, 4000, and 3500 laser points per square meter from tile 2175, 2177, and 2304 respectively. A histogram of the standard deviation of elevation values was used to identify the quartile threshold values. Using these values, the number of laser points in a tile was reduced to the threshold of 7500 laser points per square meter if the standard deviation of elevation was $\geq$ 0.03617, to 4000 laser points per square meter if the standard deviation of elevation was <0.03617 and $\geq$ 0.0106, and to 3500 laser points per square meter if the standard deviation of elevation was ≤ 0.0106. A total of 25 tiles were reduced.

The two point clouds, original and reduced, were converted to Triangular Irregular Network (TIN) format to facilitate volume computations. TIN format was chosen over the conversion of the point cloud into a raster grid to minimize uncertainties caused by interpolation methods. A third TIN, with artificially filled channel, was created by manually digitizing the edges of the gully channel to form a polygon and then subsequent removal of all the laser points within the channel polygon. Through the use of differencing technique, the original and reduced TINs were subtracted from the artificially filled channel TIN yielding volumes estimate of 18.154 m^3 and 18.146 m^3 respectively. There is a difference of less than 0.04% between the two estimates. Additionally, visual comparison of the thalweg profiles for both datasets confirms the agreement between the original and reduced dataset (**Figure 12**). In multi-temporal research efforts, it is important to obtain accurate horizontal and vertical characterization of the gully's thalweg in order to precisely characterize gully changes over time leading to improved understanding of gully evolution.

5. Summary and Concluding Remarks

This study used the concept of semivariogram to quanti-

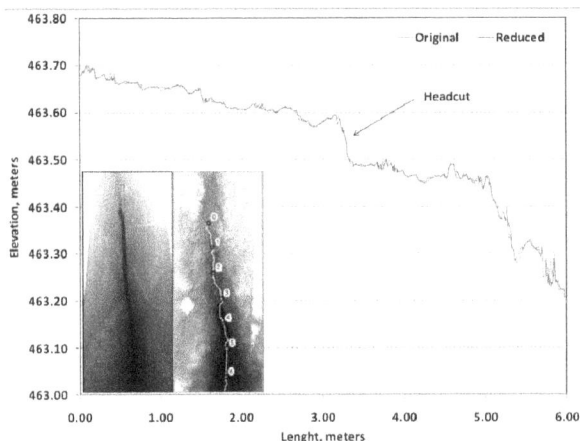

Figure 12. Thalweg topographic profile generated using ground-based LiDAR. Two frofiles are compared: blue profile was generated using all the surveyed points and red profile was generated using a reduced dataset.

tatively investigate the relationship between LiDAR point sampling density and topographic modeling needed to evaluate ephemeral and classic gullies in agricultural fields. The impact of gullies in agricultural fields can be studied at different scales, such as watershed, field, and individual gully scales. In this study, we addressed effects of point sampling density on the topographic information at the individual gully scale.

The gully investigated was partitioned into square meter tiles and the sampling density of each tile was computed by counting the number of laser points in each tile. This experiment revealed a large variation in LiDAR point sampling density throughout the gully. Tiles were ranked by standard deviation of elevation values and partitioned into three groups based on quartile of the histogram elevation values (representing three different topographic characteristics). The tile with the highest number of points in each group was selected for the sensitivity analysis. Multiple realizations of subsets of randomly selected points at pre-defined percentages were used to identify the minimum point sampling density in which the data set retains the original spatial characteristic. Using the minimum number of points per square meter thresholds, a reduced point cloud dataset was developed and compared to the original dataset yielding not significant discrepancy. This indicates that data could be collected with smaller sampling density while retaining the original spatial characteristics.

At the fine sampling density required to proper characterize ephemeral gully evolution in agricultural fields, results indicate that well planned surveys could be designed to collect between 3500 to 7500 points per square meter based on the local terrain topographic variability. Such surveys, could significantly expedite data collection without loss of topographic information. It is also important to note that, although results indicated that the re-

duced dataset did not significantly differ from the original dataset in terms of topographic information, and thus these tiles could be considered over-sampled, the reduced tiles represent only a small percentage of the entire dataset. Out of 2085 tiles containing laser points, only 25 were reduced because they had originally more laser points than the defined thresholds. And, out of the 25 reduced tiles only 14 were located in and around the gully channel. Despite the oversampling of 14 tiles in and around the gully channel, still there are 175 tiles (out of 189) located in and around the gully channel that contained fewer points per square meter than the threshold values obtained as result of this study. This is an inherent consequence of the large variation in sampling density.

Although the ideal situation would be to survey gullies with the highest possible sampling density, this is often not practical because sampling density varies with factors such as resolution of the instrument, vertical scan angle, number of overlapping scans, and land coverage. Furthermore, scientific investigation to quantify and to understand the development of ephemeral and classic gullies in agricultural fields over time often requires multitemporal surveys of multiple locations throughout the watershed.

Based on the findings of this study, future field campaigns can be designed to generate consistent datasets with minimum point sampling density considering the different topographic characteristic (3500, 4000, 7500 laser points per square meter). During the field collection the laser scanner is mounted on a tripod that can be elevated allowing the possibility of collecting data far away from the nadir situation (large vertical angles). Collection of data with such large vertical angles leads to lower sampling densities and shadowing when investigating gullies with deep channels. One possible alternative would be to survey the same location using multiple overlapping scans each with lower point density. Although the instrument would be set to collect data at a lower point density, the combined set of scans would yield higher point density. Additionally, the overlapping dataset could be used to evaluate the point cloud by identifying pairs of points with high elevation difference what could be a potential cue to remove anomalies from the data cloud.

The use of ground-based LiDAR for ephemeral and classical gully investigations in agricultural fields is relatively new and research in this field is expected to continue to grow as technology becomes less expensive and new applications are developed. The use of such technology can help in collecting detailed micro-topography information that can be used in many different research areas such as ephemeral and classical gully modeling, soil water depressional storage capacity, terrain roughness measurements, and many others. Continuation of this work will investigate the influence of vegetation canopy and standing crop residue on the laser point sampling density and the derived topographic information.

6. Acknowledgements

The authors would like to acknowledge Don Seale for his indispensible assistance during data collection. Thanks are also due to Howard Miller and Lisa French, at Cheney Lake Watershed, Inc., for contacting and coordinating with local landowners and providing logistic support.

REFERENCES

[1] W. H. Wischmeier and D. D. Smith, "Predicting Rainfall Erosion Losses," Agricultural Research Service, Washington DC, 1978, 58 p.

[2] "CREAMS: A Field-Scale Model for Chemicals, Runoff, and Erosion from Agricultural Management Systems," In: W. G. Knisel, Ed., USDA Conservation Research Report, No. 26, 1980, p. 640.

[3] K. G. Renard, G. R. Foster, G. A. Weesies and J. P. Porter, "RUSLE: Revised Universal Soil Loss Equation," Journal of Soil and Water Conservation, Vol. 46, No. 1, 1991, pp. 30-33.

[4] R. L. Bingner, R. R. Wells, H. G. Momm, F. D. Theurer and L. D. Frees, "Development and Application of Gully Erosion Components within the USDA AnnAGNPS Watershed Model for Precision Conservation," Proceedings of the 10th International Conference on Precision Agriculture, Denver, 18-21 July 2010, pp. 18-21.

[5] Soil Science Society of America, "Glossary of Soil Terms," Soil Science Society of America, Madison, 2001, p. 134.

[6] R. M. B. Quadros, C. Wisneiwski and E. Passos, "Modelagem de Canais Incisivos—Revisao," RA'EGA, Vol. 8, pp. 69-81. (Portuguese)

[7] J. Casali, J. Lopez and J. Giráldez, "Ephemeral Gully Erosion in Southern Navarra," CATENA, Vol. 36, No. 1-2, 1999, pp. 65-84.

[8] D. Woodward, "Method to Predict Cropland Ephemeral Gully Erosion," CATENA, Vol. 37, No. 3-4, 1999, pp. 393-399.

[9] O. Cerdan, V. Souchère, V. Lecomte, A. Couturier and Y. Le Bissonnais, "Incorporating Soil Surface Crusting Processes in an Expert-Based Runoff Model: Sealing and Transfer by Runoff and Erosion Related to Agricultural Management," CATENA, Vol. 46, No. 2-3, 2002, pp. 189-205.

[10] C. Parker, C. Thorne, R. Bingner, R. Wells and D. Wilcox, "Automated Mapping of Potential for Ephemeral Gully Formation in Agricultural Watersheds," National Sedimentation Laboratory, Oxford, 2007.

[11] T. Schmid, H. Schack-Kirchner and E. Hildebrand, "A Case Study of Terrestrial Laser-Scanning in Erosion Research: Calculation of Roughness Indices and Volume Balance at a Logged Forest Site," International Archives of Photogrammetry, Remote Sensing and Spatial Information Sciences, Vol. 36, No. 8, 2004, pp. 114-118.

[12] R. Perroy, B. Bookhagen, G. Asner and O. Chadwick, "Comparison of Gully Erosion Estimates Using Airborne and Ground-Based LiDAR on Santa Cruz Island, California," Geomorphology, Vol. 118, No. 3-4, 2010, pp. 288-300.

[13] L. A. James, D. G. Watson and W. F. Hansen, "Using LiDAR Data to Map Gullies and Headwater Streams under Forest Canopy: South Carolina," CATENA, Vol. 71, No. 1, 2007, pp. 132-144.

[14] B. D. Collins, K. M. Brown and H. C. Fairley, "Evaluation of Terrestrial LIDAR for Monitoring Geomorphic Change at Archeological Sites in Grand Canyon National Park, Arizona," US Geological Survey, Open-File Report, 2008, 60 p. http://pubs.usgs.gov/of/2008/1384/

[15] K. H. Hsiao, J. K. Liu, M. F. Yu and Y. H. Tseng, "Change Detection of Landslide Terrains Using Ground-Based LiDAR Data," 20th ISPRS Congress, Istanbul, 12-23 July 2004.

[16] J. J. Roering, L. L. Stimely, B. H. Mackey and D. A. Schmidt, "Using DInSAR, Airborne LiDAR, and Archival Air Photos to Quantify Landsliding and Sediment Transport," Geophysical Research Letters, Vol. 36, No. 19, 2009, Article ID: L19402.

[17] A. Wehr and U. Lohr, "Airborne Laser Scanning—An Introduction and Overview," ISPRS Journal of Photogrammetry and Remote Sensing, Vol. 54, No. 2-3, 1999, pp. 68-82.

[18] R. M. Measures, "Laser Remote Sensing Fundamentals and Applications," John Wiley & Sons, Inc., Hoboken, 1984.

[19] S. Wu, J. Li and G. H. Huang, "A Study on DEM-Derived Primary Topography Attributes for Hydrologic Applications: Sensitivity to Elevation Data Resolution," Applied Geography, Vol. 28, No. 3, 2008, pp. 210-233.

[20] Q. Guo, W. Li, H. Yu and O. Alvarez, "Effects of Topographic Variability and Lidar Sampling Density on Several DEM Interpolation Methods," Photogrammetric Engineering and Remote Sensing, Vol. 76, No. 6, 2010, pp. 701-712.

[21] P. I. Curran and P. M. Atkinson, "Geostatistis and Remote Sensing," Progress in Physical Geography, Vol. 22, No. 1, 1998, pp. 61-78.

[22] P. K. Kitanidis, "Introduction to Geostatistics: Application in Hydrogeology," Cambridge University Press, Cambridge, 1997.

[23] I. Clark, "Practical Geostatistics," Applied Science, New York, 1979.

[24] C. E. Woodcock, A. H. Strahler and D. L. B. Jupp, "The Use of Variograms in Remote Sensing: II. Real Digital Images," Remote Sensing of Environment, Vol. 25, No. 3, 1988, pp. 349-379.

[25] J. C. Davis, "Statistics and Data Analysis in Geology," John Wiley and Sons, New York, 2002.

[26] E. H. Isaaks and R. M. Srivastava, "Applied Geostatistics," Oxford University Press, New York, 1989.

[27] S. Soudarissanane, R. Lindenbergh, M. Menenti, P. Te-

unissen and T. Delft, "Incidence Angle Influence on the Quality of Terrestrial Laser Scanning Points," *Proceedings ISPRS Workshop Laser Scanning*, 1-2 September 2009, Paris, pp. 183-188.

[28] C. V. Deutsch and A. G. Journel, "GSLIB: Geostatistical Software Library and User's Guide," Oxford University Press, New York, 1998, 369 p.

[29] K. Levenberg, "A Method for the Solution of Certain Non-Linear Problems in Least Squares," *The Quarterly of Applied Mathematics*, Vol. 2, No. 2, 1944, pp. 164-168.

9

Remote Monitoring of Surfaces Wetted for Dust Control on the Dry Owens Lakebed, California

David P. Groeneveld, David D. Barz

HydroBio Advanced Remote Sensing, Santa Fe, USA.

ABSTRACT

Extensive dust control on the dry Owens Lake mainly uses constructed basins that are flooded with shallow depths of fresh water. This dust control is mandated by law as a minimum percent of the area of each individual wetting basin. Wetted surfaces are evaluated for area and degree of wetness using the shortwave infrared (SWIR) band of Landsat TM, or similar earth observation satellite sensor. The SWIR region appropriate for these measurements lies within the electromagnetic spectrum between about 1.5 and 1.8 μm wavelengths. A threshold value for Landsat TM5 band 5 reflectance of 0.19 was found to conform with surfaces having a threshold for adequate wetting at a nascent point where rapid drying would occur following loss of capillary connection with groundwater. This threshold is robust and requires no atmospheric correction for the effects of aerosol scatter and attenuation as long as the features on the image appear clear. Monthly monitoring of surface wetting has proven accurate, verifiable and repeatable using these methods. This threshold can be calibrated for any Earth observation satellite that records the appropriate SWIR region. The monitoring program is expected to provide major input for the final phase of the dust control program that will have a focus to conserve water and resources.

Keywords: Remote Sensing; Monitoring; Surface Wetness; Dust Control; Owens Dry Lake; California

1. Introduction

The dry Owens Lakebed was formerly the largest single source of anthropogenic PM_{10} (known to be a health hazard) in the western hemisphere [1] (**Figure 1**). Owens Lake is the terminus of the Owens River that desiccated after nearly complete diversion of its surface water supply to the City of Los Angeles, first drying completely during 1927 [2].

Like other terminal lakes, Owens became the depository for salts that accumulated during the several thousand years since it last spilled [3,4]. The dominant cation is sodium and the dominant anions, in decreasing order of their presence are carbonate, bicarbonate, sulfate and chloride, with about 10% of other elements [5].

Salt chemistry and temperature-controlled efflorescence have been identified as being the causal factor for extreme levels of windborne dust through a temperature-controlled process that occurs during the winter [6,7]. Wetting of the lakebed surface by rain or snow during wintertime low temperatures induces salt dissolution and subsequent re-precipitation as decahydrates of sodium sulfate and carbonate that swell four to five times their volume. Subsequent drying during warm clear winter days cause these salt crystals to shed the majority of their hydration leaving void spaces and destroying soil structure and grain to grain contact. The resulting dry and powdery carbonate and sulfate rich surface is ablated by winds of only 7.5 m/s (15 kts) [6]. Active dust storms during frontal passage have entrained millions of metric tons of particulates each year before dust control [8].

As the party primarily responsible for lakebed desiccation, the Los Angeles Department of Water and Power (Department) is mandated by California State Health and Safety Code Section 42,316 (SB270) to perform dust control at the lake. Three dust control measures were identified by the agency responsible for enforcing SB270, the Great Basin Unified Air Pollution Control District [9]. These are (1) covering emissive portions of the lakebed with gravel, (2) covering the lakebed with vegetation, and (3) covering the lakebed with shallow flooding by water. These methods are recognized within the State Implementation Plan [9], the legal order for wetting

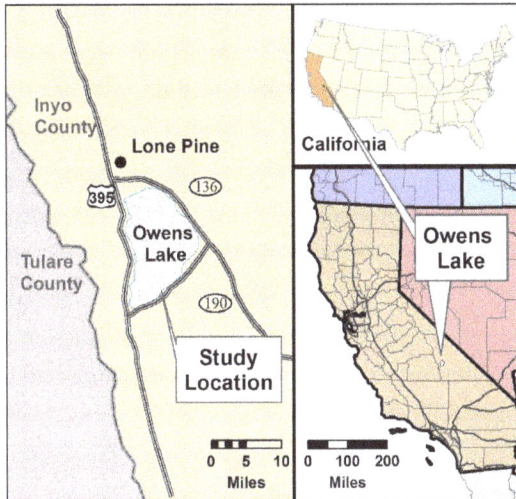

Figure 1. Location map for Owens dry lake.

mitigation. In order to respond to an aggressive schedule for dust control, the Department opted to use shallow flooding as its main dust control measure.

The purpose of this paper is to describe the development and application of a monitoring program to evaluate the degree of wetting achieved for dust control on Owens Lake. Aspects of this program were tested during its development to ensure that they yielded robust, repeatable and verifiable results. The methods that worked were used to support the monitoring program. This work was performed for, and in cooperation with, the Great Basin Unified Air Pollution Control District (District) who has responsibility for measuring air quality and enforcing standards in the Owens Lake airshed.

Eventually, over 100 km^2 of the lakebed will be covered by shallow flooding [1]. Following completion of all planned shallow flooding basins projected for the near future, the dust problem is expected to be completely controlled. However, as designed, the fresh water usage each year is enormous, and with many additional locations that need treatment, could exceed 10^8 m$^3 \cdot$yr^{-1}. Hence, one potential beneficial aspect of the monitoring system for wetness can be its application to assist water conservation.

This paper is the first of two companion papers that describe monitoring with Earth Observation Satellite (EOS) data such as Landsat Thematic Mapper (TM). The paper that follows in this journal, entitled "Remote Monitoring of Vegetation Managed for Dust Control on the dry Owens Lakebed, California", provides a similar heuristic program to develop methods to measure compliance for managed vegetation cover on about 9.06 km^2.

1.1. Wetting Standards for Shallow Flooding Dust Control

A series of dust control standards were generated for

shallow flooding in a controlled experiment called the Flooding Irrigation Project, or FIP, illustrated on **Figure 2**. The standards adopted within the State Implementation Plan [9], are that 75% of a surface receiving shallow flooding treatment for dust control must be at, or near, saturation, *i.e.*, substantially wet. These tests were conducted by the District by flooding water from a pipe equipped with emitters along its length that was laid along the contour. The water that was released flowed down the gently sloping lakebed filling void spaces in the unsaturated zone to bring the local water table to the surface. Wetting of the surface was judged visually, by close inspection, and by annotations made in the field. The resulting classes for surface condition were mapped on aerial photos that were obtained to coincide with the field work.

Compared to ponded water surfaces, flow down many small channels provides greater water conservation because wetted soil surfaces between the flow channels evaporate only a fraction of the water. Capillary evaporation from the soil is complex because it is dependent upon depth to water, soil type and soil layering [10-12]. The effectiveness of the wetting treatment showed that if 75% of the surface is wetted, over 99% dust control is achieved as determined by measurements of the surface sand flux and airborne particulates (**Figure 3**). Using modeling methods that predicted concentrations of dust at the lake shoreline, various levels of required dust control efficiency have been determined from surfaces that were less than the 75% wet. Such variable control efficiency thresholds are being applied during the last phases of the dust control project.

During FIP testing, field inspection of surface wetting

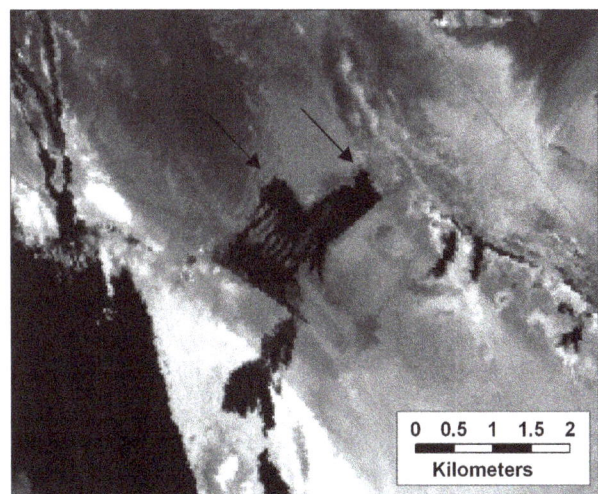

Figure 2. Landsat TM5 April 4, 1994 band 5 image of the FIP (wet surfaces image darkly). Two lengths of emitter equipped pipe (arrows) were used to direct water across the lakebed. The delta of the Owens River entering at the north end of the lakebed, is visible to the upper left.

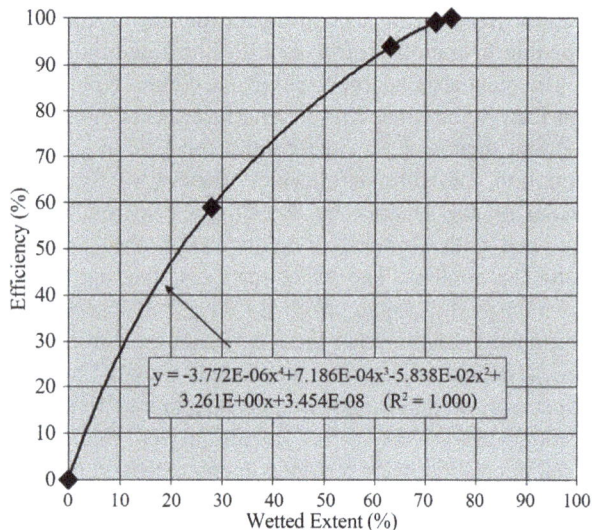

Figure 3. Dust control efficiency curve for wetted surfaces.

required some suitable metric for what was wet enough to prevent dust emissions. The District adopted saturation as a standard and used a method called the "foot tap test", whereby the observer repeatedly taps the surface of the lakebed in question with the sole of the shoe. If this agitation causes the surface to exhibit a thixotropic response, *i.e.*, changing from gel to fluid, then the test is regarded as a positive indication that the soil surface is at, or very close to saturation and is sufficiently wet to prevent the release of dust.

Although the foot tap test is qualitative, it proved to be the only non-remote sensing method that was reliable and repeatable to assess whether a surface was sufficiently wet to avoid dust emission. Other quantitative ground-based methods were evaluated during extensive testing by the District. These quantitative methods were rejected as inaccurate due to the characteristics of the Owens Lakebed soils: 1) substrate textures are highly variable, ranging from pure sands to pure clays that have markedly different water holding properties, thus, precluding the use of gravimetric or volumetric measurements, 2) the very high salinity of the substrate and groundwater (often exceeding 50 dS/m) prevents use of in situ measurements of wetting such as lysimeters, neutron probe, electrical resistance blocks, or geophysical methods such as ground penetrating radar or electrical conductivity, and 3) the lakebed substrate is highly variable in locations that were affected by wave erosion during prehistoric low stands of the lake. In addition, field crews need a fast method for determining whether a surface is sufficiently wet, or not, since they may need to assess large areas on foot.

1.2. Implemented Shallow Flooding Dust Control

Of the three accepted dust control measures for the Owens

Lake bed, shallow flooding was chosen by the Department as the default, despite the requirement to use large amounts of water in a climate where the total open water evaporation exceeds six feet per year [13]. Though gravel was judged to be a permanent fix, the required permitting and environmental review for the appropriate magnitude of the gravel source, plus the cost of extraction, crushing, transport and spreading are extremely high for the area to be treated (~100 square kilometers). In addition, saturated subsurface soils, very low bearing capacities and the weight of gravel would require significant creativity and heroic effort to move the millions of metric tons into place that would be needed. Likewise, establishing vegetation on the harsh saline playa was found to require an immense expense, metering infrastructure, and thousands of kilometers of drip irrigation tubing for an area of slightly over 900 hectares. Despite its high evaporative losses, shallow flooding was the method chosen to meet the aggressive schedule established for phased implementation (**Figure 4**).

Figure 4 illustrates the progression of shallow flood implementation through time that is not yet complete as of this paper. This process was focused to fix the portions of the lakebed identified as the most problematic for dust release first, then, subsequently fixing the next worse areas as identified by the District. The revolving process of identification, then control, will culminate with over 100 km^2 of shallow flood treatments. Once monitoring of fugitive dust from the lakebed shows reduction 99%, the requirement to have 75% of the surface substantially wet can be relaxed in search of water conservation for all

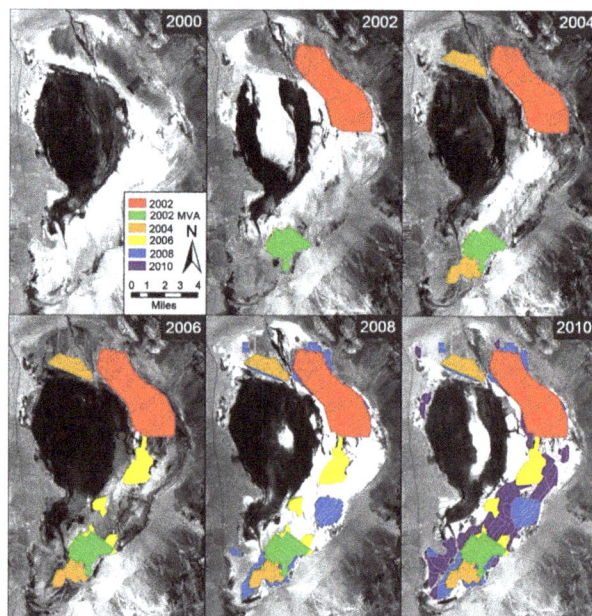

Figure 4. Progression for installation of wetting basins on Owens Lake bed displayed over Landsat TM5, Band 5 images from the same year.

wetting basins.

The basins constructed for shallow flooding are typically bordered on at least three sides by reinforced berms suitable for vehicular traffic. Interior berms are built in many of the basins for the purpose of ponding, diverting and moving water. The basins range in size from 0.06 to 4.60 km^2 and are of two basic designs that can be called "flowing" and "ponded". The flowing design mimics the FIP and is served by hosing, emitters and bubblers that provide a steady flow of recirculated water that spreads in a pattern over the surface as it runs down the gentle slope of the lakebed (as in **Figure 2**).

The drawback to the flowing design is the requirement to frequently control patterns of water drainage across variable substrates that may lead to erosion or creation of raised dry islands. By comparison, most (but not all) of the basins using the ponded design contain a minimum of, or lack completely, plumbing and recirculation capacity (**Figure 5**). Water is released into the ponded basins so that nearly the entire basin remains flooded throughout the 9-month dust season (October through June). Thus, from a standpoint of dust control and required management ef-

Figure 5. Wetting basins designed for flowing water (top) and ponded water (bottom).

fort, the ponded design is superior to the flowing design. In terms of water conservation, however, the ponded design greatly enhances the loss of water through evaporation.

Water evaporation from the wetted lakebed sediment, as in the flowing design is only a fraction of standing water. The flowing design wets the surface in patterns and flow lines rather than flooding it, therefore, providing potential water savings over the flooded condition of the ponded design because reduced area of free water surface is exposed to the atmosphere. Because it is designed only to contain open water, the flooded design offers virtually no potential for water conservation unless converted to some other dust control method.

The flowing design achieves water conservation by spreading the water and wetting the surfaces within the dust-control wetting basin—less free water surface per given area provides the savings. Shallow flooding also provides significant wetland habitat for migrating shorebirds. The habitat inadvertently produced through the Owens Lake dust control effort has been recognized as a significant birding area by the Audubon Society [14]. The flowing design produces an environment that contains a mix of ponded and wetted areas with transition zones that may provide valuable "edge" habitat; a mix that is better for wildlife than would be achieved by a single habitat type. The completely flooded condition of the ponded design provides mostly flooded terrain and therefore provides less range of habitat for birds and the food web that supports them. Much to its credit, the Department has developed a habitat management plan in cooperation with the Audubon Society and the California Department of Fish and Game that will enhance wildlife use while also meeting dust control and water conservation objectives.

Because the ponded design remains flooded throughout the dust season, virtually the only way to achieve conservation when operated with fresh water is to reduce the area of open water which would decrease its effectiveness for dust suppression. Thus, the Department is largely prevented from achieving meaningful conservation with the ponded design. This design, however, is amenable to conversion from fresh water to a managed salt and brine deposits, described in Groeneveld et al. [13] that can achieve virtually 100% water conservation. Also, a great deal of effort and cost would be needed to retrofit the flooded design to a flowing design, and this could also be done, if desired.

2. Heuristic Development of Monitoring

The dust season is recognized as lasting 9 months, from October 1 through June 30 [1]. Evaluation of wetting compliance was desired at approximately monthly intervals. Because of the large scale of the shallow flooding system and difficulty accessing all areas due to low weight bearing capacity or flooded conditions within wetting basins, remote sensing was chosen as the method to judge compliance. Once a wetting basin is judged to be out of compliance according to remote sensing results, it is visited in the field by District staff that then confirms whether the system is not sufficiently wet to prevent dust release and to confirm conclusions drawn through remote sensing. In this way, field verification is built into the program.

2.1. Initial Digital Photographic Method

The first iteration (2000-2002) for remotely-sensed monitoring used color-infrared imagery obtained by aircraft. A combination of four bands—blue, green, red and near infrared—provided a fair evaluation of surface wetness due to light absorption by water in all bands. There were significant problems with this method including long lead times for processing individual shots into a single image, relatively high expense and only fair accuracy. The many seamed and geocorrected images required to represent conditions for each snapshot in time were problematic because scaling of the images varied and the tilt and roll of the aircraft during acquisition frequently captured bright patches of specular reflectance.

Noting the shortcomings for an aircraft-based system, Landsat TM5 was evaluated for its potential as a monitoring platform during 2003-2005. This period followed the scan line correction failure aboard Landsat ETM+7 that rendered data with missing data stripes, so ETM+7 was not chosen for the testing and development. The 30 m pixels and 16-day repeat time of TM5 were judged to be adequate for the intended application while lacking the problematic issues noted for air photography. An additional enhancement over the aircraft-based photographic system, TM output included two short wave infrared (SWIR) bands with potential for direct determination of surface wetting.

2.2. Criteria for Selecting a Remotely-Sensed Metric for Surface Wetness

A series of studies were undertaken to determine how to apply Landsat TM5 for the monitoring task. Initially, this required observation of low altitude oblique photography obtained to capture various wetting states; dry, partially flooded and completely flooded surfaces. These photographs were compared to coincidental values of the various TM bands for images acquired. Of the Landsat bands, one band consistently showed good agreement with the environments visible on the air photographs, the Landsat shortwave infrared (SWIR) band 5 (TMB5).

Other researchers had used TMB5 to assess surface wetness [15,16] or combinations of all six reflected bands including TMB5 in a principle components-based method called "tasseled cap wetness" [17,18]. TMB5 measures reflected light within the window between 1.55 and 1.75 nm. In this spectral region, light is highly absorbed by water, thus, zero reflectance indicates open, clear water, while high reflectance, for example a reflectance of 0.25 to 0.6 indicates dry surfaces. TMB5 may saturate on the Owens Lakebed at a reflectance from 0.55 to 0.6.

From comparison of paired air photos and TM5 data, it was apparent that TMB5 did an acceptable job for identification of wet and dry areas. Thus, either alone or in combination with other bands such as within the tasseled cap wetness algorithm, TMB5 appeared to hold great promise for determining the surface wetting condition pursuant to determining shallow flood compliance.

Practicality was the main factor for choosing a metric to judge wetting compliance. Six criteria were chosen that focused upon reproducibility, determination of whether a surface was sufficiently wet, or not, with a simple yes/no dichotomy, and a built in safety margin between what is adequately wet and protected and what is potentially dry and emissive (**Table 1**).

2.3. Selecting Between Two Methods That Use SWIR Reflectance

A study was conducted to determine how to use TMB5 to measure system wetness and to set thresholds for adequate surface wetting to control dust. At the outset, using actual EOS data paired with ground truth was rejected for a number of reasons including the need to identify large homogeneous surfaces that contain multiple 30-m TM pixels). Collecting ground truth data was judged to be expensive and impractical because it required that the field work was in place in advance, with no guarantee of a cloud free data take. Surface wetness within the wetting basins can change markedly within a day due to addition of water for the ponded design, windblown movement of shallow water surfaces, and manipulation of the hosing and emitters supplying the flowing design (**Figure 5**). Thus, post hoc sampling following an overpass that was clear of clouds was also rejected as potentially error prone.

Table 1. Six criteria for selection of a remotely sensed metric of surface wetting.

1	Provides a progressive ranking from wet to dry.
2	Consistently places flooded surfaces at one end of the distribution and dry emissive surfaces at the other.
3	Provides for non-overlapping separation of surfaces that are moist (non-emmissive) from dry surfaces at risk for dust emission.
4	Provides significant separation between moist, non-emissive surfaces from dry and emissive surfaces as a built-in safety net.
5	Provides clear yes/no dichotomy for the compliance of each pixel.
6	Does not require groundtruth calibration to provide accurate results.

Since accurate pairing of wetted surfaces with reflectance data was crucial for developing the thresholds to evaluate surface wetting, the testing was conducted in highly controlled trials using spectrometry, photography, physical measurements and observations of the surface reported with more detail in Groeneveld *et al.* [19]. An ASD Fieldspec Pro™ spectrometer was mounted on a tripod for a downward look that covered 30 cm diameter areas (0.07 m^2) of homogeneous surfaces of variable material and wetness. The spectrometer was frequently recalibrated to reflectance using a Spectralon™ panel. All measurements of the surfaces, including gravimetric samples were obtained after completion of photography and spectrometry. The foot tap test was conducted last. These data were used to select between the two competing remote sensing methods and also as a means to establish a threshold to distinguish what was acceptably wet from what was not.

Literature review indicated two remote sensing methods with potential for monitoring surface water content. Energy in the short wave infrared infrared (SWIR) region of the spectrum measured by Landsat TM band 5 (TMB5) was used by Johnson and Barson [15] to identify wetted surfaces and Lunetta and Balogh [20] for identification of wetlands. Tasseled Cap uses all six reflectance bands of TM, including TMB5 [17,18]. A tasseled cap transformation, often available in remote sensing software packages, can be readily applied to various datasets.

TMB5, alone, has been used to identify and assess surface wetting. For wetlands in Australia, Johnston and Barson [15] found that a simple density slice of TMB5 with an empirically derived threshold was the most useful approach for identification of wet surfaces, providing 95% accuracy for identification of permanent open water or marshes. Frazier and Page [16] found that density slicing of TMB5 was as successful as multispectral classification in delineating water bodies, having an accuracy of 96.9%.

The spectrometer reflectance data for the 30 cm diameter surfaces were converted to TM5 equivalence using published band sensitivity curves in ENVI Version 4.5 software. These data were then used for comparison of the two methods: TMB5, used alone as an indicator of wetting versus tasseled cap wetness (TCW). The TC transform uses the weighted sums of the six TM reflectance bands to develop a series of components labeled Brightness, Greenness, and Wetness [17,18]. The point of these analyses was to select an index according to the criteria listed in **Table 1**, so only the TCW component was compared.

The resulting analysis presented in Groeneveld *et al.* [19], found that tasseled cap failed with regard to all of the criteria in **Table 1**, thus being an unreliable measure

of surface wetting that could not be calibrated. Use of TMB5 data, however, provided consistent and verifiable measures of surface wetness that excelled in all six of the criteria in **Table 1** (**Figure 6**). TMB5 values from flooded to dry-and-emissive were correctly ordered. A threshold value of 0.19 TMB5 reflectance was chosen as a representation between what was determined as acceptably wet (*i.e.*, not emissive, and likely to remain so for at least days) and what was not. There was a large range between dry, emissive surfaces and the 0.19 chosen threshold (46% of the range of values measured). The saturated samples, determined by foot tap test, lay below the 0.19 threshold. Thus, the use of TMB5 provided a means to adequately protect the surface from emissions while enabling relaxation of the requirement for saturation that was originally written into the State Implementation Plan [9].

Further analysis of TMB5 data has indicated that the 0.19 threshold represents a natural break between surfaces maintained in a state of wetness through capillarity and surfaces where that contact for resupply is lost causing the surface to undergo drying (**Figure 7**). Thus, 0.19 can also be characterized as occurring in the spectral region between active wetting and active drying for Owens Lake substrate.

3. Monitoring Using Shortwave Infrared

Landsat TM5 was chosen for the monitoring program due to continuing high quality of data output, suitable repetition frequency, and because it was delivered free of charge. This satellite was operating well past twice its expected useful life [21] and onboard failures lead to decommissioning TM5 in December, 2012.

Even though it produces data with missing data stripes, Landsat ETM+7 (TM7) can be used for the Owens Lake for the monitoring program because it lies near the center of the image swath and is, therefore, mostly devoid of the

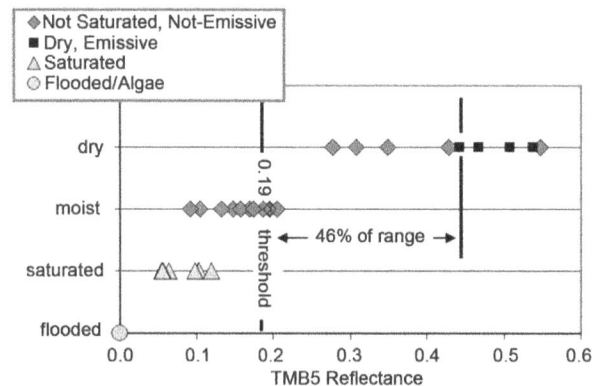

Figure 6. TMB5 results from the spectrometer study. The interval between the 0.19 threshold and lowest TMB5 for dry and emissive surfaces is 46% of the measured range of values (from Groeneveld, *et al.* 2010b).

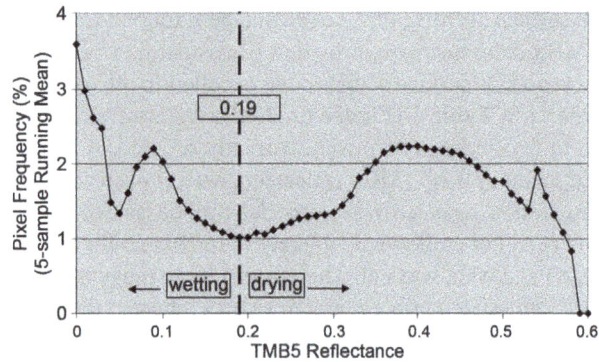

Figure 7. Histogram of TMB5 extracted from the entire Owens Lake area for the March 31, 2009 TM5 image. The 0.19 threshold is a natural break between surfaces being wetted by capillarity and those that are drying or dry.

striping. If the missing data stripes are treated as random, they can be used as a statistical sample because a set pattern of regular spacing (as presented by the stripes) imposed upon the irregular wetting basins, provides an acceptably random sample, especially if only a small percentage of any wetting basin is missed. Thus, for data analysis, the missing data of the stripes do not enter the calculation and the remaining portions of the affected wetting basin are treated as the full statistical sample. If the resulting statistics enable the basin to pass, then the credit is applied to the entire area of the wetting basin, including the portions excluded by missing data. TM7 data were also used as a stopgap if cloudy weather occluded the lakebed through more than a month. Fortunately, in the arid climate of the Owens Lake, located in the rain shadow of the Sierra Nevada, this happens infrequently.

For other EOS data that have a suitable SWIR band in the region of the water absorption feature for light, that is the region of 1.55 - 1.75 μm. Appropriate SWIR platforms include Landsat TM7, SPO5, and TM8 that has just entered service. Calibration to TMB5 equivalence is a necessary step because the 0.19 standard threshold was determined with Landsat TM5 equivalent data and band sensitivities within this window are markedly different among the EOS sensors.

For Owens Lake monitoring, cross calibration between sensors was performed using the approximately 100-km^2 natural salt deposit in the lake's topographic low. This target provided large areas with gradations of wetness that permit pairing many pixels of variable size to enable calculating relatively tight regression relationships between the two sensors. **Figure 8** presents the calibration curve for SPOT5 collected on the same day as an equivalent TM5 overpass that yields TMB5-equivalent results. Pixel values were extracted along transects that were selected to capture a combination of broad areas of flooded and dry surfaces that, in-combination, provide a continuous distribution. The choice of a wide gradation

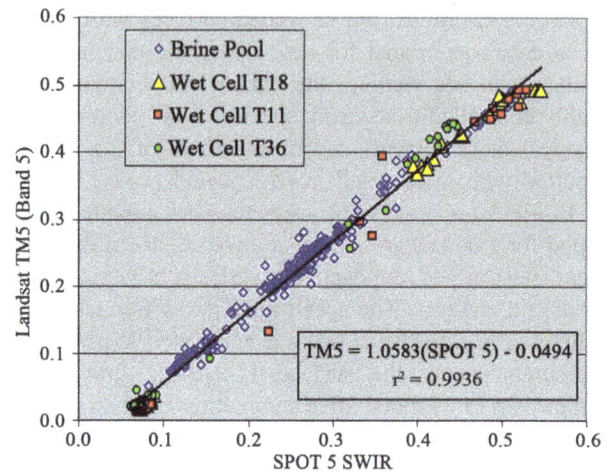

Figure 8. Regression relationship to predict TMB5-equivalent reflectance from coincidental SPOT5 SWIR data (July 2, 2008). These data were extracted from pixels along four transects across the Owens Lake.

of measurements across broad, slowly changing content of surface water and wetness minimized the scatter induced by different pixel sizes and geolocation error (generally taken to be at least one-half of a pixel dimension). Such comparisons contain large amounts of scatter if they are made for target surfaces with comparatively great spatial variability (for example, the flowing design shown in **Figure 5**).

The scale of the two sensors, 20 m pixels for SPOT5 versus 30 m TM pixels, plus geocorrection error induces scatter in the calibration line, however, this scatter should not bias the predicted line fitted by regression (**Figure 8**). With TM5 no longer available, cross platform analysis will need to use historic data for existing satellites in order to achieve TMB5 equivalency. For example, at the time of this paper, both the Landsat continuity mission (TM8) and SPOT7 that contain SWIR bands will be coming on line soon. Both of these satellites can be calibrated to yield TMB5 equivalency through interrelationship with SPOT5 or Landsat 7 that operated during the Landsat TM5's tenure.

Landsat TM8 was calibrated using extracted pixels from paired transects of TM7 that were in turn calibrated to TM5. This additional step was necessary because of the lack of coincidence for TM8 that began service over a year after decommission of TM5. The TM8 SWIR band 6 has a window (1.57 - 1.65) that is much smaller than the TM5 and TM7 SWIR windows (1.55 - 1.75). **Table 2** provides the calibrated relationships for calculating TM5 equivalency for TM7, TM8 and SPOT data for this SWIR band.

3.1. Effect of Ice on Monitoring Results

Rare periods of cold winter weather may potentially form

Table 2. Calibration results yielding TM5 Band 5 equivalency.

Relationship for TM Band 5 Equivalency	Compliance Threshold
TM5 = 0.9837 (TM7) − 0.00170	0.195
TM5 = 0.9611 (TM8) − 0.00140	0.199
TM5 = 1.0583 (SPOT5) − 0.0494	0.229

ice on the fresh water that is supplied for Owens Lake dust control. For this reason, the spectral properties of ice and fresh water were investigated using an ASD Field-Spec Pro™ spectrometer to measure the effect of ice formation upon reflectance. The question to answer was whether the formation of ice changes the interpretation of TMB5 reflectance compared to the same system not covered by ice.

The analysis used plastic cups whose insides were spray painted flat black to reduce the potential for specular reflectance. These cups were then partially filled with plaster of Paris well diluted with water and placed on a horizontal plate in order to achieve a flat, level, hardened surface. Bare plaster of Paris has about the same TMB5 reflectance as dry lakebed and the coverings of water and ice mimic actual conditions on the Owens Lakebed. Five treatments were investigated: a dry plaster surface as a control, a plaster surface covered by 2 mm of water and a plaster surface covered by 5 mm of water. The fourth and fifth treatments were the two wetted surfaces that were frozen on a leveled surface. Difficulty in obtaining a frozen surface that did not buckle into irregular surfaces during freezing limited the treatments to the two shallow depths tested.

Spectrometer measurements were made during late morning under clear skies and freezing outside temperatures, first with the control and ice covered treatments. The frozen samples were then placed at room temperature, allowed to melt, and then measured again with the spectrometer under the clear sky conditions. Calibration using a Spectralon™ panel was made before each measurement in order to enable calculation to reflectance. The data, presented in **Figure 9**, indicate that freezing does induce a slight change in water reflectance. This change is likely due to entrained air bubbles, however, the degree of the effect was so slight that it would not influence compliance determination. Hence, no adjustment need be made for frozen wetting basins.

3.2. Spectral Distortion from Atmospheric Aerosols

Monitoring of surface wetness may be needed for images that were taken through an atmosphere that contains visible amounts of water droplets, ice crystals, smoke or dust, *i.e.*, a haze, that could potentially influence monitoring

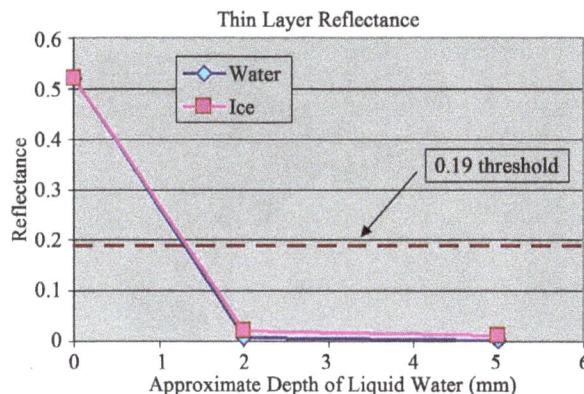

Figures 9. Attenuation of TMB5 through water and ice. Any effects due to freezing are so slight as to not warrant different methods for frozen surfaces.

results due to scatter or attenuation of light as it passes to and from the earth's surface. Urgency to evaluate an image could arise through the need to monitor some time-sensitive effect of management or change of management, or following an extended period of cloud coverage occluding previous overpasses. Such conditions would elevate the need for timely monitoring and exert pressure to accept images affected by atmospheric aerosol content. Would this alter the results obtained?

Dark object subtraction (DOS) is an accepted method for reducing the effects of atmospheric scatter [22,23]. The method assumes that the backscatter of light measured over clear, open water elevates the reflectance above zero, the expected condition with no atmospheric scatter. For DOS, this measured value is then subtracted from all values of reflectance in the region of the open water target. Were DOS used for TMB5 evaluation of wetting, this would raise the value of the 0.19 threshold to some higher value. A test was devised to determine whether DOS was warranted within the wetting basin monitoring program.

Considering that thin clouds would be an acceptable proxy for reflectance distortion due to atmospheric aerosols, a study was made of paired images taken over the Owens Lake. Thin clouds were accepted for this analysis based upon recent results that found that scatter produced by atmospheric aerosols affected red and near infrared light in the same proportion, regardless of aerosol type or density [24]. Thin cloud cover could therefore be used as a general representation for reflectance behavior as affected by any atmospheric aerosol, whether ice in cirriform clouds, droplets in cumulus, mist near the ground, or smoke or dust at intermediate altitudes.

Pixel values for TMB5 were extracted from a polygon of thin cloud cover chosen for homogeneity over targets with a range of wetness from extremes of dry to flooded (**Figure 10**). The two March, 2009 images coincide with a period of relatively high evaporative demand as the

Figure 10. Paired Landsat TM5 images that were first clear and then cloud covered. TMB5 values were extracted from both images within the polygon shown on the inset cloudy image.

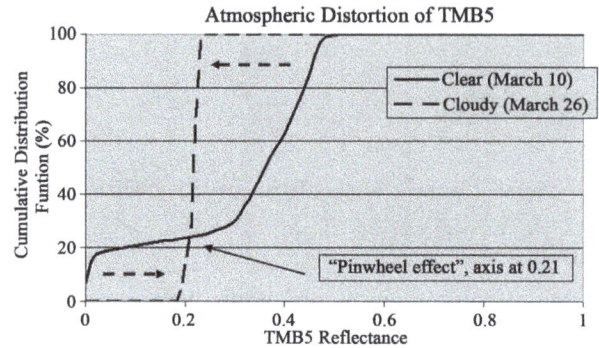

Figure 11. Results from the extracted TMB5 values for clear and cloudy conditions within the Figure 10 polygon. The distorted values moved in a pinwheel fashion around 0.21 that remained unchanged.

days grew longer and the temperatures warmer during transition from winter to spring. Under these conditions, wetted areas would wet and dry areas dry in both images. The published reference ET, expressed as ET_0 [25], was about 30 mm·d^{-1} for March 26, the time of the latter, cloud-affected overpass [26]. High evaporative demand is important to note because about 0.8 mm of rain fell four days before the March 26 image at a station located within 4 kilometers. However, given the relatively high evaporative demand, this additional water was negligible and would have been depleted within a day. **Figure 10** provides a visual comparison of both images.

Figure 11 presents cumulative distribution functions for the extracted TMB5 values within the **Figure 10** polygon for the clear and cloudy conditions. These data show that the distortion effect produced by atmospheric aerosols causes backscatter of light which increases the lower reflectance values. Rather than the DOS algorithm that would shift all values rightward, higher reflectance values were affected by signal attenuation that reduced the light returning to the sensor. Scattering and attenuation created a pinwheel effect that moved high reflectance values leftward and low reflectance values right-

ward. The axis of this pinwheel was a reflectance of 0.21. Fortunately, for the sake of simplicity, 0.21 is not far removed from the 0.19 threshold. Thus, DOS or any similar type of correction was rejected in favor of simply accepting the TMB5 values when the image in question has haze and the observer can differentiate features on the ground.

The degree of thin cloud cover visible on **Figure 10** is not acceptable for monitoring. Concentrations of aerosols approaching the opacity represented by the test data are clearly beyond the limit for reading the wetness signal necessary for monitoring. This can be confirmed by observing the near vertical line for March 26 on **Figure 11** that shows nearly unresponsive data despite the wide difference in surface types visible on the March 10. These image features were overcome by the back scatter created by the aerosol in the cloud. The gross changes, however, are useful for interpreting how reflected light in this spectral region is distorted by aerosols and that such distortion should have minimal effect upon the quality of the monitoring determination even if slight aerosol-induced spectral distortion occurs. Hence, a hazy image would be acceptable for monitoring, but thin clouds are not.

3.3. Structure of the Monitoring Program

The remote sensing aspects described above have been used in a successful monitoring program for surface wetness at Owens Lake since 2007. Monitoring has been performed on a monthly basis during this period with generally 9 - 10 evaluations per dust season from October 1 through June 30. No instances of false positives, where the monitoring says too wet but the system is actually dry, have occurred. Thus, the validity of this remote sensing monitoring system to safeguard air quality has been confirmed.

Roadways and berms for accessing the wetting basins are surfaces managed to prevent dust release and so, are given credit as compliant areas. Likewise, since all of the wetting basins are generally surrounded by dry lakebed or roadways, pixels are given credit within the wetting basin that may contain spectral mixing with pixels outside of the basins. This is accomplished using a buffer that extends the hypotenuse of each pixel to the interior (42.4 m for a 30 m TM pixel). All pixels within this buffer are counted as compliant.

The TMB5 monitoring procedure for surface wetness is standardized with reporting occurring within live spreadsheets that are summarized in a second document that provides visual verification of any wetting basin failures. **Figure 12** is the monitoring result from April 19, 2010 that was included in the documentation.

4. Challenges for the Future

The TMB5 monitoring system has proven to be a robust and dependable means to classify and track surface wetting on Owens Lake. Although the primary role of the monitoring program is to determine compliance to wetting requirements that protect air quality, the future challenge is to be able to reduce water lost to evaporation while maintaining surface protection. This challenge can be met with EOS data such as Landsat TM, the monitoring program outlined in this paper, additional data that includes

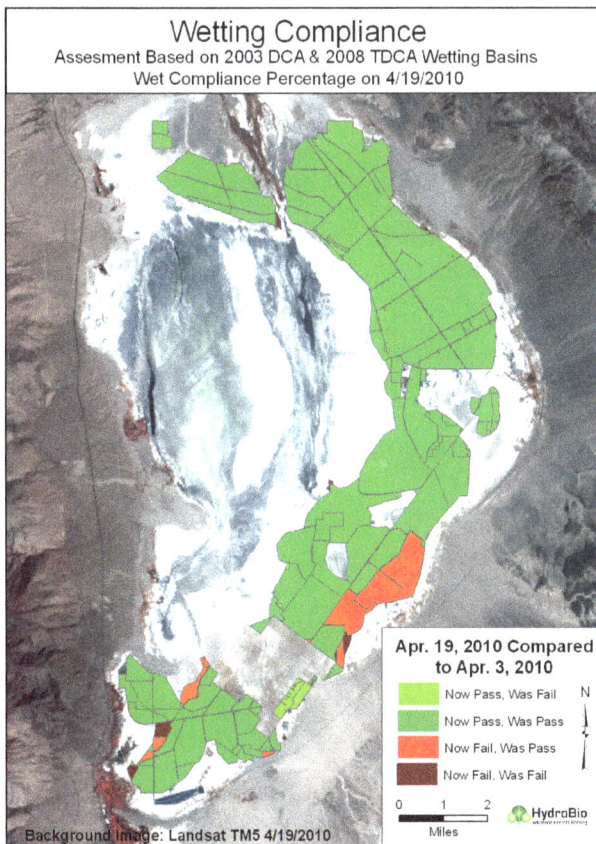

Figure 12. Monitoring result from April 19, 2010.

gage records for water releases to each wetting basin and the evaporative demand that varies with weather, for example, as characterized by the ASCE Penman ET_0 [27]. As written within the State Implementation Plan [1] once the dust control system has achieved 99% control efficiency, the prescription for wetting the majority of the surfaces, currently 75% of the area in most wetting basins must be identified as wet, will gradually be relaxed. TMB5 monitoring can be adjusted to evaluate surface wetting as the wetting basins receive reduction in supply. Hence, the last role for TMB5 monitoring is to optimize water conservation.

Although false positives have not occurred, a false negative, has been identified. As noted earlier, the remediation efforts at the Owens Lake are varied and occur across areas of differing soil hydrology, for example, areas that are wetted by sprinkler. Sprinkler-served zones are one of the few areas that fall outside the norm captured by the 0.19 TMB5 threshold. Sprinkler systems produce circular wetted areas interspersed with lakebed that is not wetted directly. Repeated operation of the sprinklers creates zones of recharge that leach salts from within the sprinkler arcs, while between these wetted zones the resulting high groundwater and capillarity induces white salt efflorescence on the soil surface. The white salt surfaces, though stable, reflect highly within TMB5 and so, the averaged values that result from pixels much larger than the circular wetted areas, are driven above the 0.19 threshold by about 0.05 TMB5 reflectance. One of the zones indicated on **Figure 12** failed for this reason. The 0.19 threshold is too stringent and raising the threshold to 0.24 TMB5 reflectance is a logical and expedient solution for that location.

Potential also exists to replace fresh water used for dust control in wetting basins with salt deposits as described in Groeneveld *et al.* [13]. The salt can be moved as brine from a deposit in the topographic low of the lakebed that contains approximately 70 million m^3 of salt, quantified by several mining companies [5]. The intent of such replacement will be to maintain dust control while keeping the salt mass wet, as occurs naturally within the source deposit. This replacement is a potential solution that can be used to reduce evaporation from wetting basins that are designed to be maintained full of water. This appears to be an ideal solution for the lower elevation regions of the lakebed where the water table and fine textures combine to reduce infiltration to near zero. When annualized, the measured evaporation rates through salt crusts that float atop the brine were found to be lower than the average annual rain and snowfall on the Owens Lake, hence, brine covered surfaces should remain perpetually wet and nonemissive [13]. Fortunately, the wetting of the replacement salt mass surfaces of brine surfaces can be confirmed using the same methods that are presented here, however, a new threshold will need to be developed specifically to detect wetted conditions.

The monitoring for wetness at Owens Lake is a living program that will be changed to enable water conservation through intensive observation and varied management. For example, effects of prolonged drought have forced a revisit to the 75% wetting standards established for wetting compliance in the spring of 2013 prompted by insufficient water present in the Owens River watershed to meet all obligations. The District and the Department allowed the wetness to be reduced in each of the wetting basins combined with more intensive and efficient usage. **Figure 13** provides a look at the percentage wet (*i.e.*, area wetter than the threshold) provided by TM8 data for April 11, 2013 judged by the TMB5-equivalent threshold listed in **Table 2**). The initial results from this program are encouraging since no dust releases have been recorded from the wetting cells with reduced water content.

5. Acknowledgements

The authors thank the Great Basin Unified Air Pollution Control District and staff for supporting this work, especially Ted Schade, Air Pollution Control Officer and Grace McCarley-Holder, Geologist. The authors are greatly indebted to the NASA/USGS Landsat Program for the gratis excellent data needed to develop the methods described in this paper.

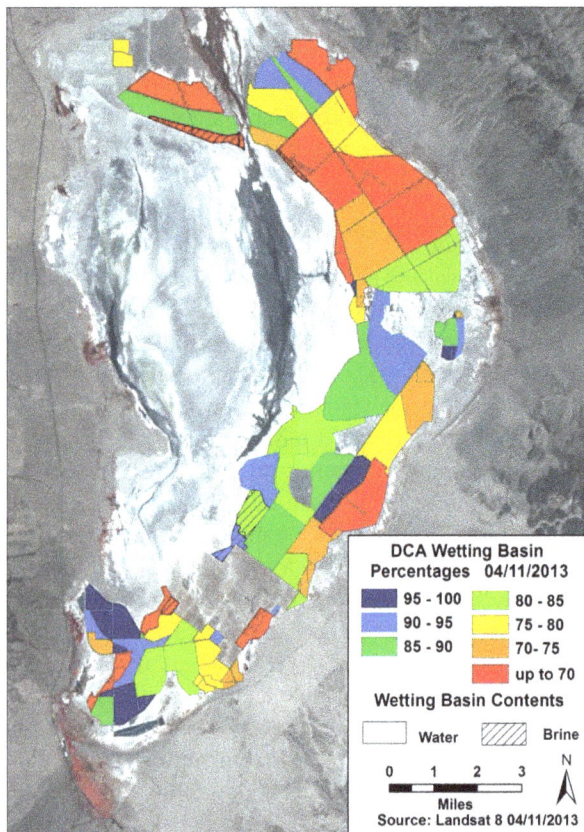

Figure 13. Monitoring result for wetness from April 11, 2013 as percent of area above the wet threshold. The system was allowed to dry as a test of the wetting criteria while accommodating the need for water conservation during a profound regional drought.

REFERENCES

[1]   Great Basin Unified Air Pollution Control District, "2008 Owens Valley PM$_{10}$ Planning Area Demonstration of Attainment State Implementation Plan. Great Basin Unified Air Pollution Control District," 2008. http://www.gbuapcd.org/Air%20Quality%20Plans/2008S IPfinal/2008%20SIP%20-%20FINAL.pdf

[2]   W. L. Kahrl, "Water and Power," University of California Press, Berkeley, 1982.

[3]   H. S. Gale, "Salines in the Owens, Searles and Panamint Basins, Southeastern California," US Geological Survey Bulletin 580, 1915, pp. 251-323.

[4]   A. S. Jayko and S. N. Bacon, "Late Quaternary MIS 6-8 Shoreline Features of Pluvial Owens Lake, Owens Valley, Eastern California," In: M. C. Reheis, R. Hershler and D. M. Miller, Eds., *Late Cenozoic Drainage History of the Southwestern Great Basin and Lower Colorado River Region*: *Geologic and Biotic Perspectives*, *Geol Soc Am Spec Pap.* 439, 2008, pp. 185-206.

[5]   S. S. Alderman, "Geology of the Owens Lake Evaporite Deposit," *6th Int Symp Salt*, 1983, pp. 75-83.

[6]   P. Saint-Amand, L. A. Mathews, C. Gaines and R. Reinking, "Dust Storms from Owens and Mono Valleys, California," US Naval Weapons Center, California, 1986.

[7]   P. Saint-Amand, P. C. Gaines and D. Saint-Amand, "Owens Lake, an Ionic Soap Operas Staged on a Natric Playa," *Centennial Field Guide. Cordilleran Section of the Geological Society of America*, Vol. 1, 1987, pp. 145-150.

[8]   T. E. Gill and T. A. Cahill, "Playa-Generated Dust Storms from Owens Lake," In: C. A. Hall Jr., V. Doyle-Jones and B. Widawski, Eds., *White Mountain Research Station Symposium IV Proceedings*: *The History of Water*: *Eastern Sierra, Owens Valley, White-Inyo Range*, 1991, pp. 63-73.

[9]   Great Basin Unified Air Pollution Control District, "2003 Owens Valley PM10 Planning Area Demonstration of Attainment State Implementation Plan. Great Basin Unified Air Pollution Control District," 2003. http://www.gbuapcd.org/Air%20Quality%20Plans/2008S IPfinal/2008%20SIP%20-%20FINAL.pdf

[10]   W. R. Gardner and M. Fireman, "Laboratory Studies of Evaporation from Soil Columns in the Presence of a Water Table," *Soil Science*, Vol. 85, No. 4, 1957, pp. 244-249.

[11]   W. R. Gardner, "Some Steady-State Solutions of the Unsaturated Moisture Flow Equation with Application to Evaporation from a Water Table," *Soil Science*, Vol. 85, No. 4, 1958, pp. 228-232.

[12] A. Hadas and D. Hillel, "Steady-State Evaporation through Non-Homogeneous Soil Columns in the Presence of a Water Table," *Soil Science*, Vol. 113, 1972. pp. 63-73.

[13] D. P. Groeneveld, J. L. Huntington and D. D. Barz, "Floating Brine Crusts, Reduction of Evaporation and Possible Replacement of Fresh Water to Control Dust from Owens Lakebed, California," *Journal of Hydrology*, Vol. 392, No. 3-4, 2010, pp. 211-218

[14] Audubon, "New Opportunities for Birds," 2013. http://ca.audubon.org/new-opportunities-birds-owens-lake, undated

[15] R. M. Johnston and M. M. Barson, "Remote Sensing of Australian Wetlands: An Evaluation of Landsat TM Data for Inventory and Classification," *Marine & Freshwater Research*, Vol. 44, No. 2, 1993, pp. 235-252.

[16] P. S. Frazier and K. J. Page, "Water Body Detection and Delineation with Landsat TM Data," *Photogrammetric Engineering & Remote Sensing*, Vol. 66, No. 12, 2000, pp. 1461-1467.

[17] E. P. Crist and R. C. Cicone, "A Physically-Based Transformation of Thematic Mapper Data—The TM Tasseled Cap," *IEEE Transactions on Geoscience and Remote Sensing*, Vol. 22, No. 3, 1984, pp. 256-263.

[18] E. P. Crist and R. J. Kauth, "The Tasseled Cap Demystified," *Photogrammetric Engineering & Remote Sensing*, Vol. 52, No. 1, 1986, pp. 81-86.

[19] D. P. Groeneveld, R. P.Watson, D. D. Barz, J. R. Silverman and W. M. Baugh, "Assessment of Two Methods to Monitor Wetness to Control Dust Emissions, Owens Lake, California," *International Journal of Remote Sensing*, Vol. 31, No. 11, 2010, pp. 3019-3035.

[20] R. S. Lunetta and M. E. Balogh, "Application of Multitemporal Landsat 5 TM Imagery for Wetland Identification," *Photogrammetric Engineering & Remote Sensing*, Vol. 65, No. 11, 1999, pp. 1303-1310.

[21] K. Hansen, "Earth-Observing Landsat 5 Turns 25," 2009. http://landsat.gsfc.nasa.gov/news/news-archive/dyk_0013.html

[22] P. S. Chavez Jr., "An Improved Dark-Object Subtraction Technique for Atmospheric Scattering Correction of Multispectral Data," *Remote Sensing of Environment*, Vol. 24, No. 3, 1988, pp. 459-479.

[23] P. S. Chavez Jr., "Image-Based Atmospheric Corrections Revisited and Improved," *Photogrammetric Engineering & Remote Sensing*, Vol. 62, No. 9, 1996, pp. 1025-1036.

[24] D. P. Groeneveld and D. D. Barz, "A Robust Empirical Relationship for Atmospheric Scatter between Red and Near-Infrared Bands," *Remote Sensing Letters*, Vol. 1, No. 4, 2010, pp. 65-74.

[25] R. G. Allen, L. S. Pereira, D. Raes and M. Smith, "Crop Evapotranspiration," FAO Irrigation and Drainage Paper No. 56, 1998.

[26] California Irrigation Management Information System, "Monitoring Data from Owens Lake," 2013. http://wwwcimis.water.ca.gov/cimis/welcome.jsp

[27] R. G. Allen, "The ASCE Standardized Reference Evapotranspiration Equation," American Society of Civil Engineers, 2005.

10

Development of Pedotransfer Functions for Saturated Hydraulic Conductivity

John P. O. Obiero[1], Lawrence O. Gumbe[1], Christian T. Omuto[1], Mohammed A. Hassan[2], Januarius O. Agullo[1]

[1]Department of Environmental and Biosystems Engineering, University of Nairobi, Nairobi, Kenya; [2]Department of Earth and Environmental Science and Technology, Technical University of Kenya, Nairobi, Kenya.

ABSTRACT

The purpose of the study was to develop pedotransfer functions for determining saturated hydraulic conductivity (K_s). Pedotransfer functions (PTFs) for predicting soil physical properties used in determining saturated hydraulic conductivity, based on moisture retention characteristics, were developed. The van Genuchten moisture retention equation was fitted to measured moisture retention properties obtained from International Soil Reference and Information Centre (ISRIC) soils data base in order to determine parameters in the equation *i.e.* saturated soil moisture content (θ_s), residual soil moisture (θ_r), air entry parameter (α) and the pore size distribution parameter (n). 457 samples drawn from the data base were used to be the maximum possible sample size that contained the measured soils characteristics data required. Using statistical regression, mathematical relationships were developed between moisture retention parameters (response variables) and appropriately selected transformed basic soil properties (predictor variables). The developed PTFs were evaluated for accuracy and reliability. It was found that pedotransfer functions developed for θ_s produced the best performance in reliability compared to the remaining parameters yielding a correlation coefficient value of coefficient of determination ($R^2 = 0.76$), $RMSE = 2.09$, $NSE = 0.75$ and $RSR = 0.5$ indicating good performance. Relatively poorest performance was obtained from the pedotransfer function developed for α which yielded a correlation coefficient, $R^2 = 0.06$, $RMSE = 0.85$ and a NSE of 0.02 reflecting the best possible equation derived for the parameter for use in predicting hydraulic conductivity. Out of the pedotransfer functions developed for each of the moisture retention parameters, the best performing PTF was identified for each parameter. The accuracy of the pedotransfer functions assessed based on R^2 were for θ_s ($R^2 = 0.80$), θ_r ($R^2 = 0.42$), α ($R^2 = 0.04$) and for n ($R^2 = 0.30$), when the variables were expressed directly in terms of the selected transformations of the basic soil properties.

Keywords: Pedotransfer Functions; Development; Evaluation

1. Introduction

Hydraulic conductivity is an important parameter that influences hydrological processes which affect flow in rivers. For instance, ground water flow is determined by saturated hydraulic conductivity (K_s). A number of formulations have been developed over the years to predict soil hydraulic conductivity from readily measurable soil properties. The need for mathematical modeling of K_s arises from the fact that insitu or laboratory measurements of hydraulic conductivity are time consuming, labour intensive and expensive as noted by [1], making it practically unlikely in reality to collect permeability data. Besides, direct measurements are also unreliable for site specific applications. The use of pe-

dotransfer functions is an alternative to determining hydraulic conductivity and involves relationships that enables K_s to be predicted using measurable soil physical and hydraulic properties. K_s serves as an input parameter in many hydrological models. There is no specific method considered most accurate in determining K_s as the method used depends on soil and environmental conditions. Furthermore, data on K_s is not readily available in most local soil survey data bases and even where they are available, the reliability is not guaranteed. There is uncertainty in the prediction of K_s using existing pedotransfer functions. Methods to develop PTFs are also unknown and not well understood especially in Africa. Well known moisture characteristic functions are based on parameters that can be determined using moisture retention characteristics. Such data is not readily available in most soil data bases. It is essential to estimate such parameters using basic and easily available soil properties like bulk density, texture etc. Attempts to develop such equations is rare.

[2] notes that modern simulation and analysis of hydrologic processes relies heavily on the accurate and reliable description of soil water holding and transmission characteristics. Predicting K_s from basic soil properties using developed pedotransfer functions is based on analysis of the existing data sets, an approach applied in agricultural hydrology and watershed management. This research demonstrates how pedotransfer functions can be developed from a data base, with limited soils data, in predicting van Genuchten moisture retention parameters, a step in predicting K_s from readily available data on basic soil properties available from soil survey data. K_s can be indirectly predicted from soil properties using moisture retention equations like van Genuchten, Brooks Corey etc. where data on moisture retention curves is available. This technique was used in the study. Readily available data on measured K_s is rare and characterized with uncertainty on their accuracy as different methods in different environments yield varied results as observed by [3] therefore necessitating use of the indirect methods.

Three types of pedotransfer functions are recognized [4]. One category predicts soil moisture retention from basic soil properties. The other is based on point prediction of water retention characteristic as used by [5] to predict points along the moisture retention curve on Iranian soil. Another category involves prediction of parameters in models describing the θ-h-k relationship. This latter approach used in the study involved pedotransfer functions that relate the simple and easy to measure soil properties to van Genutchen moisture retention parameters, a notable gap in the science of development of pedotransfer functions. Most established pedotransfer functions for predicting soil hydraulic characteristics from continuous soil properties are based on statistical regression [6] in which the response variable, y, predicted from a number of n predictor variables, x_i yields the statistical multiple linear regression tool expressed as

$$y = a + \sum_{i=1}^{n} b_i x_i + \varepsilon, \quad i = 1, \cdots, n \quad (1)$$

where "a" is the intercept, b_i is the regression coefficient and ε is the error. The procedure to develop the pedotransfer functions involves steps which include:

The first step of statistical analysis is intended to test if each response variable distribution may be considered as normal distribution [7]. The normality of a distribution is tested by certain statistics like the Shapiro-Wilk (W-value), Skewness coefficient and Kurtosis. The shape of the distribution is determined by a measure of its Skewness which is given by the equation,

$$S = \left[\frac{n}{(n-1)(n-2)} \right] \sum_{i=1}^{n} \frac{(x_i - \bar{x})^3}{s^3} \quad (2)$$

where S is Skewness coefficient, s the standard deviation, x_i the observation value, $\bar{x}$ the mean value and n the number of observations. For a normally distributed data, the skewness = 0, hence the closer the skewness coefficient is to zero (0), the better the normality of the distribution. The level of Kurtosis also provides information about the extent of the normality of the distribution and is given by,

$$K = \left[\frac{n(n+1)}{(n-1)(n-2)(n-3)} \right]$$
$$\times \sum_{i=1}^{n} \left[\frac{(x_i - \bar{x})^4}{\left[s^4 - (3n-1)^2 \right] / (n-2)(n-3)} \right] \quad (3)$$

The measure of Kurtosis is zero (0) for a normally distributed population. Graphical aids used to check normality includes the normal probability plot.

Mathematical relationships between potential predictor variables and the response variables are checked by detecting exponential, logarithmic, or square root tendencies. Important to eliminate is the problem of redundant information in the predictor set brought about by linear dependence between predictor variables [6]. This problem of multicollinearity can be examined by means of correlation matrix. Common statistics used in evaluating pedotransfer functions include [4]:

Multiple determination coefficient given by

$$R^2 = 1 - \frac{\sum_{1}^{N} (y_i - \hat{y}_i)^2}{\sum_{1}^{N} (y_i - \bar{y})^2} \quad (4)$$

Root mean square error given by

$$RMSE = \sqrt{\frac{\sum_{1}^{N} (y_i - \hat{y})}{N}} \quad (5)$$

Mean Error given by

$$ME = \sum_{1}^{N} \frac{(y_i - \hat{y})}{N} \quad (6)$$

Mean Absolute Error given by

$$ME = \sum_{1}^{N} \frac{|y_i - \hat{y}_i|}{N} \quad (7)$$

where y_i denotes the actual value, $\hat{y}_i$ the predicted value, and $\bar{y}$ the average of the actual value and N is the total number of observations. [8] evaluated performance of different pedotransfer functions for estimating certain soil hydraulic properties that include Available Water Capacity (AWC) using R^2, $RMSE$ and ME. [9]

used only *RMSE* in assessing pedotransfer functions developed for predicting hydraulic conductivity and water retention.

The van Genuchten water retention equation which relates moisture content (θ) to the soil matric suction (h) and is expressed as follows;

$$h(\theta) = \frac{1}{\alpha}\left[S_e^{n-1} - 1\right]^{\frac{1}{n}} \qquad (8)$$

where

$$S_e = \left[\frac{\theta - \theta_r}{\theta_s - \theta_r}\right] \qquad (9)$$

To determine K_s, the hydraulic conductivity function used is expressed as

$$K(h) = K_s \left[\frac{\left(1 - (\alpha h)^{n-1}\left(1 + (\alpha h)^n\right)^{-m}\right)^2}{\left(1 + (\alpha h)^n\right)^{m/2}}\right] \qquad (10)$$

To determine K_s using the above equation, a relationship between hydraulic conductivity and matric suction, h is used and can be obtained through equations that relate hydraulic conductivity to permeability [10].

2. Methodology

2.1. Data Collection and Analysis

[4] noted that data from existing international soil data bases having measured soil data can be analyzed to enable prediction of hydraulic characteristics from measured soil data. Availability of measured soil hydraulic characteristics for a wide range of soils and from a large and reliable international data bases are considered prerequisite for development of pedotransfer functions. The ISRIC soils database was used to obtain measured data on soil hydraulic characteristics and basic soil properties. It was possible to obtain all the required measured data on 457 soil samples from various parts of the world. The data contained measured moisture retention curves as well as data on texture (Sand%, Clay% and Silt%). The ISRIC dataset [11] was used for this test. This dataset contains Water Retention Curve (WRC) for both topsoil (between 0 and 20 cm depth) and subsoil (between 20 and 100 cm) from many countries in the world. It has 8 levels of measured suction potential (at 0, 0.1, 0.312, 1.0, 1.99, 5.01, 2.5, and 15.8 m) (**Figure 1**). The lowest suction potential was taken as 0.01 to avoid errors with logarithmic transformations. The observed moisture retention curve obtained from measured data on moisture retention characteristics was fitted to the van Genuchten moisture retention Equation (11)

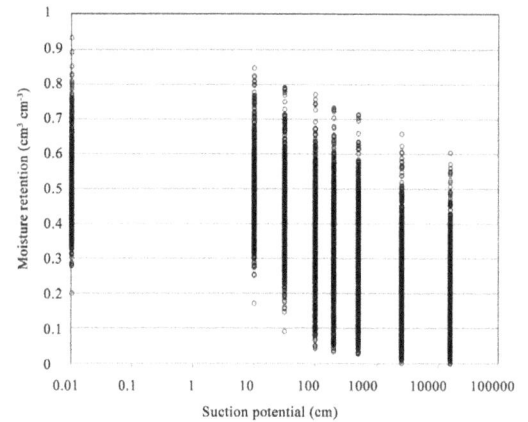

Figure 1. Measured water retention characteristics from ISRIC database.

$$\theta(h) = \theta_r + (\theta_s - \theta_r)\left[1 + (\alpha h)^n\right]^{-m} \qquad (11)$$

where $\theta(h)$ is the soil moisture content at h suction potential, θ_r is the residual moisture content, θ_s is the saturated moisture content, α is the inverse of air-entry potential, and n and m are the shape parameters for the water retention curve representing pore-size distribution index [12]. HydroMe computer programme was used to determine parameters of the van Genuchten equation (http://cran.r-project.org/web/packages/HydroMe/index.html) and is executable in R programme The pedotransfer functions were then developed to estimate these parameters from readily available measured basic soil properties. Relating the moisture retention parameters to the basic soil properties then makes it possible to predict the moisture retention characteristics from the basic properties and hence enabling prediction of the saturated hydraulic conductivity. It would therefore not be necessary to have measured data on moisture retention characteristics to determine K_s but instead, the basic soil properties would be used in determining the equations for moisture retention which can then be used to estimate K_s. Measurement of moisture retention characteristics is quite elaborate, tedious, time consuming and also expensive especially if a large number of sites is involved. From the sample data set of 457 samples, a sub dataset consisting of 342 (75%) samples was randomly selected for use in calibration of the pedotransfer functions to be developed while the remainder portion of 115 samples (25%) of the data was used in the validation process. In development of the pedotransfer functions, the following procedures were undertaken.

2.1.1. Evaluation of the Distribution of the Moisture Retention Parameters

A requirement in developing pedotransfer functions is that the response (dependent) variables should be normally distributed. As a result, the moisture retention parameter of the van Genuchten equation and their possible transformations were evaluated to establish which would best approximate normal distribution. To check the extent to which the response variables would be normally distributed, statistical measures of Shapiro Wilk (W-value), Skewness Coefficient, and measures of Kurtosis were used in the assessment.

2.1.2. Statistical Regression Analysis

Statistical regression was performed between each transformed response variable and each of the basic properties and their transformations so as to establish the goodness of fit in each case based on the measure of the coefficient of determination (R^2). The purpose of this was to determine, for each response variable, the predictor variable or its transformation that gives the best quality of fit between the two when a regression is performed. This would give an indication of which predictor variable or its transformation correlates best with the response variable or it transformation.

2.1.3. Analysis of Cross Correlations

To establish the level of dependence between the selected predictor variables for each response variable best correlated to them, a cross correlation was performed between the independent variables by determining the correlation coefficient (r) between each pair of the variables. This kind of analysis is performed to determine if there is any correlation between two predictor variables among the set chosen for determining the multiple regression equations. It is preferable that the independent variables in a multiple linear regression equation be independent among themselves.

2.1.4. Multiple Linear Regression and Validation of Developed Equations

This was carried out between each selected transformed response variable and selected combination of predictor variables/transformation consisting of the independent variables for which the response variable best correlated as determined by the measure of coefficients of determination. The multiple coefficient of determination was determined for each equation developed to assess the accuracy of the equation. The multiple regression was carried out between response variables and the predictor variables or transformations consisting of independent variables that are not themselves significantly correlated ($r < 0.5$). After development of the pedotransfer functions using the given data set, their reliability is then tested by comparing the values of predicted parameters using the developed equations with the observed data using an independent data set. An independent data set consisting of 115 samples was used in the validation process. The

evaluation statistic used included the correlation coefficient among others.

3. Results and Discussion

3.1. Evaluation of the Distribution of the Moisture Retention Parameters

Table 1 shows the values of statistical parameters used for evaluating normality of distribution for the response variables being the van Genuchten moisture retention parameters and their proposed transformations for the data set of 457 samples. The transformations of van Genuchten parameters (response variables) that produced the best approximation to normal distribution are θ_s^{-1}, e^{θ_r}, $\sqrt{\alpha}$ and $\sqrt{n}$. This was based on measures of the Shapiro-Wilk (W-value), Skewness coefficient and Kurtosis. For illustration, **Figure 2** shows the histogram and normality plot in the normal distribution check for the saturated soil moisture content transformation (θ_s).The transformations of van Genutchen parameters (response variables) that produce the best approximation to normal distribution are θ_s^{-1}, e^{θ_r}, $\sqrt{\alpha}$ and $\sqrt{n}$. This was based on measures of the Shapiro-Wilk (W-value), Skewness coefficient and Kurtosis.

3.2. Statistical Linear Regression Analysis

The quality of fit for the linear regression between each

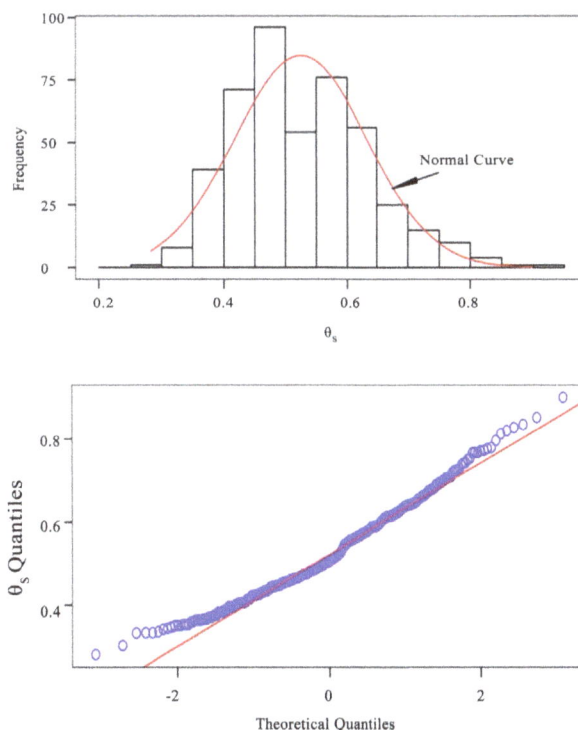

Figure 2. Check for normality of distribution of the saturated soil moisture content (θ_s) showing the histogram and normality plot.

Table 1. Normality of distribution check for parameters in the van Genuchten water retention characteristic and their transformations.

	Kurtosis	W-value	Skewness
θ_r	0.99	0.09	−0.21
α	0.67	4.08	29.47
n	0.78	2.65	11.13
$\ln(\theta_s)$	0.99	0.04	−0.39
$\sqrt{n}$	0.99	0.28	−0.29
θ_s^{-1}	0.98	0.06	0.06
e^{θ_s}	0.96	0.82	0.66
θ_s^2	0.93	1.05	1.25
$\ln(\theta_r)$	0.75	−4.16	38.34
$\sqrt{\theta_r}$	0.93	−1.09	1.77
$1/\theta_r$	0.03	21.09	444.94
θ_r^2	0.89	1.34	2.29
e^{θ_r}	0.99	0.39	0.12
$\theta_r^{1/3}$	0.83	−2.02	5.78
$\ln(\alpha)$	0.72	−2.94	11.8
$1/\alpha$	0.03	21.17	450.93
e^{α}	0.62	5.01	44.15
α^2	0.20	13.89	238.13
$\sqrt{\alpha}$	0.95	0.91	2.27
$\text{Log}(n)$	0.98	0.49	0.94
$\sqrt{n}$	0.91	1.45	3.67
n^2	0.46	6.23	56.04
$1/n$	0.93	1.32	3.93
e^{n}	0.24	12.69	201.48

selected transformed dependent variables (moisture retention parameters) and the transformed independent variables (basic soil properties) was notably different for each predictor variable transformation. For instance, the linear regression performed between θ_s and sand yielded a quality of fit, measured by the value of R^2, to be 0.10 while that between θ_s and e^{sand} yielded $R^2 = 0.01$ (**Table 2**). The same observation was made in the case of regresson performed with silt, clay, bulk density and organic carbon. Also each selected transformed variable (*i.e.* $\ln(\theta_s)$, e^{θ_r}, $\sqrt{\alpha}$, etc.) produced different values of R^2 when a linear regression was performed between each of the response variables and a transformed predictor varile for each of the variables sand, silt, clay bulk density and organic carbon. For example, a regression performed between the independent variable $sand^2$ and θ_s^{-1} yielded $R^2 = 0.14$ while that between $sand^2$ and e^{θ_r} produced $R^2 = 0.20$, hence no particular transformed variable could be said to generally give the best quality of fit with all the dependent variables whether it is sand, silt, clay, bulk density or organic carbon. Also, there is no particular response variable that could be said to yield the best quality of fit compared to others for all the independent variables or transformations to which regression was performed. The purpose of regression analysis was to identify the pair of transformed dependent and independent variables that yield the best quality of fit between them as measured by coefficient of determination (R^2) so that for each dependent variable, one is able to identify which transformations of each of the dependent variables would be used in developing the pedotransfer functions during multiple regression equation to relate each response variable and selected predictor variables and appropriate transformations. This would help establish the best possible mathematical relationship between the moisture retention parameters (response variables) and selected

Table 2. Measurement of the quality of fit based on coefficient of determination (R^2) between the moisture retention parameters (response variable) and sand with its transformation (predictor variables) in the linear regression.

	Sand	$Sand^2$	$Sand^3$	$Sand^5$	$Sand^{10}$	$Sand^{15}$	$\sqrt{Sand}$	$\ln(sand)$	$sand^{-1}$	e^{Sand}
θ_s	$R^2 = 0.10$	$R^2 = 0.10$	$R^2 = 0.08$	$R^2 = 0.06$	$R^2 = 0.03$	$R^2 = 0.03$	$R^2 = 0.09$	$R^2 = 0.06$	$R^2 = 0.01$	$R^2 = 0.01$
θ_s^{-1}	$R^2 = 0.14$	$R^2 = 0.14$	$R^2 = 0.12$	$R^2 = 0.08$	$R^2 = 0.05$	$R^2 = 0.04$	$R^2 = 0.12$	$R^2 = 0.08$	$R^2 = 0.01$	$R^2 = 0.02$
$\ln(\theta_s)$	$R^2 = 0.12$	$R^2 = 0.12$	$R^2 = 0.10$	$R^2 = 0.10$	$R^2 = 0.07$	$R^2 = 0.04$	$R^2 = 0.12$	$R^2 = 0.08$	$R^2 = 0.01$	$R^2 = 0.02$
θ_R	$R^2 = 0.21$	$R^2 = 0.21$	$R^2 = 0.19$	$R^2 = 0.15$	$R^2 = 0.04$	$R^2 = 0.03$	$R^2 = 0.11$	$R^2 = 0.07$	$R^2 = 0.01$	$R^2 = 0.02$
e^{θ_r}	$R^2 = 0.20$	$R^2 = 0.20$	$R^2 = 0.18$	$R^2 = 0.13$	$R^2 = 0.07$	$R^2 = 0.05$	$R^2 = 0.18$	$R^2 = 0.13$	$R^2 = 0.02$	$R^2 = 0.03$
$\sqrt{\alpha}$	$R^2 = 0.01$	$R^2 = 0.01$	$R^2 = 0.02$	$R^2 = 0.02$	$R^2 = 0.01$	$R^2 = 0.01$	$R^2 = 0.00$	$R^2 = 0.00$	$R^2 = 0.00$	$R^2 = 0.02$
$\ln(\alpha)$	$R^2 = 0.01$	$R^2 = 0.03$	$R^2 = 0.05$	$R^2 = 0.07$	$R^2 = 0.09$	$R^2 = 0.09$	$R^2 = 0.01$	$R^2 = 0.00$	$R^2 = 0.00$	$R^2 = 0.08$
$\text{Log}(n)$	$R^2 = 0.01$	$R^2 = 0.13$	$R^2 = 0.15$	$R^2 = 0.15$	$R^2 = 0.16$	$R^2 = 0.15$	$R^2 = 0.15$	$R^2 = 0.08$	$R^2 = 0.02$	$R^2 = 0.08$
$\sqrt{n}$	$R^2 = 0.11$	$R^2 = 0.15$	$R^2 = 0.18$	$R^2 = 0.21$	$R^2 = 0.20$	$R^2 = 0.12$	$R^2 = 0.08$	$R^2 = 0.05$	$R^2 = 0.01$	$R^2 = 0.10$
n^{-1}	$R^2 = 0.07$	$R^2 = 0.08$	$R^2 = 0.09$	$R^2 = 0.08$	$R^2 = 0.09$	$R^2 = 0.06$	$R^2 = 0.06$	$R^2 = 0.04$	$R^2 = 0.01$	$R^2 = 0.03$

basic soil properties (independent variables) in the pedotransfer function. A number of observations on the regression relationships can be made for each response variable and how it relates with each of the predictor variables and their transformations. The transformed response variable θ_s^{-1} yielded the best quality of fit ($R^2 = 0.14$) with the independent variables $sand$ or $sand^2$ compared to the other transformations when a regression analysis was performed. In the case of regression with silt, the best possible quality of fit was obtained with the transformation $silt^{-1}$ ($R^2 = 0.04$) and so was with bulk density, BD^3 ($R^2 = 0.60$), $\sqrt{clay}$ or $clay^2$ ($R^2 = 0.12$), organic carbon, $\ln(orgC)$ ($R^2 = 0.27$). In general, the linear regression performed between θ_s^{-1} and the bulk density transformations produced best quality of fit as compared to those of silt, sand, organic carbon and clay, showing that the best relationship was obtained with bulk density followed by organic carbon, sand, clay and silt in

that order. In the case of residual moisture content, θ_r, the selected transformation variable e^{θ_r} related best with clay with a quality of fit $R^2 = 0.32$ obtained when regression was performed between e^{θ_r} and $\sqrt{clay}$. The poorest fit was obtained when the regression was performed with transformations of organic carbon e.g. $\ln(orgC)$ with $R^2 = 0.01$. α yielded the best possible quality of fit when a regression was performed with $\sqrt{clay}$ or $1/clay$. $\sqrt{\alpha}$ generally showed poor quality of fit with most of the independent variables or transformations. The transformation n^{-1} also showed poor quality fit with most of the selected response variables when regression analysis was performed. The best quality fit was obtained with the transformation of sand when the linear regression was performed between n^{-1} and $sand^3$ ($R^2 = 0.09$). **Figures 3(a)-(d)** illustrates the regression between selected transformations of response variables and the predictor variables to which they best fit.

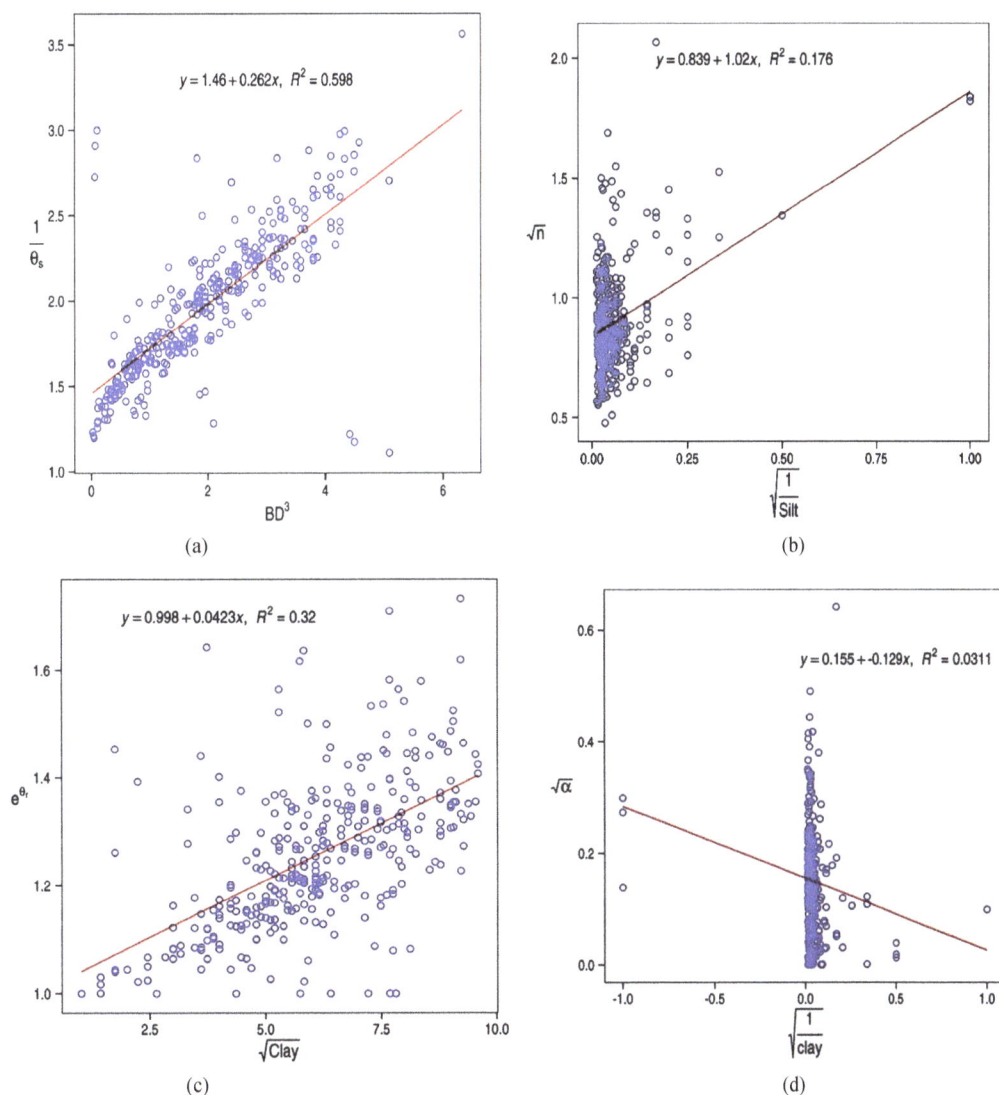

(a)

(b)

(c)

(d)

Figure 3. Response versus predictor variables correlations with corresponding best possible quality fit.

After performing linear regression between two variables involving the selected transformed response variables, associated with the moisture retention parameters, and the predictor variables associated with the basic soil properties, the following conclusions were arrived at. For the transformed response variable θ_s^{-1} the predictor variables that yielded the best possible quality of fit with it when linear regression was performed were sand, silt^{-1}, BD^3, clay^2, and ln(orgC). These variables would then be used in developing the multiple linear regression between the said response variable and the predictor variables. In the case of e^{θ_r}, the best possible fit (measured by the value of R^2) was obtained from linear regression with sand, silt^{10}, BD^2, $\sqrt{clay}$, and ln(orgC), while in the case of $\sqrt{\alpha}$ the best possible fits were obtained with e^{sand}, Silt^{-1}, BD, Clay^{-1}, and ln(orgC). In the case of $\sqrt{n}$, the best possible quality of fit was obtained with the predictor variables Sand^5, silt^{-1}, ln(clay), (orgC^2)$^{-1}$.

3.3. Multiple Regression Analysis and Accuracy of the Pedotransfer Functions Developed

Table 3 shows the quality of fit obtained when multiple linear regression was performed between each selected response variable and the transformed predictor variables to which it relates best e.g. a multiple linear regression performed between the response variable θ_s^{-1} and sand, silt^{-1}, BD^3, clay^2, ln(orgC), yielded the quality of fit measured by $R^2 = 0.63$ which is considered acceptable and therefore accurate. The resulting equation is illustrated below Equation (12). The predictor variables indicated above that related best in linear regression with the selected response variables were the ones considered in developing the pedotransfer functions using multiple linear regression.

$$\theta_s^{-1} = 1.447 + 0.003045\text{sand} - 0.1505\text{silt}^{-1}$$
$$+ 0.2126 BD^3 - 006\text{clay}^2 - 0.04542\ln(\text{orgC}) \quad (12)$$

where θ_s is the saturated soil moisture content (cm^3/cm^3), silt the % silt, sand the % sand, BD the bulk density (g/cm^3), clay the % clay, and orgC the Organic carbon content (g/kg). The qualities of fit for the remaining transformed response variables are illustrated in **Table 3** for each of e^{θ_r}, $\sqrt{\alpha}$ and $\sqrt{n}$ when the simple linear regression was done with the selected predictor variable transformations.

3.4. Analysis of Cross Correlation and Correlation Matrix

Cross correlation was done for the independent variables (transformations) used in developing multiple linear regression for each of the response variables θ_s^{-1}, e^{θ_r}, $\sqrt{\alpha}$ and $\sqrt{n}$. **Tables A1-A4** (Appendix) shows the relationships, based on correlation coefficient, between the various predictor variables used for each of the response variable transformations. In the case of θ_s^{-1} the predictor variables *sand* and $\sqrt{clay}$ appear to be highly negatively correlated with $r = -0.71$ while BD^3 and ln(orgC) are also fairly correlated in a negative sense ($r = -0.59$). For the dependent variable e^{θ_r}, the response variables BD^3 and ln(orgC) are also closely correlated in the negative sense with $r = -0.62$. Other reasonable correlations include that between e^{sand} and silt^{-2} ($r = 0.89$), BD and ln(orgC) ($r = -0.64$) in the case of predictor variables for $\sqrt{\alpha}$. The other significant correlations observed is that between sand^5 to (silt)$^{-1}$ ($r = 0.74$) and sand^5 to ln(Clay) ($r = -0.69$) in the predictor variables for $\sqrt{n}$.

3.5. Multiple Linear Regression Results

For each of the selected response variables, **Table 3** shows the various possible combinations of the predictor variables that were used in performing the multiple linear regression and also indicates the qualities of fit obtained based on the coefficient of determination R^2.

Table 3. Measurement of Quality of fit based on coefficient of determination (R^2) between the moisture retention parameters (response variables) and selected predictor variable transformations in the multiple linear regression.

	sand, silt^{-1}, BD^3, clay^2, ln(orgC)	sand, e^{BD}, sand	silt^{-1}, BD^3	$\sqrt{clay}$, BD^2, silt^{-1}	BD^3, ln(orgC)	BD^3, $\sqrt{clay}$	sand, BD^3
θ_r^{-1}	$R^2 = 0.63$	$R^2 = 0.63$	$R^2 = 0.63$	$R^2 = 0.61$	$R^2 = 0.59$	$R^2 = 0.61$	$R^2 = 0.63$
	sand, silt^{10}, BD^2, $\sqrt{clay}$, ln(orgC)	sand, $\sqrt{clay}$	$\sqrt{clay}$, silt^{10}, BD^2	clay, silt^{10}, BD^2	$\sqrt{clay}$, Silt^{10}, ln(orgC)	$\sqrt{clay}$, BD^2	$\sqrt{clay}$, ln(orgC)
e^{θ_r}	$R^2 = 0.37$	$R^2 = 0.33$	$R^2 = 0.35$	$R^2 = 0.32$	$R^2 = 0.32$	$R^2 = 0.35$	$R^2 = 0.32$
	e^{sand}, silt^{-2}, BD, clay^{-1}, ln(orgC)	e^{sand}, BD, clay^{-1}	e^{sand}, clay^{-1}, ln(orgC)	clay^{-1}, silt^{-2}, ln(orgC)	silt^{-2}, BD, clay^{-1}	BD, clay^{-1}	silt^{-1}, clay^{-1}
$\sqrt{\alpha}$	0.05	0.05	0.04	0.04	0.05	0.04	0.04
	sand^5, 1/silt, ln(clay), orgC^{-2}	sand^5, orgC^{-2}	silt^{-1}, ln(clay)	ln(clay), or GC^{-2}	orgC^{-2}, BD^3	ln(clay), BD^3	sand^5, BD^3
$\sqrt{n}$	$R^2 = 0.24$	$R^2 = 0.21$	$R^2 = 0.23$	$R^2 = 0.12$	$R^2 = 0.04$	$R^2 = 0.11$	$R^2 = 0.20$

The various possible multiple linear regression equations obtained for the various independent variables are underlisted (Equations (13)-(27)). The choice of the independent variables or their transformations was based on combinations that would yield the highest possible quality of fit *i.e.* the predictor variable combinations chosen include the variable that relate very well with the response variable, hence the equations are the best possible equations that could be obtained using the variables indicated. Equations developed involving θ_s^{-1} are as follows with the measures of accuracy based on coefficient of determination (R^2) indicated:

$$\frac{1}{\theta_s} = 0.813 + 0.003534 \text{sand} \tag{13}$$
$$+ 0.3099 e^{BD} \quad \left(R^2 = 0.63\right)$$

$$\frac{1}{\theta_s} = 1.397 + 0.00317 \text{sand} \tag{14}$$
$$+ 0.2451 BD^3 \quad \left(R^2 = 0.63\right)$$

$$\frac{1}{\theta_s} = 1.397 + 0.003385 \text{sand} \tag{15}$$
$$- 0.1433 \frac{1}{\text{silt}} + 0.246 BD^3 \quad \left(R^2 = 0.63\right)$$

$$\frac{1}{\theta_s} = 1.477 + 0.003045 \text{sand}$$
$$- 0.1505 \frac{1}{\text{silt}} + 0.212 BD^3 - 9.3857E \tag{16}$$
$$- 0.006 \text{clay}^2 - 0.0452 \ln(\text{orgC}) \quad \left(R^2 = 0.63\right)$$

The set of possible multiple linear regression equations developed involving e^{θ_r} are listed as follows:

$$e^{\theta_r} = 1.083 + 0.0382\sqrt{\text{clay}} - 5.0290E$$
$$- 021 \text{silt}^{10} - 00392 BD^2 \quad \left(R^2 = 0.35\right) \tag{17}$$

$$e^{\theta_r} = 1.062 - 0.0007465 \text{sand}$$
$$+ 0.03532\sqrt{\text{clay}} \quad \left(R^2 = 0.33\right) \tag{18}$$

$$e^{\theta_r} = 1.009 + 0.04099\sqrt{\text{clay}} - 4.9035E - 021 \text{silt}^{10}$$
$$+ 0.004877 \ln(\text{orgC}) \quad \left(R^2 = 0.32\right) \tag{19}$$

$$e^{\theta_r} = 1.007 + 0.03896\sqrt{\text{clay}}$$
$$- 0.03928 BD^2 \quad \left(R^2 = 0.35\right) \tag{20}$$

The set of possible multiple linear regression equations developed involving $\sqrt{\alpha}$ are listed as follows:

$$\sqrt{\alpha} = 0.18 - 1.0440E - 043 e^{\text{sand}}$$
$$- \frac{0.071381}{\text{silt}^2} - 0.02068 BD - \frac{0.088661}{\text{clay}} \tag{21}$$
$$+ 0.0008447 \ln(\text{orgC}) \quad \left(R^2 = 0.05\right)$$

$$\sqrt{\alpha} = 0.1847 - 2.0502E - 043 \exp(\text{sand})$$
$$- 0.02556 BD - 0.09316 \frac{1}{\text{clay}} \quad \left(R^2 = 0.05\right) \tag{22}$$

$$\sqrt{\alpha} = 0.1564 - 2.0817E - 043 e^{\text{sand}}$$
$$- 0.08877 \frac{1}{\text{clay}} + 0.004372 \ln(\text{orgC}) \quad \left(R^2 = 0.04\right) \tag{23}$$

The set of possible multiple linear regression equations developed involving $\sqrt{n}$ are listed as follows:

$$\sqrt{n} = 0.9871 + 4.0051E - 011 \text{sand}^5 + 0.623 \frac{1}{\text{silt}}$$
$$- 0.03977 \ln(\text{Clay}) - \frac{0.00006391}{\text{orgC}^2} \quad \left(R^2 = 0.24\right) \tag{24}$$

$$\sqrt{n} = 1.101 + 0.8702 \frac{1}{\text{silt}} - 0.07228 \ln(\text{clay}) \left(R^2 = 0.24\right) \tag{25}$$

$$\sqrt{n} = 0.8617 + 8.4062E - 011 \text{sand}^5$$
$$+ 2.2259E - 005 \left(\frac{1}{\text{orgC}^2}\right) \quad \left(R^2 = 0.21\right) \tag{26}$$

$$\sqrt{n} = 1.22 - 0.09247 \ln(\text{clay}) + \frac{0.00011681}{\text{orgC}^2} \quad \left(R^2 = 0.12\right) \tag{27}$$

Based on the above indicated equations that produced the best quality of fit, equations relating the moisture retention parameters that are not transformed to the trans formed basic properties were also compared to the measured (fitted) parameters and yielded the values of R^2 indicated (Equations (28) to (31)) alongside the equations.

$$\theta_s = \left(0.8103 + 0.003534 \text{sand} + 0.3099 e^{BD}\right)^{-1} \left(R^2 = 0.77\right) \tag{28}$$

3.6. Validation of the Developed Pedotransfer Functions to Evaluate Reliability

Table 4 shows regression equations developed for each of

$$\theta_r = \ln\left(1.083 + 0.0382\sqrt{\text{clay}} - 5.029E - 0.021 \text{silt}^{10} - 0.0329 BD^2\right) \quad \left(R^2 = 0.43\right) \tag{29}$$

$$n = \left(0.8617 + 8.4062E - 011 \text{sand}^5 + 2.2259E - 0.005 \frac{1}{\text{orgC}^2}\right)^2 \quad \left(R^2 = 0.28\right) \tag{30}$$

Table 4. Summary of equations for transformations of van Genuchten parameters and their performance in reliability.

Response variable	Equation	R^2	RMSE	RSR	NSE	ME	PBIAS
θ_s^{-1}	$\theta_s^{-1} = 0.8103 + 0.003534\text{sand} + 0.3099e^{(BD)}$	0.76	2.09	0.50	0.75	0.001	0.07
θ_s^{-1}	$\theta_s^{-1} = 1.397 + 0.003176\text{sand} + 0.2451BD^3$	0.76	2.12	0.50	0.75	0.004	0.19
θ_s^{-1}	$\theta_s^{-1} = 1.477 + 0.003045\text{SAND} - 0.15051/\text{silt} + 0.2126BD^3$ $-9.3857E - 006\text{clay}^2 - 0.04542\ln(\text{orgC})$	0.74	2.11	0.51	0.74	0.009	0.44
θ_s^{-1}	$\theta_s^{-1} = 1.397 + 0.003385\text{sand} - 0.1433\text{silt}^{-1} + 0.246BD^3$	0.76	2.13	0.51	0.74	0.006	0.31
e^{θ_r}	$e^{\theta_r} = 1.083 + 0.0382\sqrt{\text{clay}} - 5.0290E - 021\text{silt}^{10} - 0.0392BD^2$	0.41	1.10	0.77	0.41	0.011	0.90
e^{θ_r}	$e^{\theta_r} = 1.062 - 0.0007465\text{sand} + 0.03532\sqrt{\text{clay}}$	0.36	1.14	0.80	0.36	0.008	0.63
e^{θ_r}	$e^{\theta_r} = 1.009 + 0.04099\sqrt{\text{clay}} - 4.9035E - 021\text{silt}^{10}$ $+0.004877\ln(\text{orgC})$	0.40	1.11	0.78	0.39	0.007	0.55
e^{θ_r}	$e^{\theta_r} = 1.007 + 0.03896\sqrt{\text{clay}} - 0.03928BD^2$	0.41	1.10	0.77	0.40	0.011	0.90
e^{θ_r}	$e^{\theta_r} = 1.003 + 0.0417\sqrt{\text{clay}} + 0.005286\ln(\text{orgC})$	0.38	1.12	0.79	0.38	0.007	0.57
$\sqrt{\alpha}$	$\sqrt{\alpha} = 0.18 - 1.0440E - 043\exp(\text{sand}) - 0.07138\text{silt}^{-1}$ $-0.02068BD - 0.088661/\text{clay} + 0.0008447\ln(\text{orgC})$	0.04	0.85	1.00	0.01	−0.012	−8.82
$\sqrt{\alpha}$	$\sqrt{\alpha} = 0.1847 - 2.0502E - 043e^{\text{sand}} - 0.02556BD - 0.09316\text{clay}^{-1}$	0.04	0.85	0.99	0.01	−0.011	−8.29
$\sqrt{\alpha}$	$\sqrt{\alpha} = 0.1564 - 2.0817E - 043e^{\text{sand}} - 0.088771/\text{clay}$ $+0.004372\ln(\text{orgC}).$	0.06	0.85	0.99	0.02	−0.013	−9.39
$\sqrt{n}$	$\sqrt{n} = 0.9871 + 4.0051E - 011\text{sand}^5 + 0.623\text{silt}^{-1}$ $-0.03977\ln(\text{Clay}) - 0.00006391/\text{orgC}^2$	0.20	1.89	0.90	0.20	−0.006	−0.62
$\sqrt{n}$	$\sqrt{n} = 1.101 + 0.87021/\text{silt} - 0.07228\ln(\text{Clay})$	0.18	1.91	0.91	0.17	−0.002	0.17
$\sqrt{n}$	$\sqrt{n} = 0.8617 + 8.4062E - 011\text{sand}^5 + 2.2259E - 0051/\text{orgC}^2$	0.20	1.88	0.89	0.20	−0.005	−0.59
$\sqrt{n}$	$\sqrt{n} = 1.22 - 0.09247\ln(\text{Clay}) + 0.00011681/\text{orgC}^2$	0.08	2.02	0.96	0.08	−0.002	0.19

$$\alpha = \left(0.1564 - 2.0817E - 043e^{\text{sand}} - 0.08877\frac{1}{\text{clay}} + 0.004372\text{Ln}\left(\text{orgC}\right)\right)^2 \quad \left(R^2 = 0.04\right) \qquad (31)$$

the selected transformations of the moisture retention parameters of the van Genutchen moisture retention equation and measures of their performance in validation based on the statistical measures indicated. The transformed model parameter that yielded the best reliability is the saturated soil moisture content θ_s^{-1} (considering the measure of coefficient of determination ($R^2 = 0.76$). This indicates that θ_s is reasonably well predicted by the pedotransfer function developed. Prediction of transformed variable $\sqrt{\alpha}$ yielded the lowest possible value for the selected measure for the quality of fit ($r = 0.25$) reflecting the lowest level of reliability compared to the other pedotransfer functions developed for moisture retention parameters under consideration. The transformed variable for residual moisture content e^{θ_r} yielded modest performance with the best possible value of $R^2 = 0.41$.

[4] notes that continuous pedotransfer functions can be applied in the case of more site specific applications, where measured data is available, since they do not provide site specific information. The pedotransfer functions predict soil hydraulic characteristics for broadly defined textural classes. **Figures 4-7** illustrates the performance of the developed pedotransfer functions for the transformed response variables based on analysis by linear regression.

4. Conclusion and Recommendation

The performance of developed pedotransfer functions for estimating moisture retention parameters in the van Genuchten moisture retention equation varied for the individual moisture retention parameters. Equations deve-

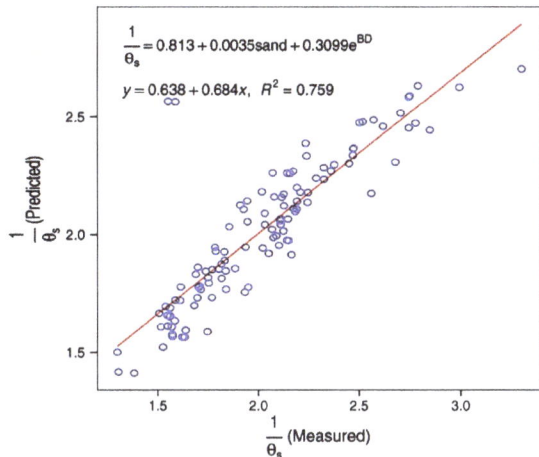

$$\frac{1}{\theta_s} = 0.813 + 0.0035\,sand + 0.3099\,e^{BD}$$

$$y = 0.638 + 0.684x, \quad R^2 = 0.759$$

Figure 4. Predicted value of transformed variable θ_s^{-1} versus measured by correlation for a developed equation that yielded the best possible quality of fit.

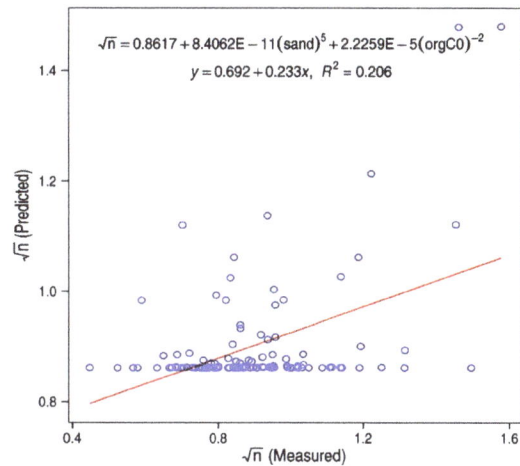

$$e^{\theta_r} = 1.083 + 0.0382(clay)^{\frac{1}{2}} - 5.029E - 0.21(silt)^{10} - 0.0392(BD)^2$$

$$y = 0.753 + 0.397x, \quad R^2 = 0.413$$

Figure 5. Predicted value of transformed variable e^{θ_r} versus measured by correlation for a developed equation that yielded the best possible quality of fit.

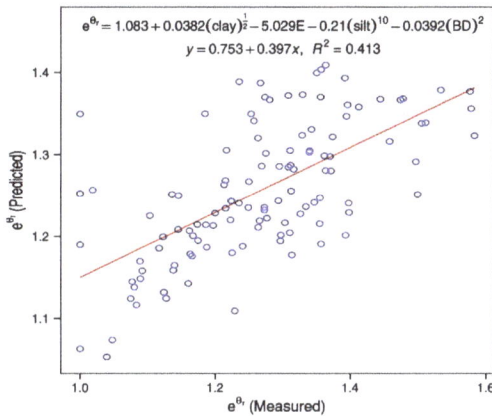

$$\sqrt{\alpha} = 0.156 - 2.0817E - 43e^{sand} - 0.0888(clay)^{-1} + 0.00437\,LN(orgC)$$

$$y = 0.146 + 0.0289x, \quad R^2 = 0.0635$$

Figure 6. Predicted value of transformed variable $\sqrt{\alpha}$ versus measured by correlation for a developed equation that yielded the best possible quality of fit.

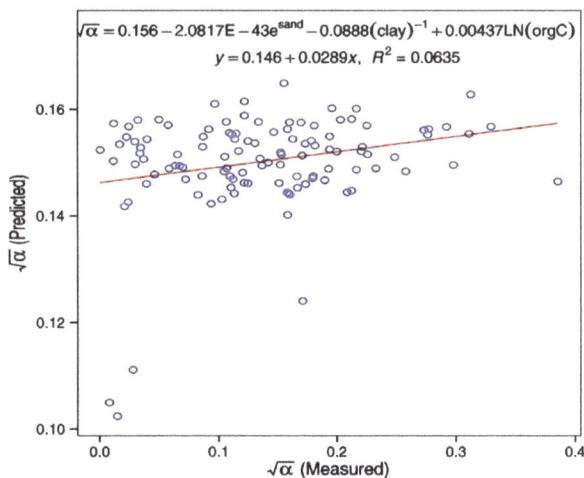

$$\sqrt{n} = 0.8617 + 8.4062E - 11(sand)^5 + 2.2259E - 5(orgC0)^{-2}$$

$$y = 0.692 + 0.233x, \quad R^2 = 0.206$$

Figure 7. Predicted value of transformed variable $\sqrt{n}$ versus measured by correlation for a developed equation that yielded the best possible quality of fit.

loped for predicting saturated soil moisture content (θ_s) yielded the best performance in model validation with the highest possible value of coefficient of determination $R^2 = 0.76$, $RMSE = 2.09$ and a NSE of 0.75. The relatively poorest performance was observed with the parameter "α" with the best possible value of R^2 being 0.06 indicating unsatisfactory performance and a $RMSE$ value of 0.85. The remaining parameters (θ_r and "n") registered performance intermediate between the aforementioned two extremes. It is therefore evidently not possible to obtain excellent performance in pedotransfer functions for all the parameters. The equations derived from these parameters would be the best possible obtainable for estimating hydraulic conductivity. Continued effort should be made in developing more pedotransfer functions to provide users with more alternative equations and especially those PTFs that use the very basic soil properties available in many local soils data bases.

5. Acknowledgements

I wish to acknowledge the support provided by the Deans Committee Research Grant, University of Nairobi (UoN) who funded the research project, The Principal's office College of Architecture and Engineering (UoN) who funded journal publication fees. I also wish to acknowledge support given by staff of the department of Environmental and Biosystems Engineering (UoN) who contributed useful academic ideas that enriched this document.

REFERENCES

[1] H. R. Fooladmad, "Pedotransfer Functions for Point Estimation of Soil Moisture Characteristic Curve in Some Iranian Soils," *African Journal of Agricultural Research*,

Vol. 6, No. 6, 2011, pp. 1586-1591.

[2] J. P. O. Obiero, "Evaluation of Infiltration Using Green Ampt Model and Rainfall Runoff Data for Lagan and Sambret Catchments, Kericho, Kenya," M.Sc. Thesis, Department of Agricultural Engineering, University of Nairobi, Nairobi, 1996.

[3] A. Rasoulzadeh, "Estimating Hydraulic Conductivity Using Pedotransfer Functions," In: L. Elango, Ed., *Hydraulic Conductivity Issues, Determination, and Applications*, Intech, 2011, pp. 145-164.

[4] K. E. Saxton and W. J. Rawls, "Soil Water Characteristic Estimates by Texture and Organic Matter for Hydrologic Solutions," *Soil Science Society of America Journal*, Vol. 70, No. 5, 2006, pp. 1569-1578.

[5] M. G. Schaap and F. J. Leij, "Using Neural Networks to Predict Soil Water Retention and Soil Hydraulic Conductivity," *Soil and Tillage Research*, Vol. 47, No. 1-2, 1998, pp. 37-42.

[6] J. Tomasella and M. Hodnett, "Pedotransfer Functions in Tropical Soils," In: Y. Pachepsky and W. J. Rawls, Eds., *Development of Pedotransfer Functions in Soil Hydrology. Developments in Soil Science*, Vol. 30, Elsevier B. V., Amsterdam, 2004, pp. 415-429.

[7] H. Vereecken and M. Herbst, "Statistical Regression," In: Y. Pachepsky and W. J. Rawls, Eds., *Development of Pedotransfer Functions in Soil Hydrology. Developments in Soil Science*, Vol. 30, Elsevier B. V., Amsterdam, 2004, pp. 3-19.

[8] R. T. Walczack, F. Moreno, C. Slawinski, E. Fernandez and J. L. Arrue, "Modelling of Soil Water Retention Curve Using Soil Solid Phase Parameters," *Journal of Hydrology*, Vol. 329, No. 3-4, 2006, pp. 527-533.

[9] T. Wells, S. Fityus, D. W. Smith and H. Moe, "The Indirect Estimation of Saturated Hydraulic Conductivity of Soils, Using Measurements of Gas Permeability. I. Laboratory Testing with Dry Granular Soils," *Australian Journal of Soil Research*, CSIRO Publishing, 2006. http://www.highbeam.com

[10] J. H. Wosten, M. Y. A. Pachepsky and W. J. Rawls, "Pedotransfer Functions: Bridging the Gap between Available Basic Soil Data and Missing Soil Hydraulic Characteristics," *Journal of Hydrology*, Vol. 251, No. 3-4, 2001, pp. 123-150.

[11] N. H. Batjes, Ed., "A Homogenized Soil Data File for Global Environmental Research: A Subset of FAO, ISRIC, and NRCS Profiles (Version 1.0)," Working Paper and Preprint 95/10b, International Soil Reference and Information Centre, Wageningen, 1995. http://www.isric.org/NR/exeres/545B0669-6743-402B-B79A-DBF57E9FA67F.htm

[12] M. van Genutchen, "A Closed Form Equation for Predicting the Hydraulic Conductivity of Unsaturated Soils," *Soil Science Society of America Journal*, Vol. 44, No. 5, 1980, pp. 892-898.

Appendix

Analysis of Cross Correlations

Table A1. Cross correlation tests for transformed predictor variables for θ_s^{-1}.

	Sand	Silt^{-1}	BD^3	$\sqrt{\text{clay}}$	ln(orgC)
Sand	$r = 1$	$r = 0.42$	$r = 0.26$	$r = -0.71$	$r = -0.10$
Silt^{-1}	$r = 0.42$	$r = 1$	$r = 0.19$	$r = -0.13$	$r = -0.25$
BD^3	$r = 0.26$	$r = 0.19$	$r = 1$	$r = -0.28$	$r = -0.59$
$\sqrt{\text{clay}}$	$r = -0.71$	$r = -0.13$	$r = -0.28$	$r = 1$	$r = 0.08$
ln(orgC)	$r = -0.10$	$r = -0.25$	$r = -0.59$	$r = 0.08$	$r = 1$

Table A2. Cross correlation tests for transformed predictor variables in e^{θ_r}.

	Sand	Silt^{10}	BD^2	$\sqrt{\text{clay}}$	ln(orgC)
Sand	$r = 1$	$r = -0.25$	$r = 0.23$	$r = -0.71$	$r = -0.10$
Silt^{10}	$r = -0.25$	$r = 1$	$r = 0.06$	$r = -0.21$	$r = -0.09$
BD^2	$r = 0.23$	$r = 0.06$	$r = 1$	$r = -0.24$	$r = -0.09$
$\sqrt{\text{clay}}$	$r = -0.71$	$r = -0.21$	$r = -0.24$	$r = 1$	$r = 0.08$
ln(orgC)	$r = -0.10$	$r = -0.09$	$r = -0.62$	$r = 0.08$	$r = 1$

Table A3. Cross correlation tests for transformed predictor variables in $\sqrt{\alpha}$.

	e^{sand}	silt^{-2}	BD	clay^{-1}	ln(orgC)
e^{sand}	$r = 1$	$r = 0.89$	$r = 0.13$	$r = 0.3$	$r = -0.21$
silt^{-2}	$r = 0.89$	$r = 1$	$r = 0.14$	$r = 0.23$	$r = -0.24$
BD	$r = 0.13$	$r = 0.14$	$r = 1$	$r = 0.22$	$r = -0.64$
clay^{-1}	$r = 0.30$	$r = 0.23$	$r = 0.22$	$r = 1$	$r = 0.08$
ln(orgC)	$r = -0.21$	$r = -0.24$	$r = -0.64$	$r = 0.08$	$r = 1$

Table A4. Cross correlation tests for transformed predictor variables in $\sqrt{n}$.

	sand^5	Silt^{-1}	ln(clay)	orgC^{-1}
sand^5	$r = 1$	$r = 0.74$	$r = -0.69$	$r = 0.34$
Silt^{-1}	$r = 0.74$	$r = 1$	$r = -0.25$	$r = 0.52$
ln(clay)	$r = -0.69$	$r = -0.25$	$r = 1$	$r = -0.21$
orgC^{-1}	$r = 0.34$	$r = 0.52$	$r = -0.21$	$r = 1$

11

Application of MODIS-Based Monthly Evapotranspiration Rates in Runoff Modeling: A Case Study in Nebraska, USA

Jozsef Szilagyi[1,2]

[1]Department of Hydraulic and Water Resources Engineering, Budapest University of Technology and Economics, Budapest, Hungary; [2]School of Natural Resources, University of Nebraska-Lincoln, Lincoln, USA.

ABSTRACT

Daily and monthly flow-rates of the Little Nemaha River in Nebraska were simulated by the lumped-parameter Jakeman-Hornberger as well as a distributed-parameter water-balance accounting procedure for the 2003-2008 and 2000-2009 periods, respectively, with and without the help of the MODIS-based monthly estimates of evapotranspiration (ET) rates. While the daily lumped-parameter model simulation accuracy remained practically unchanged with the inclusion of the monthly MODIS-based ET rates interpolated into daily values (R^2 of 0.66 vs 0.68, simulated to measured runoff ratio remaining the same 96%), the monthly water-balance accounting model outcomes did improve to some extent (from an R^2 of 0.67 to 0.7 with simulated to measured runoff ratio of 72% vs 115%). In both cases the models had to be slightly modified for accommodation of the ET rates as predefined input values, not present in the original model setups. These results indicate the potential practical usefulness of satellite-derived ET estimates (CREMAP values in the present case) in monthly water-balance modeling. CREMAP is a calibration-free ET estimation method based on MODIS-derived daytime surface temperature values in combination of basic climatic variables, such as air temperature, humidity and solar radiation within a Complementary Relationship framework of evaporation.

Keywords: Remotely Sensed Evapotranspiration; Complementary Relationship; Runoff Modeling

1. Introduction

With the free public availability of the ~1-km, global, Moderate Resolution Spectroradiometer (MODIS, available at https://lpdaac.usgs.gov/lpdaac/products/modis_products_table) data since 2000, the number of remote-sensing based, basin-scale evapotranspiration (ET) estimation methods have seen an unprecedented growth. For a review of the available techniques, see [1]. In general, the different approaches are based on the application of a vegetation index and/or the land surface temperature to solve the energy balance equation at the ground. Common to all these methods is that they are based on simplifying assumptions and require some sort of parameter calibration, typically aided by precipitation and runoff data at the basin scale. Probably the simplest and totally calibration-free MODIS-based ET estimation method (called CREMAP) was proposed by Szilagyi et al. [2] and Szilagyi [3] with subsequent demonstration of its

effectiveness and practical usefulness in different studies (see below), mostly involving groundwater recharge estimation.

The CREMAP method makes use of the scale of the MODIS pixel of about 1-km, by assuming that each pixel would contain a mixture of vegetation, thus differences in albedo, surface roughness (also assumed to be as much influenced by elevation changes within and around a given MODIS pixel as by vegetation type) and net radiation among neighboring cells are considered negligible over a flat or rolling vegetated terrain and over the typical computational time interval of a month. The method estimates the regional monthly evapotranspiration rate for each MODIS pixel employing the Complementary Relationship (CR) of Bouchet [4] by using the WREVAP program of Morton et al. [5] with inputs of monthly minimum, maximum air and dew-point temperatures (Prism Climate Group [6]), combined with the incident solar radiation data from GCIP/SRB [7]. The regional ET rate is then related to the regional mean of the daily surface temperature value (T_s) of MODIS (except in the winter months when patchy snow cover may grossly violate the assumed near constancy of the albedo values), and is assumed that deviations of the T_s values among the individual cells from the regional mean are directly proportional to fluctuations in sensible heat fluxes and, due to an assumed spatially constant net radiation term, in latent heat fluxes as well, around the CR-obtained regional mean. The resulting monthly linear transformations require another T_s vs ET pair which can be obtained by specifying the ET rate of the coldest, and thus wettest MODIS cell in the region, via the Priestley-Talor [8] equation. In the winter months each cell is assigned the regional ET rate.

While the CREMAP ET rates have proven valuable for specifying spatially varying regional recharge rates in Nebraska and Hungary [9-11], as well as defining net recharge to the groundwater as a function of vadose-zone depth in the shallow groundwater area of the Platte River in Nebraska [12], their practical value in runoff modeling has not been investigated, thus motivating the present study.

It is widely accepted among regional groundwater modelers that a groundwater model-independent estimation of the external forcing, such as recharge to the groundwater, is highly desirable for groundwater model calibration in order to reduce the number of unknown parameters as well as to separate uncertainties in the recharge estimates from those of the inherent groundwater model parameters, such as the spatially highly variable saturated hydraulic conductivity. The same should apply to runoff modeling, especially for spatially distributed watershed models: "a-priori" knowledge of the relevant ET rates should improve model accuracy. Question is whether any such "external" ET estimate is accurate enough to fulfill such expectations.

In the present study the state-wide monthly CREMAP ET rates of Szilagyi [3] are utilized for the Little Nemaha watershed in the south-eastern part of Nebraska, USA (**Figure 1**).

2. Study Site and Runoff Model Descriptions

The Little Nemaha watershed (with a drainage area of 2051 km² above Auburn, Nebraska) is an agricultural catchment, typical of the mid-western region of the US, producing mostly corn and soybean [13]. It is a catchment only negligibly affected by irrigation within the state, being situated in the south-eastern portion of it where precipitation is the most abundant, 775 mm·yr⁻¹ for 2000-2009. The climate is continental, with a May-June peak in precipitation (P) and a July peak in ET and

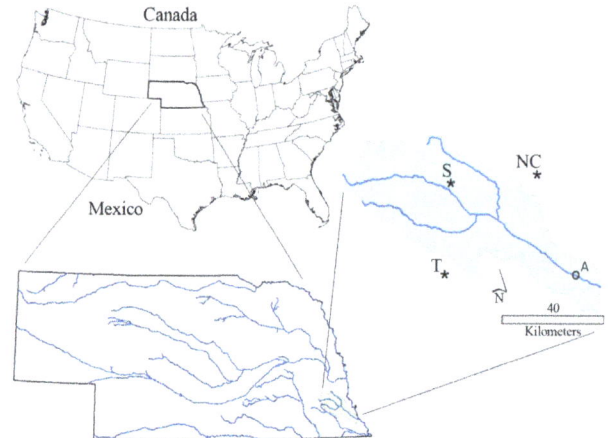

Figure 1. Location of the Little Nemaha watershed within Nebraska, USA. Daily precipitation and temperature are measured at Auburn (A), Nebraska City (NC), Syracuse (S) and Tecumseh (T). Daily streamflow rates are from USGS station at Auburn (A).

air temperature (T) (**Figuer 2(a)**). **Figure 3** displays the spatial distribution of the mean annual P and T values, beside the CREMAP ET rates, at the MODIS grid-resolution, obtained by spatial interpolation, using inverse-distance weighting of the values measured at the four climate stations of **Figure 1**. The area is part of the dissected plains region of Nebraska, with a gentle topography underlain by glacial till deposits, with a mean depth to the groundwater of about 15 m from the surface. The predominant physical soil texture at the land surface (generalized from STATSGO data of USDA [14]) is clayey loam turning into loam in the stream valleys. As **Figure 2(b)** demonstrates, inter-annual variability of precipitation is significant, the 2002-2006 period having had experienced a significant drought.

Daily runoff was simulated by a lumped-parameter conceptual model of Jakeman and Hornberger [15]—JH model for further reference—over the 2003-2008 period, having a complete daily precipitation (as well as temperature, snow depth and snow accumulation) record, and where the starting and ending years denote hydrologic years, beginning in November (the driest non-winter month in Nebraska, see **Figure 2(a)**), the previous year. The model first transforms the daily precipitation rate, r_i (mm·d⁻¹), of day i into "excess rainfall" through the definition of an antecedent precipitation index, s_i (−) as

$$s_i = c\left[r_i + \left(1 - \tau^{-1}\right) r_{i-1} + \left(1 - \tau^{-1}\right)^2 r_{i-2} + ... \right]. \quad (1)$$

Here c (d mm⁻¹) is just an adjustment factor ensuring that the volume of "excess rainfall" equals the volume of observed runoff over the calibration period. The order of the polynomial in (1) was set to seven through "trial and error". The variable, τ (−), represents the rate at which

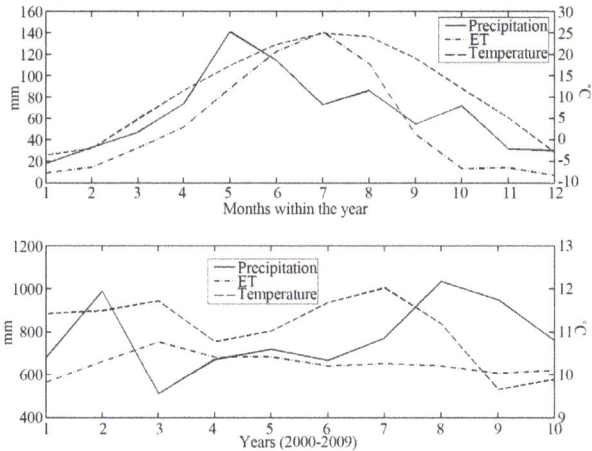

Figure 2. Catchment-averaged values: (a) mean monthly P (mm), CREMAP-derived ET (mm) and T (°C); (b) annual P (mm), CREMAP-derived ET (mm) and mean T (°C) of the Little Nemaha watershed for the 2000-2009 period.

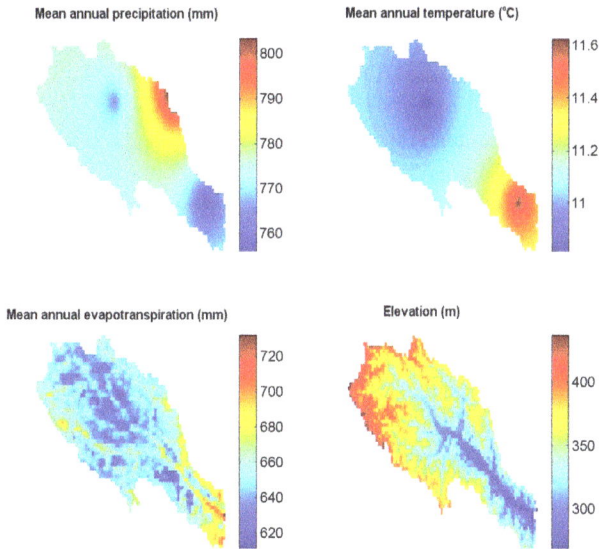

Figure 3. Distribution of the P, T and CREMAP-derived *ET* (from Szilagyi, *et al.*, 2013a) values over the Little Nemaha catchment for the 2000-2009 period. The elevation (a.m.s.l) distribution at the MODIS resolution is also shown. The catchment- and period-averaged values of P, T and ET are: 775 mm·yr^{-1}, 11°C, and 650 mm·yr^{-1}, respectively.

catchment wetness declines, a function of the seasons, expressed in the form as

$$\tau\left(T_m\right) = \tau_0 e^{g(30-T_m)}. \quad (2)$$

T_m is the catchment-averaged mean air temperature (°C) in month m (= 1,...,72), g (°C^{-1}) is a temperature modulation factor, and τ_0 is the reference value of τ. Excess rainfall, u_i (mm d^{-1}), then obtained as the product of the actual s_i and r_i values, subsequently transformed into runoff via the help of two parallel linear reservoirs, with storage coefficients k_q and k_s (d^{-1}) for quick- and slow-

flow responses, with simultaneous inputs to them as αu_i and $(1 - \alpha)u_i$, respectively, where α (−) is a constant multiplier from the (0, 1) range. The original JH model described thus far was amended by considerations of snow. During winter months excess rainfall is taken to be zero on days with reported snow accumulations and is augmented by melt-water of snow, whenever reported snow depth values decrease between consecutive days. By "trial and error" during model calibration (see below) the factor that best transformed reported snow-depth values in mm to water depth in mm was 0.01. This value is a magnitude smaller than the typically applied snow-water equivalent, but it also accounts for any sublimation and evaporation of snow across the watershed before the snow appears as runoff. The resulting model has five parameters to calibrate, *i.e.*, τ_0, g, α, k_q and k_s, similar to the original JH model.

Runoff at the monthly time-scale was simulated by a modified version of a simple distributed-parameter water-balance model of Vorosmarty *et al.* [16] and Szilagyi and Vorosmarty [17] over the 2000-2009 calendar years. The model tracks the soil moisture, *SM* (mm), of the root zone of the vegetation with rooting depth, *RD* (m), and field capacity, *FC* (−), values specified at each MODIS cell, derived from knowledge of the vegetation type and physical soil texture. **Table 1** lists the values employed in this study. Open water surfaces and wetlands did not contribute to runoff, which involved only two cells. The assumed maximum rate of cell *ET* (*ET$_{max}$* in mm·mo^{-1}) each month was first estimated by the Jansen-Haise [18] equation as

$$ET_{\max} = 1.6742 \cdot 10^{-2} R_s \left[0.014\left(1.8T_m + 32\right) - 0.37\right]n. \quad (3)$$

R_s (cal·cm^{-2}·d^{-1}) is the incident solar radiation, n is the number of days within the month, and T_m (°C) is the monthly mean air temperature of the cell. As an alternative, *ET$_{max}$* was also estimated by the Priestley-Taylor [8] equation as

$$ET_{\max} = 1.26\delta\left(\delta + \gamma\right)^{-1} R_n. \quad (4)$$

δ (hPa·K^{-1}) is the slope of the saturation vapor pressure curve at the monthly mean air temperature, γ (hPa·K^{-1}) is the psychrometric constant and R_n is net radiation (expressed in water-depth equivalent of mm·mo^{-1}) at the surface. The monthly R_n estimates came from the WREVAP model as watershed representative values. The monthly changes in *SM* (*i.e.*, ΔSM) are calculated as follows

$$\Delta SM = \left(P - ET\right)\Delta t, \, P > ET_{\max} \, \& \, SM < FC \quad (5)$$

$$\Delta SM = 0, \, P > ET_{\max} \, \& \, SM = FC \quad (6)$$

$$\Delta SM = -\beta\left(ET_{\max} - P\right)SM\Delta t, \, P < ET_{\max}. \quad (7)$$

Table 1. Field capacity and rooting depth values employed in the monthly runoff model as functions of physical soil type and/ or vegetation (after Vorosmarty *et al.* [16]; Szilagyi and Vorosmarty [17]).

Soil texture	Field capacity (−)	Rooting depth (m)			
		Forest	Agricultural crop	Urban, barren	Open water, wetlands
Clay	0.35	1.17	0.67	0.34	0
Clayey loam	0.32	1.6	1	0.5	0
Loam	0.29	2	1.25	0.63	0

β (mm^{-1}), the slope of the moisture-retention curve, empirically estimated as

$$\beta = \ln\left(FC\right)\left(1.1282FC\right)^{-1.2756}. \qquad (8)$$

FC is specified in mm (*i.e.*, *FC* [mm] = 1000 *FC*[−]× *RD*). *ET* is estimated as ET_{max} whenever $P > ET_{max}$, and as the difference in precipitation rate and (negative) soil moisture change, provided $P < ET_{max}$.

Excess monthly rainfall occurs in the model as the difference in estimated *SM* and *FC* (mm) in each cell whenever *SM* > *FC* (mm), thus reducing *SM* to *FC* in such months. The excess rainfall values of the individual cells are subsequently averaged over the cells to obtain one u_m ($m = 1,..., 120$) value for each month. A part of excess rainfall [*i.e.*, $(1 - \alpha)u_m$] then is routed through a single linear reservoir, representing slow-flow response of the entire catchment (not individual MODIS cells), by expressing α with the help of a power function as

$$\alpha = \left(\omega u_m / u_m^{max}\right)^{\lambda}. \qquad (9)$$

u_m^{max} is the largest simulated monthly excess rainfall value, and ω (−) and λ (−) are parameters to be calibrated. Note that quick-flow response, *i.e.* αu_m, is not routed in the monthly time-scale. To account for high intensity rains and the ensuing runoff even at the monthly time-scale, a constant threshold value, P_{th} (mm·d^{-1}), was applied above which daily precipitation rates contributed directly (*i.e.*, this portion of the daily precipitation did not go through the soil moisture accounting procedure) to monthly quick-flow response. Note that this way even at the monthly time-step the daily precipitation values were utilized, not typical in modeling at a monthly time-scale. The monthly model has four parameters to calibrate: P_{th}, k_s (mo^{-1}), ω and λ.

Calibration of both models was performed by setting intervals for the individual parameters and trying out different parameter values within those intervals via systematically increasing the parameter values by predefined increments starting at the minimum value of the parameter ranges. The value of the explained variance (R^2) in runoff was set as the objective function. As calibration progresses, both the intervals and the increments can successively be decreased once optimal values are

reached for each parameter at a given increment value, until a predefined accuracy is reached in the parameter values. Because of the relative shortness of the available data (6 and 10 years, respectively) as well as the low number of flood events, the data periods were not further subdivided into calibration and verification periods, rather, we trained the models with all the available data. Also, the aim was not to use the models for future simulations with the calibrated parameter values but rather to compare their calibrated performances with and without the monthly CREMAP *ET* data.

The monthly model with the Jensen-Haise equation seriously underestimated runoff by about 60% due to an overestimation of ET_{max} in general, in comparison with the maximum CREMAP *ET* values each month. As a result, there could always be found a combination of ω and λ values that improved the R^2 and at the same time worsened the ratio of simulated and observed runoff (*RR*) values ever more slightly, due to a near-flat objective function section in the ω vs λ space. The problem has been avoided by replacing the Jensen-Haise equation with the Priestley-Taylor one. The below results are therefore for this modified version (*i.e.*, employing the Priestley-Taylor equation) of the original model of Vorosmarty *et al.* [16].

3. Results and Discussion

After calibrating the models (**Table 2**), both models had to be slightly modified for inclusion of the external *ET* data and be recalibrated. For the JH model Equation (2) became

$$\tau = \tau_0 \exp\left\{g\left[P_m / \left(T_m^+ + 1\right)ET_{CR}^{-1}\right]\right\}. \qquad (10)$$

ET_{CR} (mm·mo^{-1}) is the watershed-averaged CREMAP *ET* rate linearly interpolated from the monthly values to the given day; T_m^+ is the monthly mean air temperature, replaced by zero for months with negative values, and g now is in degree centigrade. By Equation (10) catchment wetness declines fast when it is hot [as with Equation (2)] and humid, *i.e.*, *ET* is high as well. This last condition is the extra information that comes from the CREMAP *ET* values. By containing the product of the T_m and *ET* val-

Table 2. Calibrated model-parameter values with and without the inclusion of the lumped or spatially distributed CREMAP ET rates.

Daily model parameters	Value without CREMAP ET	Value with CREMAP ET, Lumped
τ_0 (−)	1	1
g[°C^{-1} (°C with CREMAP ET)]	0.4	1540
α (−)	0.75	0.76
k_q (d^{-1})	1.25	1.25
k_s (d^{-1})	0.033	0.033
Monthly model parameters		Lumped Distributed
P_{th} (mm d^{-1})	35	35 35
ω (−)	0.34	0.58 0.57
λ (−)	0.29	0.46 0.46
k_s (mo^{-1})	2.19	1.36 1.49

ues, and not just the latter, Equation (10) in a way corrects for incorrect ET rates coming from a linear interpolation between consecutive monthly ET values, since in a colder day it can in general be expected that wetness declines in a slower pace (and actual ET is smaller than the interpolated value), and vice versa, on a hot day faster, than what results from the interpolation, derived from average monthly conditions.

In case of the monthly model, ET_{max} is replaced by the CREMAP cell-ET rate and Equation (7) becomes

$$\Delta SM = (P - ET)\Delta t \qquad (11)$$

utilizing CREMAP ET values, with runoff occurring whenever the updated SM is larger than FC of the given MODIS cell. In the rare occasions when negative SM would occur, the SM value is set to zero. Note that this cannot happen with the analytical solution of Equation (7), describing a linear reservoir with exponentially decaying outflow as $SM(t + \Delta t) = SM(t)\exp\{\beta[P(t) - ET_{max}(t)]\}$. **Table 3** lists the different measures of model performance. For the daily model, inclusion of the watershed-averaged monthly CREMAP ET values left model efficiency practically the same, since the gain is only a few percent at most in any of those measures. This is in accordance with the findings of Oudin et al. [19], who did a similar experiment, since the CREMAP ET values, when averaged over an area (in this case the Little Nemaha watershed) approximate the CR-obtained regional ET rate, the same Oudin et al. [19] applied in their study. The explanation for the current study is that the monthly ET rates cannot fully explain the daily variability of the different physical factors that influence runoff, especially if the physical processes involved are non-linear. Note that the shorter

the interval the CR is applied for, the less reliable its predictions, as was noted by Morton et al. [5], which in turn may largely explain why Oudin et al. [19], utilizing the CR at a daily time-step, did not succeed with their runoff simulation improvement experiment. **Figure 4** displays the measured and simulated runoff values for a selected period.

At the monthly time-step, however, a slight model improvement is perceptible (**Figure 5**) with the inclusion of the CREMAP ET rates, either as spatially distributed or catchment-averaged values. Explained variance, R^2, increased from 67% to 70%, surpassing the 68% value of the daily model. The root-mean-square-error (RMSE) of the monthly model with lumped or distributed CREMAP

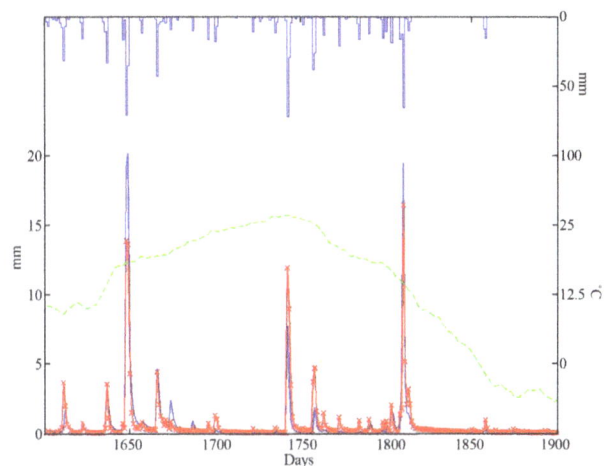

Figure 4. Sample measured (blue line) and simulated (red crosses) daily runoff values (mm) for 20 March 2007-15 January 2008, without the interpolated CREMAP ET rates. Daily precipitation [top blue line (mm)] and air temperature values [green dashes (°C)] are also shown.

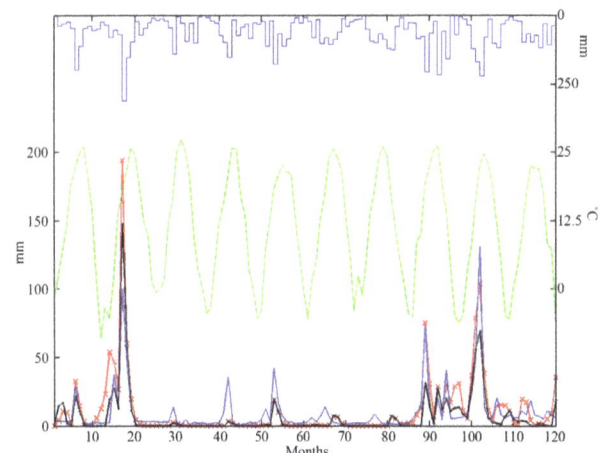

Figure 5. Measured (blue line) and simulated monthly runoff values (mm) from January 2000 to December 2009, with (red crosses) and without (black dots) the spatially distributed CREMAP ET rates. Monthly precipitation [top blue line (mm)] and air temperature values [green dashes (°C)] are also shown.

Table 3. Model performance measures: R (–) is correlation coefficient, R^2 is explained variance, $RMSE$ (mm) is root-mean-square-error, NSC (%) is the Nash-Sutcliffe performance indicator, RR (–) ratio of modeled and measured runoff over the modeling period (2003-2008 and 2000-2009, respectively).

	Daily model	Daily model with lumped CREMAP ET	Monthly model		Monthly model with lumped CREMAP ET		Monthly model with CREMAP ET	
	2003-2008	2003-2008	2000-2009	2003-2008	2000-2009	2003-2008	2000-2009	2003-2008
R	0.81	0.82	0.82	0.84	0.83	0.86	0.84	0.86
R^2	0.66	0.68	0.67	0.71	0.7	0.74	0.7	0.74
NSC	66	67	63	62	41	72	43	72
RMSE	0.75	0.73	10.82	11.64	13.67	10.11	13.43	9.97
RR	0.96	0.96	0.72	0.59	1.14	0.91	1.15	0.93

ET over the 2003-2007 period of the daily model has become smaller (10.11 and 9.97 mm) than that of the daily one (10.27 mm) with its runoff aggregated for the months. A somewhat larger improvement can be seen in the Nash-Sutcliffe model efficiency ($NSC = 1 – RMSE^2/V$, where V is variance of the measured runoff) indicator, *i.e.*, 67% vs 72%, and an even larger one in the R^2 values of 0.68 vs 0.74 between the daily and monthly model outputs for the same period. The largest overall change due to the inclusion of CREMAP ET values can be observed in the simulated to measured runoff ratios, both in the 2000-2009 and 2003-2008 periods: from an underestimation of 28% to an overestimation of 14% - 15%, and from an underestimation of 41% to only 7 - 9 percent, respectively. Note that the daily model too has a 4% underestimation even though excess rainfall is tuned via the adjustment factor, c, to the observed runoff volume. A difference in runoff volumes is only possible due to the storage delay of the slow-flow component. The only case when a model performance indicator value listed in **Table 3** would not improve with the inclusion of the CREMAP ET values is the NSC value for the entire 2000-2009 period. The sole reason for it is the significant underestimation of watershed ET in month 17 (*i.e.*, May, 2001) with the largest daily and monthly precipitation sums of 79 and 312 mm, respectively, and the largest measured daily runoff rate of ~28 mm, within the period.

It is interesting to note that the lumped monthly CREMAP-ET values yielded practically the same model performance (and similar calibrated parameter values) as the distributed ET rates. The reason must to a certain extent lie in the lumped treatment of the individual cell runoff rates and/or in the fact that different spatial distributions of excess rainfall may indeed generate similar runoff.

Some difference can be found in the calibrated value of the slow-flow storage coefficient between the daily and monthly models. In the daily model average residence time (*i.e.*, k_s^{-1}) of water within the ground is about

a month, while in the monthly model it is about 15 - 20 days. In the daily model about 25% of the simulated runoff originates from slow-flow response (in either case, with or without CREMAP ET), while in the monthly model it is 44% and 39% (with—either lumped or distributed—or without CREMAP ET). Szilagyi *et al.* [20], employing an objective base-flow-separation algorithm with long-term (*i.e.*, longer than 30 years) daily discharge measurements at Auburn, Nebraska obtained a base-flow index (*i.e.*, slow-flow contribution to runoff) of about 40% - 45%, close to the current monthly model's result. Note that in the monthly model, distribution of excess rainfall into quick- and slow-flow responses is dependent on the rainfall amounts (Equation (9)), while in the daily model, it is a fixed fraction, set by a constant α value.

4. Conclusion

The focus of the present study was to demonstrate that remote sensing based ET estimates (more specifically the CREMAP ET values of Szilagyi [3]) can be of a quality that may improve existing simple daily rainfall-runoff transformation and/or monthly water-balance model simulation results. While the improvement in the former case was hardly perceptible, it was somewhat more significant in the latter. This is not surprising since the CREMAP ET values are at a monthly time-scale therefore they are not capable of sufficiently recreating the daily variability of ET (as a loss-term) through a simple linear interpolation between consecutive monthly values to estimated daily ET rates, necessary for daily rainfall-runoff transformations. With monthly water-balance calculations this restriction is absent. Future routine application of CREMAP-derived ET rates may boost model efficiency of existing distributed watershed models that run on a monthly basis in areas where the CREMAP method is applicable. The distributed ET rates may also help with the calibration of distributed model parameters, a task not present in the current water-balance model where such parameter values were "a-priori" fixed based on soil and vegetation cover data. The spatially distributed *ET* estimation method of CREMAP has the distinguished property of not requiring any calibration. The tradeoff is that it cannot be applied over arbitrary areas, e.g., over mountainous and/or arid regions. While remote sensing based *ET* estimation is a fast evolving arena with ever more complicated and data-intensive methods appearing, typically requiring some sort of calibration based on, e.g. the application of vegetation indices, there is no reason why not to try out and employ existing simple methods, especially if those are calibration-free and therefore add no additional burden to the necessary calibration of the groundwater and/or runoff model in question.

5. Acknowledgements

This work has been supported by the Hungarian Scientific Research Fund (OTKA, #83376) and the Agricultural Research Division of the University of Nebraska. The WREVAP FORTRAN code, with the corresponding documentation, can be downloaded from the personal website of the author (snr.unl.edu/szilagyi/szilagyi.htm).

REFERENCES

[1]   G. B. Senay, S. Leake, P. L. Nagler, G. Artan, J. Dickinson, J. T. Cordova and E. P. Glenn, "Estimating Basin Scale Evapotranspiration (ET) by Water Balance and Remote Sensing Methods," *Hydrological Processes*, Vol. 25, No. 26, 2011, pp. 4037-4049.

[2]   J. Szilagyi, A. Kovacs and J. Jozsa, "A Calibration-Free Evapotranspiration Mapping (CREMAP) Technique," In: L. Labedzki, Ed., *Evapotranspiration*, INTECH, Rijeka, 2011, pp. 257-274.
http://www.intechopen.com/books/evapotranspiration

[3]   J. Szilagyi, "Recent Updates of the Calibration-Free Evapotranspiration Mapping (CREMAP) Method," In: S. G Alexandris and R. Sticevic, Eds., *Evapotranspiration—An Overview*, INTECH, Rijeka, 2013, pp. 23-28.
http://www.intechopen.com/books/evapotranspiration-an-overview

[4]   R. J. Bouchet, "Evapotranspiration Reelle, Evapotranspiration Potentielle, et Production Agricole," *Annales Agronomae*, Vol. 14, 1963, pp. 743-824.

[5]   F. I. Morton, F. Ricard and F. Fogarasi, "Operational Estimates of Areal Evapotranspiration and Lake Evaporation—Program WREVAP," National Hydrologic Research Institute Paper No. 24, Ottawa, 1985.

[6]   PRISM Climate Group, "Climate Data," Oregon State University, Corvallis, 2004. http://prism.oregonstate.edu

[7]   National Oceanographic and Atmospheric Administration (NOAA), "Surface Radiation Budget Data," 2009.
http://www.atmos.umd.edu/~srb/gcip/cgi-bin/historic.cgi

[8]   C. H. B. Priestley and R. J. Taylor, "On the Assessment of Surface Heat Flux and Evaporation Using Large-Scale Parameters," *Monthly Weather Review*, Vol. 100, No. 2, 1972, pp. 81-92.

[9]   J. Szilagyi, V. Zlotnik, J. Gates and J. Jozsa, "Mapping Mean Annual Groundwater Recharge in the Nebraska Sand Hills, USA," *Hydrogeology Journal*, Vol. 19, No. 8, 2011, pp. 1503-1513.

[10] J. Szilagyi, A. Kovacs and J. Jozsa, "Estimation of Spatially Distributed Mean Annual Recharge Rates in the Danube-Tisza Interfluvial Region of Hungary," *Journal of Hydrology and Hydromechanics*, Vol. 60, No. 1, 2012, pp. 64-72.

[11] J. Szilagyi and J. Jozsa, "MODIS-Aided Statewide Net Groundwater-Recharge Estimation in Nebraska," *Ground Water*, Vol. 51, No. 5, 2013, pp. 735-744.

[12] J. Szilagyi, V. Zlotnik and J. Jozsa, "Net Recharge versus Depth to Groundwater Relationship in the Platte River Valley of Nebraska, USA," *Ground Water*, Vol. 52, No. 1, 2014, in Press.

[13] P. Dappen, I. Ratcliffe, C. Robbins and J. Merchant, "Map of 2005 Land Use of Nebraska," 2007.
http://www.calmit.unl.edu/2005landuse/statewide.shtml

[14] United States Department of Agriculture, "State Soil Geographic (STATSGO) Data Base," United States Department of Agriculture, Washington DC, 1991.

[15] A. J. Jakeman and G. M. Hornberger, "How Much Complexity Is Warranted in a Rainfall-Runoff Model," *Water Resources Research*, Vol. 29, No. 8, 1993, pp. 2637-2649.

[16] C. Vorosmarty, B. Moore, A. L. Grace and P. Gildea, "Continental-Scale Models of Water Balance and Fluvial Transport: An Application to South America," *Global Biogeochemical Cycles*, Vol. 3, No. 3, 1989, pp. 241-265.

[17] J. Szilagyi and C. J. Vorosmarty, "Water-balance Modelling in a Changing Environment: Reductions in Unconfined Aquifer Levels in the Area between the Danube and Tisza Rivers in Hungary," *Journal of Hydrology and Hydromechanics*, Vol. 45, No. 5, 1997, pp. 348-364.

[18] M. Jensen and H. Haise, "Estimating Evapotranspiration from Solar Radiation," *Journal of Irrigation and Drainage Engineering*, Vol. 89, No. IR-4, 1963, pp. 15-41.

[19] L. Oudin, C. Michel, V. Andreassian, F. Anctil and C. Loumagne, "Should Bouchet's Hypothesis Be Taken into Account in Rainfall-Runoff Modelling? An Assessment over 308 Catchments," *Hydrological Processes*, Vol. 19, No. 20, 2005, pp. 4093-4106.

[20] J. Szilagyi, E. F. Harvey and J. Ayers, "Regional Estimation of Base Recharge to Ground Water Using Water Balance and a Base-Flow Index," *Ground Water*, Vol. 41, No. 4, 2003, pp. 504-551.

12

A Structural Overview through GR(s) Models Characteristics for Better Yearly Runoff Simulation

Safouane Mouelhi[1], Khaoula Madani[2], Fethi Lebdi[3]

[1]National Researches Institute of Water, Forests and Rural Engineering, Tunis, Tunisia; [2]National Water Distribution Utility, Tunis, Tunisia; [3]Food and Agriculture Organisation, Addis-Abeba, Ethiopia.

ABSTRACT

In rainfall-runoff modelling, a monthly timescale and an annual one are sufficient for the management of deductions. However, to simulate the flow at a large time-step (annual), we generally precede the use of a model working for a finer time-step (daily) while aggregating the desired outputs. The finest time-steps are considered, apriori, as the most performant. By passing from one time-step to another, and in order to work in the desired time-step (annual) and calculate the potential gains or loss, this article proposed a comparative study between the aggregation method of outputs of a modal working at a finer time step, and a method in which we use a conceived model from the beginning. To ensure this comparative and empirical approach, the choice has been focused on (GRs) models to a daily time-step (GR4J), monthly time step (GR2M) and annual time step (GR1A). The modelling platform used is the same for all three models taking into account the specificities of each one: the same data sample, the same optimization method, and the same function criterion are used during the construction of these models. Due to the moving between these time steps, results show that the best way to simulate the annual flow is to use an appropriate and designed modal initially conceived to this time step. Indeed, this simulation seems to be less effective when using a model at a finer time-step (daily).

Keywords: Time Step; Runoff-Rainfall; GR(s) Models; Yearly; Monthly; Daily

1. Introduction

The hydrological literature is rich in conceptual rainfall-runoff models at different time steps especially the daily one such as Beven and Kirkby [1], Thomas [2], Milly [3], Vandewiele et Xu, 1994 [4], Bergström [5], Perrin et al. [6], Mathevet [7], Mouelhi et al. [8,9].

The approach consists in passing from a large time-step to a finer one or vice versa, in order to calibrate the parameters, optimize the functions of the modals, or test the interdependence of parameters of the same model from one time-step to another, which is increasingly adopted: Hughes [10], Nalbantis [11], Kavetski et al. [12], Haddeland et al. [13], Littlewood and Croke [14], Kling and Nachtnebel [15], Widen-Nilsson et al. [16], Clark and Kavetski [17].

However, to simulate the flows at a larger time-step (monthly or annually), we often use a rainfall-runoff model operating at a daily time-step, then, by the aggregation of outputs, we obtain the flows at the desired one (monthly, seasonal, annual).

This method results from "the received" idea of the structural superiority of modals performing at finer time-step (daily or hourly) compared to others at a larger one.

However, this summing method of models' output has not been compared to the method where appropriate modals at a desired time-step are applied. Is it always effective? Which model can we choose? And for which time-step? Passing from one time-step to another, what is the evolution of the structural morphology of the modelling conceptual globing rainfall-runoff?

While trying to encrypt the gain or loss, this article attempts to answer these questions. It is caused by the use of a model to estimate the flow rates on a time-step higher than that for which the modal is conceived to work.

This work deals with the case of the annual time step, where the flows are simulated: either by an annual model; or monthly or daily models when aggregating the outputs to the annual time step.

The first paragraph deals with the choice of the model, which is brought to GRs models (GR4J, GR2M and G1A). The second one is devoted to explain the platform of the used modelling for the three time-steps. In this section, a new method of initialization has been tested. The last paragraph explains and interprets results.

2. The Chosen Models: GR(s)

The choice of GR(s) models compared to others is based on the following reasons: the availability of a model at each time-step. Same modelling platform, (sample data and modelling methodology) used in their construction, and better performance compared to other global conceptual models.

2.1. Daily Time-Step GR4J (Perrin *et al.*, 2003)

GR4J is a global conceptual rainfall-runoff model with four free parameters. It is the result of gradual improvements of the work undertaken within the National Research Institute of Science and Technology for Environment and Agriculture "IRSTEA": Michel [18], Edijatno and Michel [19], Edijatno [20], Edijatno *et al.* [21], Rakem [22], Perrin [23], Perrin *et al.* [24].

The version used in this article is the one proposed by Perrin. It is deducted following a comparative and empirical approach among 38 global conceptual models including TopModel, HBV, etc. [24].

The structure of GR4J model is based on a reservoir of production called also "Reservoir-Soil", a reservoir of draining, and on two Unitarian hydrographs (**Figure 1**).

Pluviometry (P) and potential evapotranspiration (E), expressed in millimetres (mm), are inputs's variables of the model. The interception phase consists in calculating,

from these two inputs, the net rain (Pn) or the net evapotranspiration (En) as follows (Equation (1)):

$$\begin{cases} \text{If } P \geq E \text{ Then } Pn = P - E \text{ and } En = 0 \\ \text{If } P < E \text{ Then } En = E - P \text{ and } Pn = 0 \end{cases} \quad (1)$$

The Soil Reservoir is characterized by its maximum capacity "X1", the first parameter of GR4J model. (S) is the reservoir content responsible for realizing a follow-up of humidity in the basin.

In the case where all the rain is consumed during the interception, the amount (En) is used to evaporate the water in the reservoir-ground at a real rate (Es). Conversely, when rain remains (Pn), a part of it (Ps) is stored in the reservoir. (Es) and (Ps) are calculated as functions of filling rate of the reservoir of production

$$\begin{cases} Ps = \dfrac{A\left(1-\left(\dfrac{S}{A}\right)^2\right)W}{1+\dfrac{S}{A}W} \quad avec\ W = \tanh\left(\dfrac{Pn}{A}\right) \\ \\ Es = \dfrac{S\left(2-\dfrac{S}{A}\right)V}{1+\left(2-\dfrac{S}{A}\right)V} \quad avec\ V = \tanh\left(\dfrac{En}{A}\right) \end{cases} \quad (1)$$

The content of the reservoir of production is updated as follows

$$S' = S - Es + Ps \quad (3)$$

A percolation term, which is the reservoir output and the function of the S' level, feeds the flow as follows (Equation (4)):

$$Perc = S'\left[1-\left(1+\left(\dfrac{S'}{2.25X_1}\right)^{-4}\right)^{-\frac{1}{4}}\right] \quad (4)$$

The content of the reservoir is updated again (Equation (5))

$$S'' = S' - Perc \quad (5)$$

The supplementary part of rain ($Pn - Ps$) or effective rainfall, to which we add the percolation term, is separated into two components. A direct flow ($Q1$), which represents 10% of the effective rainfall, joins the outfall after a rooting by a Unitarian hydrograph ($SH2$). The other part ($Q9$) representing 90% of the effective rainfall is routed by another Unitarian hydrograph ($SH1$) followed by a rooting reservoir.

The Unitarian hydrograph is a function that allows the creation of a time lag between rainfalls and flows. The two used hydrographs ($SH1$ and $SH2$) depend on the same reference period. ($X3$), as a third parameter wedged from the model, characterizes the rising time of the Uni-

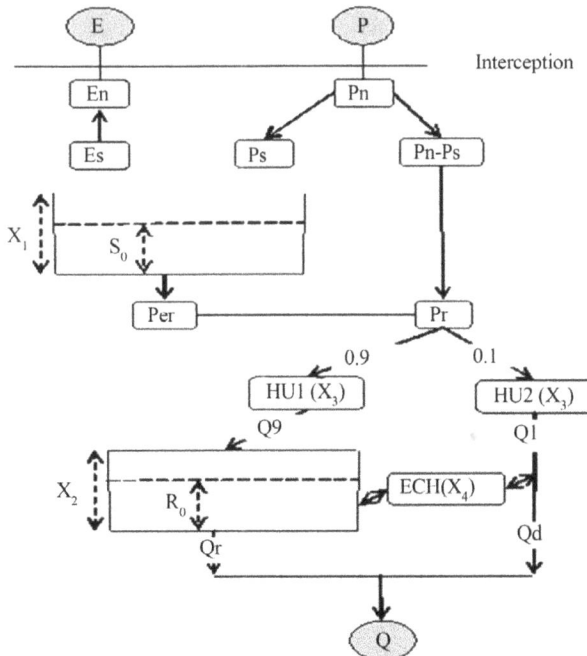

Figure 1. Architecture of GR4J model.

tarian hydrograph. Expressions of functions (SH1) and (SH2) are as follows (Equation (6)):

$$
\begin{cases}
pour\ 0 \le j \le X_3, & SH1(j) = \left(\dfrac{j}{X_3}\right)^{\frac{5}{2}} \\[2mm]
pour\ j > X_3, & SH1(j) = 1 \\[2mm]
pour\ 0 \le j < X_3, & SH2(j) = \dfrac{1}{2}\left(\dfrac{j}{X_3}\right)^{\frac{5}{2}} \\[2mm]
pour\ C \le j < 2X_3, & SH2(j) = 1 - \dfrac{1}{2}\left(2 - \dfrac{j}{X_3}\right)^{\frac{5}{2}} \\[2mm]
pour\ C \ge 2X_3, & SH2(j) = 1
\end{cases}
\tag{2}
$$

(j) can also take non-integer values. Hydographs's ordinates (UH1 and UH2) are calculated from curves in "S" representing the cumulation of Unitarian rainfall proportion by the hydrograph in function of time, noting respectively SH1 and SH2 (Equation (7)):

$$
\begin{cases}
UH1(j) = SH1(j) - SH1(j-1) \\
UH2(j) = SH2(j) - SH2(j-1)
\end{cases}
\tag{3}
$$

At each time step (i), outputs (Q9) and (Q1) of the two hydrographs are calculated by (Equation (8)):

$$
\begin{cases}
Q9(i) = 0.9 \displaystyle\sum_{k=1}^{l} UH1(k) \cdot \Pr(i-k+1) \\[2mm]
l = Int(X_3) + 1 & (Int:IntegerPart) \\[2mm]
Q1(i) = 0.1 \cdot \displaystyle\sum_{k=1}^{m} HU2(k) \cdot \Pr(i-k+1) \\[2mm]
m = Int(2X_3) + 1 & (Int:Integer\ part)
\end{cases}
\tag{8}
$$

The (Q9) quantity passes through a non-linear rooting reservoir of level (R) and of maximum capacity (X2) which is the second parameter to wedge from the model. This reservoir is characterized by its act of instant draining of this type: "$Qr(t) = K \cdot (R(t))^5$", wherein (K) is a constant. The level (R) of the rooting reservoir is affected by the relative quantity in underground exchanges, which is a function of the fourth parameter of the model (X4).

$$
ECH = X_2 \cdot \left(\dfrac{R}{X_4}\right)^{7/2}
\tag{9}
$$

The level (R) of draining reservoir is updated as in the Equation (10)

$$
R' = \max\left(0; R + Q9 + ECH\right)
\tag{10}
$$

Having transited the rooting reservoir, the quantity is deducted by the integration of the draining act as follows (Equation (11))

$$
Qr = R' \cdot \left(1 - \left(1 + \left(\dfrac{R'}{X_2}\right)^4\right)^{-\frac{1}{4}}\right)
\tag{11}
$$

The level of the reservoir is updated according to the Equation (12)

$$
R'' = R' - Qr
\tag{12}
$$

Having undergone the hydrograph action (HU2), the quantity (Q1) is subject to the same changes to give the flow component according to Equation (13):

$$
Qd = \max\left(0; Q1 + ECH\right)
\tag{13}
$$

The flow is then given by the Equation (14):

$$
Q = Qr + Qd
\tag{14}
$$

2.2. Monthly Time-Step: GR2M (Mouelhi et al., 2006b)

The (G2M) is a global conceptual model with two parameters. As for the GR4J case, it has underwent many transformations since its first construction up to the most recent version Edijatno [25], Kabouya [26], Kabouya and Michel [27], Makhlouf [28], Makhlouf and Michel [29], Nascimento [30], Mouelhi [31], Mouelhi et al. [8].

The most recent structure has been developed using a "step by step" approach or "stepwise approach". It is a question of testing a multitude of interconnected components and retaining only the combination leading to a better performance [8].

Empirically, and further to a comparative study with eight other structures of global conceptual rainfall-runoff models, the GR2M model seems to be the most performant.

This model is characterized by two functions (**Figure 2**): 1) a function of production that revolves around a reservoir-ground of a maximum capacity (X1), which is the first parameter to be wedged. Transferring a percolation of reservoir-ground is ensured by a dependent feature of the stock status "S"; 2) a transfer function represented by a quadratic draining reservoir of a capacity fixed to 60 mm.

This reservoir is modified by an underground exchange, whose coefficient (X2) is the second parameter to optimize. The runoff is calculated according to seven operations (Equations (15)-(24)): Under the action of rainfall (P), S_{m-1} level becomes S^*, where (m) refers to the month in question:

$$
S^* = \frac{S_{m-1} + \varphi \cdot X_1}{1 + \varphi \cdot \dfrac{S_{m-1}}{X_1}} \quad avec \quad \varphi = \tanh\left(\frac{P}{X_1}\right)
\tag{15}
$$

It follows a contribution to the (PR) flow:

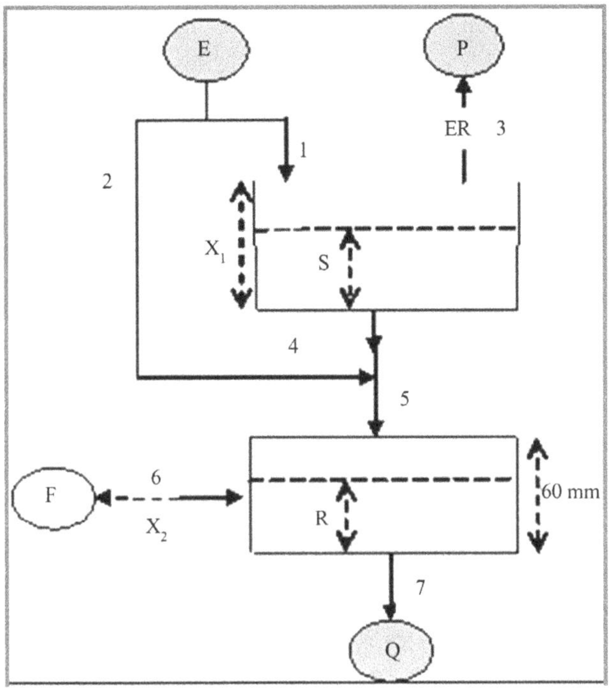

Figure 2. Architecture of GR2M model (Mouelhi, 2003; Mouelhi et al., 2006b).

$$PR = S_{m-1} + P - S^*$$ (16)

(1) Under the action of potential evapotranspiration (ET), the level (S^*) becomes (S^{**}) as follows:

$$S^{**} = \frac{S^* \cdot (1 - \psi)}{1 + \psi \cdot \left(1 - \frac{S^*}{X_1}\right)} \quad avec \quad \psi = \tanh\left(\frac{E}{X_1}\right)$$ (17)

(2) By percolation (S^{**}) becomes (S_m) at the end of the month (m):

$$S_m = \frac{S^{**}}{\left[1 + \left(\frac{S^{**}}{X_1}\right)^3\right]^{1/3}}$$ (18)

$$PS = S^{**} - S_m$$ (19)

(3) The reservoir (R), which level at the beginning of the month (R_{m-1}) becomes (R^*):

$$R^* = R_{m-1} + PR + PS$$ (20)

(4) The exchange term (ECH) is calculated as follows, where (X_2) is a positive and dimensionless parameter:

$$ECH = (X_2 - 1) \cdot R^*$$ (21)

(5) Under the action of this exchange, the reservoir level becomes:

$$R^{**} = X_2 \cdot R^*$$ (22)

(6) The rooting reservoir has a fixed capacity equal to

60 mm. It drains according to a quadratic law

$$Q = \frac{\left(R^{**}\right)^2}{R^{**} + 60}$$ (23)

(7) After draining the level of the reservoir, the m month becomes:

$$R_m = R^{**} - Q$$ (24)

2.3. Annual Time-Step: GR1A (Mouelhi *et al.*, 2006a)

The GR1A model was built by testing numerous junctions of components stemming from finer and bigger time-step: multiannual, monthly and daily, [9] and [31]. The form is found with a single parameter. It seems very simple where hydrological concepts no longer appear (Equation (25)):

$$Q_k = P_k \cdot \left(1 - \frac{1}{\left(1 + \left(\frac{0.7 P_k + 0.3 P_{k-1}}{X \cdot E_k}\right)^2\right)^{0.5}}\right)$$ (25)

Q_k: the simulated flow of the year k. P_k: the observed rainfall of the year k. P_{k-1}: the amount of the observed rainfall of the year $k-1$. E_k: the amount of potential evapotranspiration of the year k. The parameter of the model to be optimized.

3. Modelling Platform

To insure the passage from one time-step to another without biasing results, the modelling platform used during the construction of these modals is the same used in this work.

3.1. Sample Data

The sample data used in this study is 407 watersheds having served in modelling work conducted at (IRSTEA), including the development of models GR4J, GR2M and GR1A: Mouelhi *et al.* [8] and [9], Perrin [23], Perrin *et al.* [24], Makhlouf and Michel [27], Makhlouf [28], Mouelhi [31], Andréassian [32], Andréassian *et al.* [33]; Perrin *et al.* [34].

This sample collects rainfall data (P), of an evapotranspiration (E) and flow (Q) of very varied hydroclimatic conditions of 407 watersheds of: 298 French basins, 70 Americans, 26 Australians, 9 Ivorian and 4 Brazilian. (**Table 1**).

The rainfall data (P), of flow (Q) and some of the potential evapotranspiration data (E) are provided to a daily time-step. It was necessary to aggregate data at different

Table 1. Sample statistics characteristics of data.

	E (mm/year)	P (mm/year)	Q (mm/year)	S (km²)
Average	935	1010	466	972
Standard Deviation	299	342	349	3630
Minimum	633	294	0.2	0.1
Quantile 5%	666	549	6	3.5
Quantile 95%	1629	1619	1180	4250
Maxmum	2045	2299	2043	50,600

desired time-steps to meet the objectives of this work.

This sample presents quite different climatological, hydrological and anthropogenic conditions. On one hand, there exist semi-arid conditions in Australia or in the south of the United States, with watercourses knowing flows only a few days in the year. On the other hand, we find wet tropical conditions in the south of the Ivory Coast or the North of Australia.

The French basins are characterized by a quiet big climatic diversity, with Mediterranean and continental influences.

The watershed size varies over a wide range, from 0.1 to more than 50,000 km². The larger watersheds are those of La Seine in Paris (43,800 km²) and of São Francisco in the dam of Três Marias in Brazil (50,600 km²).

This sample also reflects a variety of seasonal patterns. Indeed, basins exist with very contrasted wet and dry seasons and also fairly unfirming scheme throughout the year both at the level of rainfall and flow.

3.2. Methodology of the Application of Models

3.2.1. Variable Target, Assessment Criteria and Optimization Method

The objective behind choosing the variable target is to take into account, in a balanced way, the different ranges of elapsed flows without favouring the quality of reproduction of low or strong values.

However, residues of a model are generally not homoscedastic that is to say, their variance is dependent on the value of the flow [32].

If the choice of variable target has been on the flows (Q), strong yield basins would have been privileged. The interest is proceeding by a transformation on variable flow (Q) in order to take into account, in a relatively uniforming way, all the orders of High flows.

The flow rate could be an appropriate choice of transformation. This variable tends to take into account the humid and arid basins in an equivalent manner. Yet, during the measures, the flow, as well as the pluviometry, are tainted by error. Dividing the flow by rainfall, the relative errors accumulate. Thus, the risk of biased estimates is even stronger.

A logarithmic transformation on flows, to which we add a low constant to avoid the problem of the useless flows, could be valid [35].

So, values of flows will be levelled, and errors of the model vary then in the same height order for all classes of flows.

A transformation power of flows (transformation with 0.5 power), allows to have the intermediate case between the logarithmic transformation and the solution (Q) without biasing calculations [36].

Such a choice of transformation allows, at the same time, to reduce the character of non-homoscedasticity residue models, and to keep the works having served in the construction of coherent used models in this article. Consequently, it is the variable "root of Q" that is used.

The criterion function used is based on the so-called Nash Criterion (CN) [37]. When performing the chosen transformation (Q root), for each watershed. This criterion can be written as follows (Equation (26)):

$$CN = 1 - \frac{\sum_{i=1}^{N}\left(\sqrt{Q_i} - \sqrt{\hat{Q}_i}\right)^2}{\sum_{i=1}^{N}\left(\sqrt{Q_i} - \overline{\sqrt{Q}}\right)^2} \qquad (26)$$

N: total number of calculated and observed values for each watershed. Q: observed and calculated runoff depth (in millimetres per time-step). $\hat{Q}$: runoff depth estimated by this model (in millimetres per time-step). $\sqrt{Q}$: average over N of the square root values of the observed water passed runoff depth.

(CN) will take values from ($-\infty$ to 1). The model is considered as performant when the estimated flows get closer to the observed flows, that is, when CN is close to 1.

Previous studies have demonstrated the relevance of this criterion relative to another part of the overall conceptual modelling [38].

The choice of the optimization method has been focused on the so-called "step by step" (Michel *et al.*, 1989; Nascimento, 1995). It is a direct method, which operates a local optimization (maximization or minimization) of an objective function chosen by the user regardless of the method.

This is a maximization of the (CN) function. This method adopts a displacement strategy along the axes of the parameter space, with a research step that can vary from one iteration to another. It proved a better performance compared to other methods in the context of global conceptual rainfall-runoff modelling [23].

3.2.2. Implementation and Evaluation of the Robustness

The robustness models was assessed using the technique

of the "double sample", also called "Split-Sample test" [39]. It is a question of cutting the period of observation into two sub-periods, one for the wedging and the other one for the control (or validation). Their roles will be swapped so that we have two periods of stalling and two periods of controls. Thus the model tested will be evaluated only at the control.

For each watershed, "Nash" corresponds to the sum of the squared deviations between observed and calculated values relative to the first and the second control, with regard to the sum of squares of distances between the value and average observed over all the period of observation (wedging and control). It is translated according to the following formulation (Equation (27)):

$$CNSS = 1 - \frac{\sum_{i=N_{i1}+1}^{N_2} \left(\sqrt{Q_i} - \sqrt{\hat{Q}_i}\right)^2 + \sum_{i=N_2+1}^{N} \left(\sqrt{Q_i} - \sqrt{\hat{Q}_i}\right)^2}{\sum_{i=N_1+1}^{N} \left(\sqrt{Q_i} - \sqrt{M}\right)^2}$$

(27)

$CNSS$: Nash criterion using the "Split-Sample" technique. N_1: start-up time. N_2: end of the first sub-period. N: total observation period (day for GR4J, Months for GR2M and year for GR1A for each watershed: $\sqrt{M}$ average of $\sqrt{Q}$ observed in all period's totality. Q: the observed runoff depth. $\hat{Q}$: Calculated water blade by the modal. And $\{S\}0$: the initial state of the system.

The initial state of the system consists in predefining values of departure for the parameters of the model before passing to the optimization phase. In other, words before its confrontation to the observed hydrological reality.

In the case where the model would contain reservoirs, this initialization phase would also consist in predefining the initial levels of these before the wedging.

In an empirical or conceptual modelling, this initialization phase appears as a "physical": necessity: "It is better to know from where we start to understand where we are going."

Generally, this initial state is defined arbitrarily by the modeller. However, the initial state choice can influence the optimal set of parameters of the model and its performance. To overcome this constraint, we usually choose a period of initiation, which consists in turning the system for an observation period without being taken into account in the calculation of the performance.

For better daily time-step, the starting period, associated with the best performance, is of one year, that is to say, 365 observations (Perrin, 2000; Perrin *et al.*, 2001). It is the same period used in this article for the same time-step.

At the monthly time step, a year of commissioning is equivalent to only 12 observations. To mulch this prob-

lem, a new method has been tested, here by the "starting up of cyclic system".

This latter consists in feeding the reservoir(s) by an interannual monthly rainfall. The equivalent output is the average of interannual monthly flows (**Figure 3**). The operation is repeated until the stabilization of the reservoir level. The corresponding level will be taken as the initial level of it.

The calculations showed that 20 iterations of cyclic regime with one effective year of starts are sufficient at a monthly time-step. As for the annual one, the modal chosen here is without reservoir (Equation (25)). It will be enough to fix the initial value of the parameter by avoiding the mathematically insensible zones (not near zero and infinity).

4. Results and Discussion

Given large number of tests in wedging and controlling as for a used model, 814 wedgings, 814 controls and 407 values of criterion of performance, it was interesting to be able to analyse the results of the performance of each model, by summarizing it into one or two numbers.

The quantile (30%) of performances distribution has been mainly chosen, noted (Cr3) as well as the average of the values of $CNSS$. These distributions are obtained by ranking $CNSS$ criterion values of a model in control through ascending and constructing the corresponding experimental distribution.

First of all, the GR1A model was applied on the sample data to encrypt its performance. Then, with monthly data on the same sample, GR2M monthly model was used, where annual rates were estimated by summing the outputs. Similarly, annual rates were also estimated from daily data by using GR4J model (**Figure 4** and **Table 2**).

As reported, and strongly influenced by negative values of the performance criterion, the 30% quantile (CR3) is adopted here as a criterion in addition to the average (**Table 2**). These criteria are also argued in **Figure 4** in order to get an overall idea of the performance of models over the entire sample data.

The annual model GR1A widely outclasses the daily model GR4J, and less clearly the model GR2M. These results show that, to feign annual flows, the best way is to use a model suited to this time-step, by avoiding getting lost in useless details that can weight down the task

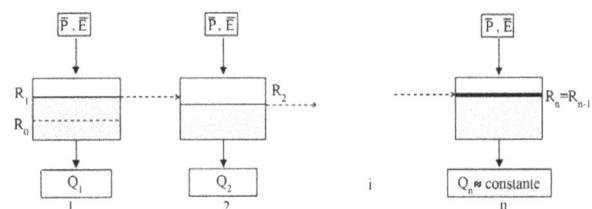

Figure 3. Cyclic started regime.

of regrouping data and the implementation of the model.

This report comes against the preconceived ideas to use a finer model and to aggregate the data of outputs for every desired time-step.

However, from the at about 60% quantile, (CR6), the performance of annual, monthly and daily models become comparable with a slight superiority of GR4J then GR2M compared to GR1A (**Figure 4**). This could be explained by the fact that the distribution of rainfall and evapotranspiration over the year is not reflected in the architecture of GR2M and GR1A models.

Indeed, for arid or semi-arid watersheds, rainy season is concentrated in few months or even few days of the year.

On the other hand, to visualise the morphology of models from one time to another, the GR4J, GR2M and GR1A architectures have been gathered on the same figure (**Figure 5**). The daily and monthly models are relatively connected to the conceptual plan.

This conceptualization disappears completely at an annual time-step. Only the underground exchange, taking the form of a linked coefficient to the evapotranspiration cinquefoil, remains. Thus, the passage from monthly time-step to yearly time step is done "brutally".

Table 2. Performances of the GR4J and GR2M models at the yearly time step.

	CNSS (%)		
	CR3	Average	*CR6*
GR4J	24.5	7.9	70.9
GR2M	38.5	36.5	69
GR1A	45.0	41.3	69.8

We believe that the prospect of building a global conceptual model at a seasonal time-step seems to be interesting. Conceptually, it will allow to see the composition of a model opening at an intermediate time step between the monthly and the annual.

Furthermore, the contrast between dry and wet seasons will be taken into account, which is not considered at the level of GR2M and GR1A architectures.

5. Conclusions

A modelling methodology was adopted to avoid biasing results and to encrypt the gain or loss caused by the use of a model, in order to estimate the flow at a time-step higher than that to which the modal is conceived to operate.

So, the choice of models is carried to GR1A, GR2M and GR4J, where the modelling platform is opted here. It also has been used to their constructions: sample data (407 watersheds), performance criterion of Nash by introducing a technique called "Split-Sample" (*CNSS*), variable target (root flows) and optimization method (step by step).

Let us note here that a new initialization technique of reservoirs' levels, called "cyclic start-up scheme" allowed the escape at their arbitrary choices.

Contrary to preconceived ideas, results show that the use of a designed model working at a large time-step (GR1A for the annual), is more accommodated than the use of a finer model (GR2M or GR4J) by aggregating the exits at an annual time-step.

Structurally, the conceptualization of models disappears at an annual time-step (GR1A) compared to daily and monthly time-steps (GR2M and GR4J), which remain relatively similar structures.

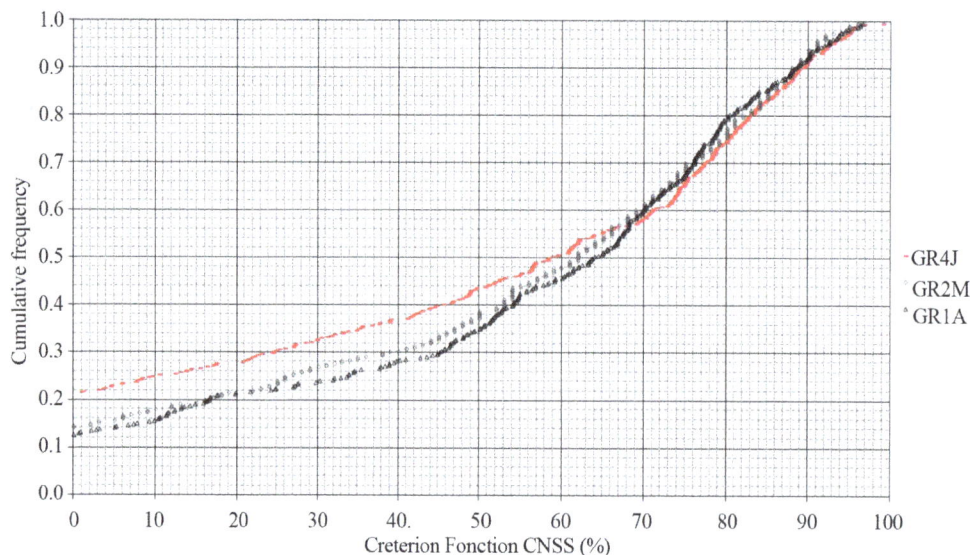

Figure 4. Comparison of G4J, GR2M and GR1A model performance at the annual time step.

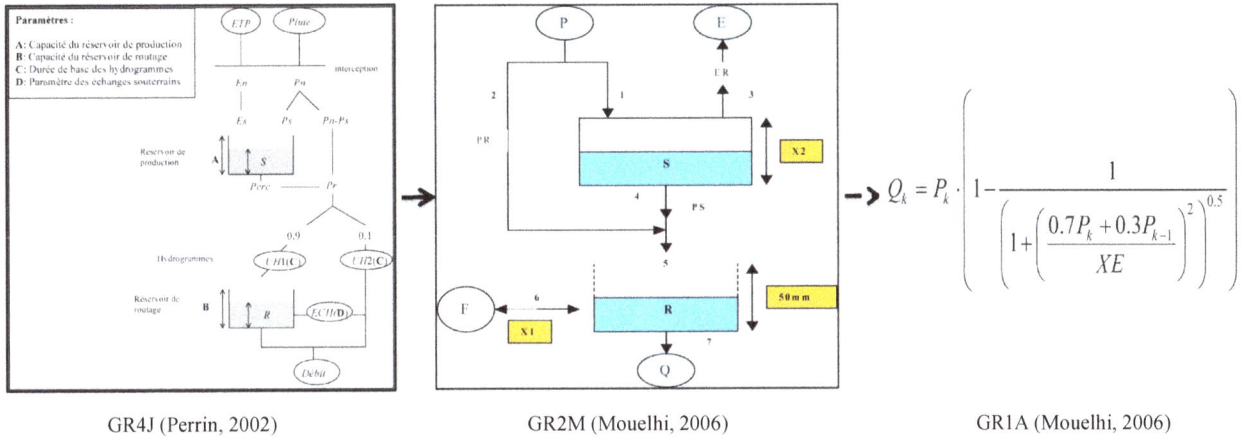

GR4J (Perrin, 2002)   GR2M (Mouelhi, 2006)   GR1A (Mouelhi, 2006)

Figure 5. The *GR*(s) models architecture overview.

It seems to be interesting then to build a model at a seasonal time-step (intermediate between the monthly and the annual), which will take into account the effect of rainfall distribution and evapotranspiration throughout the year.

REFERENCES

[1] K. J. Beven and M. J. Kirkby, "A Physically Based, Variable Contributing Area Model of Basin Hydrology," *Hydrological Sciences Bulletin*, Vol. 24, No. 1, 1979, pp. 43-69.

[2] H. A. Thomas, "Improved Methods for Rational Water Assessment," Water Resources Council, Washington DC, 1981.

[3] P. C. D. Milly, "Climate, Interseasonal Storage of Soil Water, and the Annual Water Balance," *Water Resources Research*, Vol. 17, No. 1-2, 1994, pp. 19-24.

[4] C. Y. Xu and G. L. Vandewiele, "Parsimonious Monthly Rainfall-Runoff Models for Humid Basins with Different Input Requirements," *Advances in Water Resources*, Vol. 18, No. 1, 1995, pp. 39-48.

[5] S. Bergström, "The HBV Model," In: V. P. Singh, Ed., *Computer Models in Watershed Modeling*, Water Resources Publications, Highlands Ranch, Colorado, 1995, pp. 443-476.

[6] C. Perrin, C. Michel and V. Andréassian, "Improvement of a Parsimonious Model for Streamflow Simulation," *Journal of Hydrology*, Vol. 279, No. 1, 2003, pp. 275-289.

[7] T. Mathevet, "Which Global Rainfall-Runoff Models in an Hourly Time Step? Development and Empirical Comparison of Models on a Large Sample of Watersheds," Ph.D. Thesis, AgrosParisTech (Ex. ENGREF), 2005, 463 p.

[8] S. Mouelhi, C. Michel, C. Perrin and V. Andréassian, "Stepwise Development of a Two-Parameter Monthly Water Balance Model," *Journal of Hydrology*, Vol. 318, No. 1-4, 2006, pp. 200-214.

[9] S. Mouelhi, C. Michel, C. Perrin and V. Andréassian, "Linking Stream Flow to Rainfall at the Annual Time Step: The Manabe Bucket Model Revisited," *Journal of Hydrology*, Vol. 328, No. 1-2, 2006, pp. 283-296.

[10] D. A. Hughes, "Variable Time Intervals in Deterministic Hydrological Models," *Journal of Hydrology*, Vol. 143, No. 3-4, 1993, pp. 217-232.

[11] I. Nalbantis, "Use of Multiple-Time-Step Information in Rainfall-Runoff Modeling," *Journal of Hydrology*, Vol. 165, No. 1, 1995, pp. 135-159.

[12] D. Kavetski, G. Kuczera and S. W. Franks, "Semidistributed Hydrological Modeling: A 'Saturation Path' Perspective on TOPMODEL and VIC," *Water Resources Research*, Vol. 39, No. 9, 2003, p. 1246.

[13] I. Haddleland, D. P. Lettenmaier and T. Skaugen, "Reconciling Simulated Moisture Fluxes Resulting from Alternate Hydrologic Model Time Steps and Energy Budget Closure Assumptions," *American Meteorological Society*, Vol. 7, No. 3, 2006, pp. 355-370.

[14] I. G. Littlewood and B. F. W. Croke, "Data Time-Step Dependency of Conceptual Rainfall—Stream Flow Model Parameters: An Empirical Study with Implications for Regionalization," *Hydrological Sciences Journal*, Vol. 53, No. 4, 2008, pp. 685-695.

[15] H. Kling and H. P. Nachtnebel, "A Spatiotemporal Comparison of Water Balance Modeling in an Alpine Catchment," *Hydrological Processes*, Vol. 23, No. 7, 2009, pp. 997-1009.

[16] E. Widen-Nilsson, L. Gong, S. Halldin and C.-Y. Xu, "Model Performance and Parameter Behavior for Varying Time Aggregations and Evaluation Criteria in the WASMOD-M Global Water Balance Model," *Water Resources Research*, Vol. 45, No. 5, 2009, 14 p.

[17] M. P. Clark and D. Kavetski, "Ancient Numerical Dae-

mons of Conceptual Hydrological Modeling: 1. Fidelity and Efficiency of Time Stepping Schemes," *Water Resources Research*, Vol. 46, No. 10, 2010, 23 p.

[18] C. Michel, "What Can We Do in Hydrology with a Conceptual Model with One Parameter?" *La Houille Blanche*, No. 1, 1983, pp. 39-44.

[19] Edijatno and C. Michel, "A Rainfall Runoff Model with Three Parameters," *La Houille Blanche*, No. 2, 1989, pp. 113-121.

[20] Edijatno, "Development of a Basic Rainfall-Runoff Model to Daily Time," Ph.D. Thesis, Louis Pasteur University/ENGEES, Strasbourg, 1991.

[21] Edijatno, N. O. Nascimento, X. Yang, Z. Makhlouf and C. Michel, "A Daily Watershes Model with Three Free Parameters," *Hydrological Sciences Journal*, Vol. 44, No. 2, 1999, pp. 263-277.

[22] Y. Rakem, "Critical Analysis and Mathematical Reformulation of Empirical Rainfall-Runoff (GR4J) Model," Ph.D. Thesis, Ecole Nationale des Ponts et Chaussées, Paris, 1999.

[23] C. Perrin, "To an Improvement of a Lumped Rainfall-Runoff Model through a Comparative Approach," Ph.D. Thesis, Institut National Polytechnique de Gronoble, Gronoble, 2000.

[24] C. Perrin, C. Michel and V. Andréassian, "Improvement of a Parsimonious Model for Streamflow Simulation," *Journal of Hydrology*, Vol. 279, No. 1, 2003, pp. 275-289.

[25] Edijatno, "Improvement of a Simple Rainfall Runoff Model at a Daily Time Step on Small Watersheds," Master Report, Cemagref Publication, 1987, 45 p.

[26] M. Kabouya, "Rainfall—Runoff Modeling at Monthly and Annual Time Step in Northern Algeria," Ph.D. Thesis, Université Paris Sud, Laboratoire d'Hydrologie et de Géochimie Isotopique d'ORSAY, Paris, 1990, 374 p.

[27] M. Kabouya and C. Michel, "Estimation of Surface Water Resources at Monthly and Annual Time Step, Application to a Semi-Arid Country," *Revue des Sciences de l'Eau*, Vol. 4, No. 4, 1991, pp. 569-587.

[28] Z. Makhlouf, "Supplements on the GR4J Rainfall-Runoff Model and Estimation of Its Parameters," Ph.D. Thesis, Paris XI Orsay, Cemagref, 1993.

[29] Z. Makhlouf and C. Michel, "A Two-Parameter Monthly Water Balance Model for Frensh Watersheds," *Journal of*

Hydrology, Vol. 162, No. 3-4, 1994, pp. 199-318.

[30] N. O. Nascimento, "Assessment Using an Empirical Model of Human Actions Affect on the Rainfall-Runoff Relationship in Watershed Scale," Ph.D. Thesis, ENPC, Paris, 1995.

[31] S. Mouelhi, "Towards a Coherent Chain of Global Conceptual Rainfall-Runoff Models at Multiyear, Yearly, Monthly and Daily Time Steps," Ph.D. Thesis, Ecole Nationale du Génie Rural, des Eaux et Forêts, Paris, 2003.

[32] V. Andréassian, "Impacts of Forest Cover Change on the Hydrological Behavior of Watersheds," Ph.D. Thesis, Université de Pierre et Marie Curie Paris VI, Paris, 2002.

[33] V. Andréassian, C. Perrin, C. Michel, I. Usart-Sanchez and J. Lavabre, "Impact of Imperfect Rainfall Knowledge on the Efficiency and the Parameters of Watershed Models," *Journal of Hydrology*, Vol. 250, No. 1, 2001, pp. 206-223.

[34] C. Perrin, L. oudin, V. Andreassian, C. Rojas-Serna, C. Michel and T. Mathevet, "Impact of Limited Streamflow Data on the Efficiency and the Parameters of Rainfall—Runoff Models," *Hydrological Sciences Journal*, Vol. 52, No. 1, 2007, pp. 131-151.

[35] B. Ambroise, J. L. Perrin and D. Reutenauer, "Multicriterian Validation of a Semi-Distributed Conceptual Model of the Water Cycle in the Fecht Catchment (Vosges Massif, France)," *Water Resources Research*, Vol. 31, No. 6, 1995, pp. 1467-1481.

[36] F. H. S. Chiew, M. J. Stewardson and T. A. McMahon, "Comparison of Six Rainfall-Runoff Modeling Approaches," *Journal of Hydrology*, Vol. 147, No. 1-4, 1993, pp. 1-36.

[37] J. E. Nash and J. V. Sutcliffe, "River Flow Forecasting Through Conceptual Models. Part I—A Discussion of Priciples," *Journal of Hydrology*, Vol. 27, No. 3, 1970, pp. 282-290.

[38] E. Servat, A. Dezetter and J. M. Lapetite, "Study and Selection of Calibration Criteria of Rainfall-Runoff Models," IRD (ex. ORSTOM), Note 2, Programme ERREAU, IRD Publication, Montpellier, 1989.

[39] V. Klemes, "Operational Testing of Hydrological Simulation Models," *Journal of Hydrological Sciences*, Vol. 31, No. 1, 1986, pp. 13-24.

13

Comparison of SCS and Green-Ampt Methods in Surface Runoff-Flooding Simulation for Klang Watershed in Malaysia

Reza Kabiri, Andrew Chan, Ramani Bai

Faculty of Engineering, University of Nottingham Malaysia Campus, Kajang, Malaysia.

ABSTRACT

The main aim in this research is comparison the parameters of some storm events in the watershed using two loss models in Unit hydrograph method by HEC-HMS. SCS Curve Number and Green-Ampt methods by developing loss model as a major component in runoff and flood modeling. The study is conducted in the Kuala Lumpur watershed with 674 km^2 area located in Klang basin in Malaysia. The catchment delineation is generated for the Klang watershed to get sub-watershed parameters by using HEC-GeoHMS extension in ARCGIS. Then all the necessary parameters are assigned to the models applied in this study to run the runoff and flood model. The results showed that there was no significant difference between the SCS-CN and Green-Ampt loss method applied in the Klang watershed. Estimated direct runoff and Peak discharge (r = 0.98) indicates a statistically positive correlations between the results of the study. And also it has been attempted to use objective functions in HEC-HMS (percent error peaks and percent error volume) to classify the methods. The selection of best method is on the base of considering least difference between the results of simulation to observed events in hydrographs so that it can address which model is suit for runoff-flood simulation in Klang watershed. Results showed that SCS CN and Green-Ampt methods, in three events by fitting with percent error in peak and percent error in volume had no significant difference.

Keywords: SCS Curve Number; Green-Ampt; Loss Method; GIS; HEC-Geo-HMS; HEC-HMS; Runoff; Flood Modeling

1. Introduction

Usual methods of runoff and flooding estimation are costly, time consuming along with error because of having various variables contribute in the watershed. As such, using Geographic Information System (GIS), to develop hydrology model through the sub-watershed data in water resources management and planning seem to be critical. There are various methods to simulate surface runoff and flooding by using different loss model methods in HEC-HMS which some of them consist of the SCS Curve Number model [1], CASC2D [2], TOPMODEL [3], GIUH [4], University of British Columbia Watershed Model (UBCWM) and Geomorphological Instantaneous Unit Hydrograph (GIUH). Among the methods, the SCS (Natural Resources Conservation service Curve Number method (NRCS-CN)) method is widely used. Many studies have been conducted by [5-9] who have applied the GIS tools

to estimate runoff CN value to make an empirical runoff estimation and also many researches was implemented by [10-13] to demonstrate SCS application in hydrological studies. This method is based on a rainfall-runoff model that was created to quantify direct runoff. In fact it presumes an initial abstraction according to curve number value. Curve numbers used in this study is according to USDA National Engineering Handbook [14]. To estimate the direct runoff (excess rainfall) the major components of a watershed which contribute to runoff are the data such as land use, soil data and antecedent moisture conditions (AMCs) which are designed to estimate the loss and runoff volume [15].

Green-Ampt is one of the other complicated methods which is assumed to better estimation of the impacts of land use on runoff. As stated by [16] infiltration parameters can be directly related to watershed characteristics. Green-Ampt method developed in 1911 which is an infiltration equation and requires the homogeneous soil characterizations such as hydraulic conductivity, wetting front soil suction head, moisture contents and impervious value. Some studies have been conducted on the performance of CN to Green-Ampt [17-19]. These studies demonstrate that results of direct runoff modeled are similar and state to be user friendly application of SCS-CN method compare the Green-Ampt. Wilcox *et al.* (1990) expressed that CN and Green-Ampt models leave the results close to where the scope of the study was on six small catchments in USA.

In this study, SCS Curve Number and Green-Ampt equations are applied to determining loss model as a major component in runoff and flooding modeling. The objective of this study is to compare the results of SCS-CN and Green-Ampt model to estimate runoff and flooding in Klang watershed on some rainfall event data. It is important to mention that mapping watershed modeling is done using HEC-GeoHMS extension in ArcGIS which is able

to produce the catchment delineation automatically and also acts as an interface between ArcGIS and HEC-HMS software.

2. Material

2.1. Study Area

This study was conducted in the Klang watershed, located in Kuala Lumpur, Selangor province in Malaysia given in **Figure 1**. The scope lies between 101°30' to 101°55' E Longitudes and 3°N to 3°30'N latitude. The area of Klang watershed is approximately 674 km². The elevation ranges from 10 to 1400 meter above mean sea level and the mean annual precipitation is about 2400 mm. About 50% of Klang watershed has occupied by urban area and much of it is perched on susceptible land to flooding. The **Figures 2** and **3** illustrate the major landuses and soil in the study area respectively. **Table 1** address most cover types that are commonly encountered in Klang watershed areas.

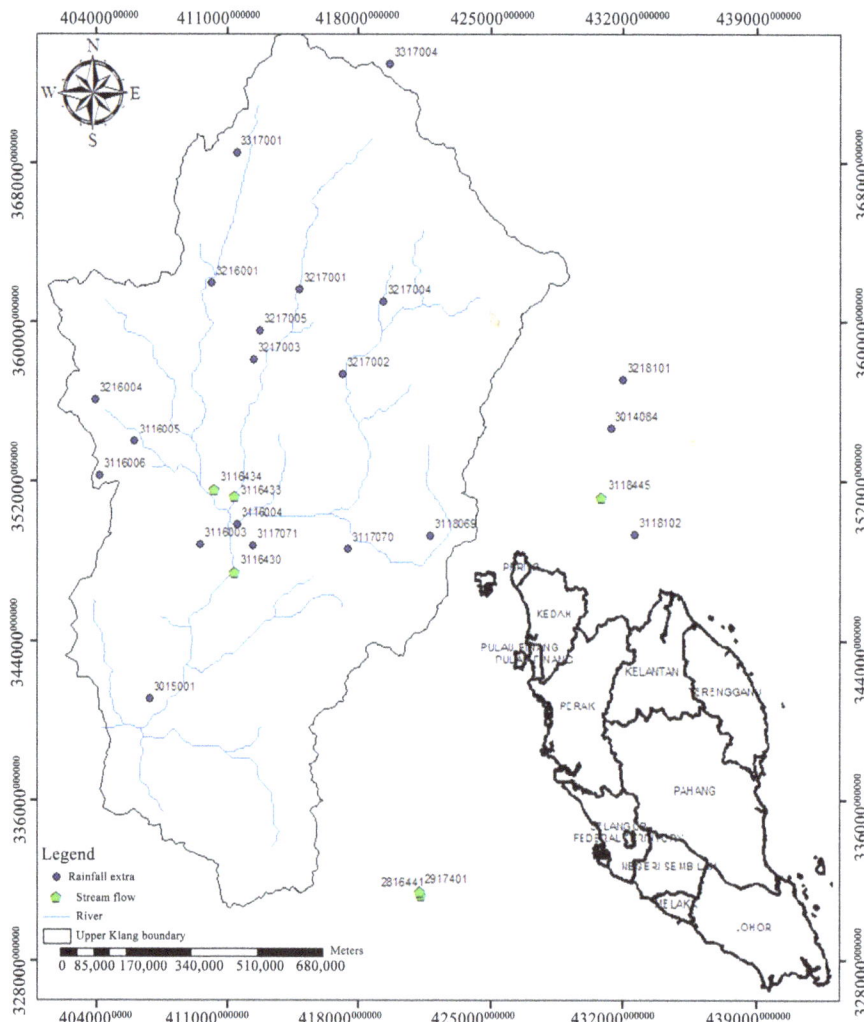

Figure 1. Location of the study area.

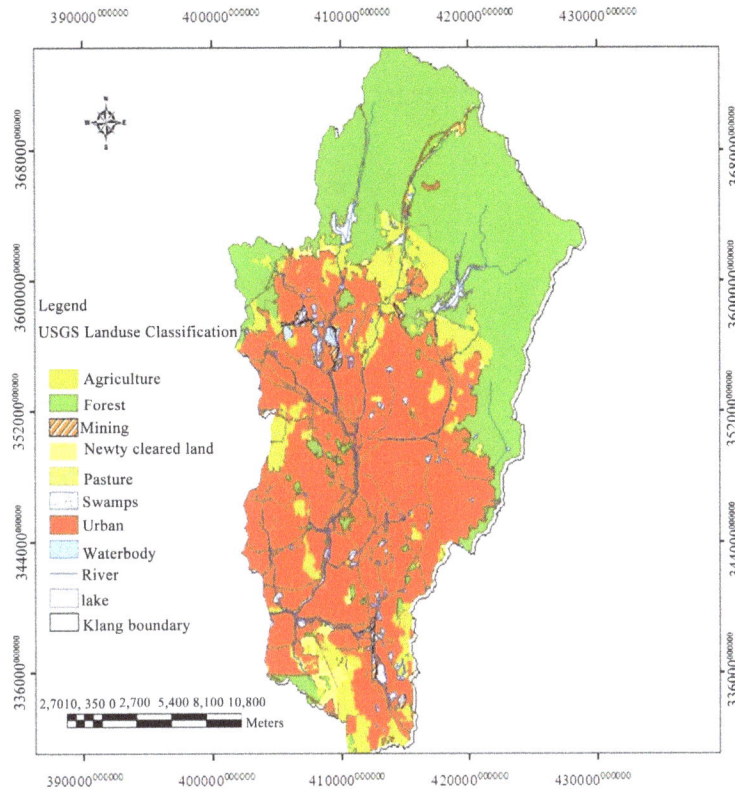

Figure 2. Land use/cover map of the Klang watershed.

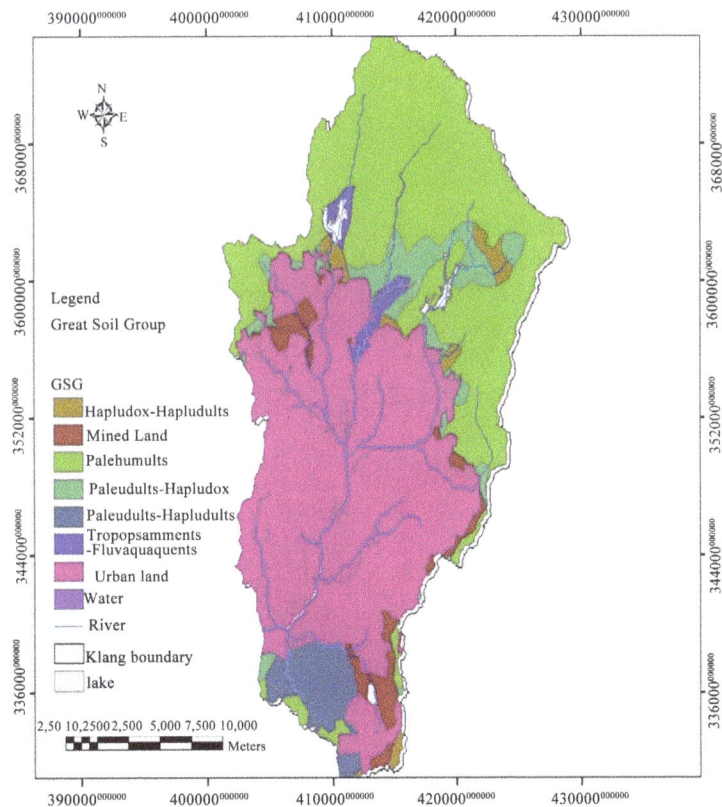

Figure 3. Soil map of the Klang watershed.

Table 1. Land use/cover classes present in the Klang watershed (from DID, 2002).

Land use	Area (Km²)	Percent of total area
Agriculture	59.45	8.82
Forest	248.28	36.83
Mining	4.1	0.61
Newly cleared land	8.58	1.27
Pasture	6.23	0.92
Swamps	0.64	0.09
Urban	334.82	49.67
Water body	11.97	1.78
Total area	**674**	**100**

2.2. Data Sources

The Landuse, Soil, rainfall data and hydrometric data (Hourly discharge) were obtained from Department of Irrigation and Drainage of Malaysia (DID). Digital Elevation Model (DEM) obtained from the Shuttle Radar Topography Mission (SRTM) with the resolution of 90 meters per pixel. 18 rainfall gage stations were selected in the scope of study which contributes to process of areal rainfall mapping. The **Table 2** given the geographical coordination of 18 rainfall gage stations located in the study area. **Figure 1** shows the spatial map of all the rainfall station.

2.3. Software Used for Data Processing

ArcGIS version 9.3.1 powerful Geographical Information System (GIS) software with the HEC-GeoHMS extension used for creating hydrological maps. The extension is a hydrological tool developed by US Army Corps of Engineers, Hydrologic Engineering Center, 2003 and also HEC-HMS software is used for Runoff and flooding analysis.

3. Methodology

According to the **Figure 4**, there has been created catchment delineation for the Klang watershed to make the sub-watershed parameters by using HEC-GeoHMS extension in ARCGIS as an input into HEC-HMS. In this regard, there has been attempt to reproduce all the spatial maps such as initial content, saturated content, suction and conductivity maps extracted from soil data for Green-Ampt method and also other necessary maps for SCS-CN method such as Hydrological soil groups (HSGs), CN and initial abstraction maps. In addition,

spatial impervious map developed by overlaying the DEM and landuse map by cross function in ArcGIS. To enter the precipitation data in HEC-HMS for each sub-watershed, there has been made an aerial rainfall data interpolation for the rainfall event used in the modeling using geostatistical extension in ArcGIS. Since the landuse map in this study is devoted to 2002, therefore relevant flood events are extracted from the year of 2002. The rainfall events with the simple hydrograph shape selected which seem to be appropriate in runoff-flooding modeling by HEC-HMS. The events of 11 June and 21 Dec. are used for validation. Muskingum method is run and finally Muskingum method has been run to enter the channel characterization for flood hydrograph setup in HEC-HMS.

To add the point, that there are two reservoirs in Klang watershed (Batu dam and Klang gate dam). According to its characterization a storage-discharge relationship was run in HEC-HMS to determine the detention impact of the reservoirs.

3.1. Loss Model to Determine Excess Precipitation (Direct Runoff)

3.1.1. SCS-Curve Number Method

The SCS-CN method is used in runoff volume calculation using the values related to landuse and soil data so that integration of these data determine CN values for the watershed to consider amount of infiltration rates of soils. The CN values for all the types of land uses and hydrologic soil groups in Klang watershed are adopted from Technical Release 55 [14]. In this regard, Soils are categorized into hydrologic soil groups (HSGs). The HSGs consist of four categories A, B, C and D, which A and D is the highest and the lowest infiltration rate respectively. To create the CN map, the hydrologic soil group and land use maps of the Klang watershed are combined by cross function in ARCGIS to get a new map integrated of both the land use and soil data.

3.1.2. Green-Ampt Method

Green and Ampt method is also used to calculate the infiltration and loss rate in runoff modeling. The Green Ampt Method is an acceptable loss model and is a simplified representation of the infiltration process in the field [20]. It is a function of the soil suction head, porosity, hydraulic conductivity and time. The general formula of Green-Ampt method is given below [21].

$$\int_o^{F(t)} \frac{1 - \Psi \Delta \theta}{F + \Psi \Delta \theta} dF = \int_o^t K dt \qquad (1)$$

where, F is the total depth of infiltration. Ψ is wetting front soil suction head, θ is water content in terms of volume ratio and K is a saturated hydraulic conductivity.

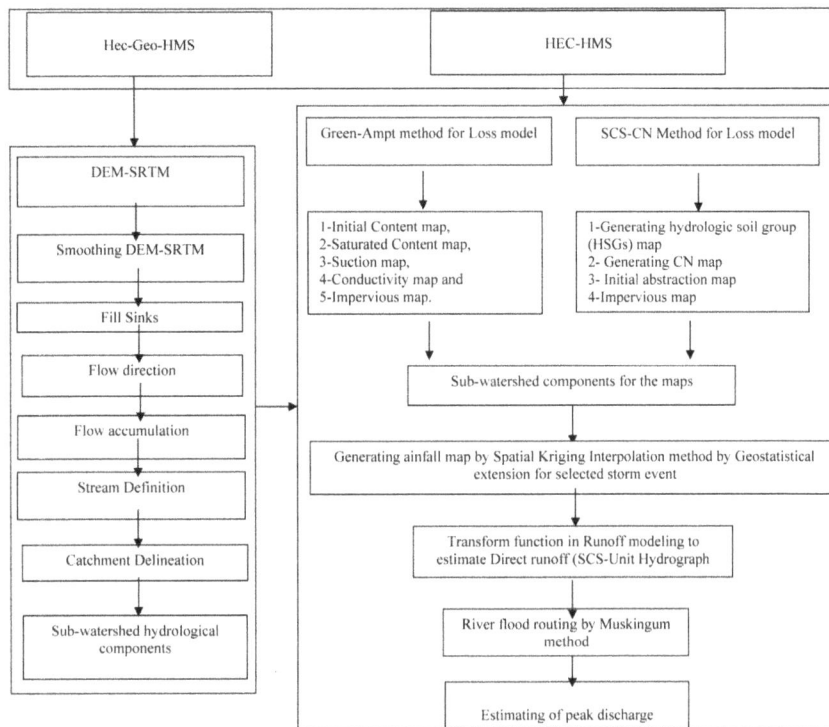

Figure 4. Flowdiagram of flood modeling using Hec-Geo-HMS.

Table 2. Rainfall station used in the study.

Number	Station ID	Longitude	Latitude	Number	Station ID	Longitude	Latitude
1	3216005	101.65	3.26	10	3317001	101.7	3.33
2	3015001	101.66	3.08	11	3117002	101.72	3.25
3	3117070	101.75	3.15	12	3217003	101.7	3.24
4	3116004	101.7	3.16	13	3016001	101.6	3.02
5	3217002	101.75	3.23	14	3216004	101.63	3.22
6	3217004	101.77	3.26	15	3317004	101.77	3.37
7	3116006	101.63	3.18	16	3116003	101.68	3.15
8	3116074	101.7	3.15	17	3117101	101.7	3.1
9	3117104	101.75	3.13	18	3016102	101.41	3.05

The soil texture is important component due to it impacts soil physical properties which are used in Green-Ampt method to calculate the loss parameters. In order to estimate soil properties in the Kland watershed it is categorized into USDA soil texture classification. Therefore, the values suggested by [22] have been adapted in soil characterizations.

3.2. SCS-Unit Hydrograph

The curve of runoff changes in terms of time is called hydrograph. It is able to prepare the maximum runoff, volume and the amount of retention of flooding in a watershed. In this study, SCS Dimensionless Hydrograph has been used to generate unit hydrograph for the selected event rainfall. This method has been by USDA on the various watersheds in US. It based on the converting time and flow axis to dimensionless hydrograph in flood hydrograph. It is implemented by dividing the real time of hydrograph by "time to peak", and also dividing the flow of hydrograph by "flow to peak. The method is based on the two assumptions which state firstly, flow at any time is proportional to the volume of runoff, and secondly, time factors affecting the hydrograph shape are constant [14]. The parameters used in SCS dimensionless unit hydrograph are Time of concentration, Lag time, Duration of the excess Rainfall, Time to peak flow, Peak flow. The relevant equations listed below:

$$S = \frac{25400}{CN} - 254 \qquad (2)$$

$$Q = \frac{(P - 0.2S)^2}{P + 0.8S} \qquad (3)$$

$$S = \frac{1000}{CN} - 10 \qquad (4)$$

where, Q is direct runoff (mm), P is accumulated rainfall (mm), S is potential maximum soil retention (mm), and CN is Curve Number.

The unit hydrograph for any regularly shaped watershed can be constructed once the values of Q_p and T_p are defined. The time to peak, time of concentration and is defined as:

$$T_p = 0.6T_c + \sqrt{T_c} \qquad (5)$$

$$T_c = \frac{L^{0.8} \left(\frac{1000}{CN} - 9 \right)^{0.7}}{1140 S^{0.5}} \qquad (6)$$

$$q_p = \frac{2.083 Q \cdot A}{t_p} \qquad (7)$$

where, T_p is Time to peak (min), T_c is Time of concentration (hr.), L is hydraulic length of watershed (ft), S is average land slope of the watershed (percent), q_p is peak flow (m^3/s), Q is direct runoff (cm), A is area of watershed (Km^2). t_p is Time to peak (hr.)

The standard lag time is defined as the length of time between the centroid of precipitation mass and the peak flow of the hydrograph. The time of concentration is defined as the length of time between the ending of excess precipitation and the first milestone on descending hydrograph.

3.3. Flow Calculation in Reach

There are some methods to consider the flow hydrograph in HEC-HMS. According the available data of the Klang watershed, Muskingum method is run to determine the effect of detention of the river on flood hydrograph. Reach element conceptually represents a segment of stream or river. The general formula of Muskingum developed by US Army Corps of Engineers.

$$S = xkQ_i^{m/d} + (1 - x)kQ_o^{m/d} \qquad (8)$$

where, S is the amount of storage (m^3), Q_i and Q_o is

inflow and outflow (m^3/s), m and d are the constant values which express the logarithmic relationship between storage and elevation.

K is called to storage coefficient having dimensions of time and expressing the ratio of storage to outflow level and can be considered as travel time through the reach element. X is a constant coefficient specifying the relative influence of inflow (Q_i) and outflow (Q_o) levels which ranges from 0.0 up to 0.5 with a value of 0 results in maximum attenuation and 0.5 results in no attenuation (HEC-HMS tutorial). In this study due to having the most urbanization areas occupied in Klang watershed, value of coefficients has been taken as 0.5.

4. Results and Discussion

4.1. Generating Hydrological Watershed Characterization

Once downloading the DEM from SRTM site, it is run some processes on it to generate the sub-watersheds and relevant hydrological characterization. The smoothing and filling function are applied by HEC-Geo-HMS to remove the null and noise of DEM. Flow direction, flow accumulation and stream definition functions are run to reproduce the drainage network of DEM. Finally "catchment delineation" function in HEC-Geo-HMS generated 33 sub-watersheds. The **Figure 5** displays generated sub-watersheds and **Table 3** presents morphological characterization of Klang watershed derived from DEM.

4.2. Generating HGSs and CN Maps

Three hydrologic groups including A, B and D were found in the Klang watershed. 32, 11.6 and 55.5 percent of soil placed in group A, B and D, respectively. **Figure 6** illustrates CN map. And also **Table 4** presents CN values obtained by overlaying the land use and soil maps. It is founded that the lowest CN value was found to be 30 in forest and industrial area with the highest CN value was found to be 93 (except the water body which CN equal to 100).

Next step is to make average for each sub-watershed which has been delineated already. The GIS Cross function is employed to generate sub-watershed CN and Green-Ampt maps using Equation (9):

$$\text{Soil } Cod_{sub} = \frac{\sum A_i \text{ Soil } cod_i}{\sum A_i} \qquad (9)$$

where: Soil Cod_{sub} is weighted average soil parameter for sub-watershed; Soil cod_i is the parameter value and A_i is area inside the specified sub-watershed.

All the values assigned to sub-watershed in Klang area are presented in **Table 5**.

Table 3. Sub-watershed parameters derived of Klang watershed.

Sub-watershed	Area (Km²)	Perimeter (Km)	Mean elevation (m)	Watershed Slope %	Slope of main channel %	Lag time (hr)
s1	49.94	52.30	4416	38.12	0.046	3.11
s2	52.44	51.76	506.2	44.26	0.079	2.9
s3	28.15	34	180.3	29.21	0.044	1.6
s4	76	60.66	365.2	30.32	0.035	3.26
s5	20.44	33.6	310.1	42.02	0.038	2.67
s6	14.98	25.9	215.2	32.55	0.04	2.9
s7	5.22	14.3	123.2	28.2	0.031	1
s8	4.49	12.19	83.6	13.5	0.019	0.75
s9	24.66	31.9	102.4	16.7	0.005	1.43
s10	16.33	29.98	185.2	25.67	0.02	1.68
s11	16.65	25.43	49	2.43	0.002	2.57
s12	21.21	35.82	73.9	10.3	0.003	1.28
s13	19.45	32.14	111.5	20.46	0.004	1
s14	19.23	29.5	102.6	17.09	0.006	1.43
s15	6.49	22.94	55.2	3.21	0.005	1.89
s16	40.11	52.18	85.9	11.43	0.019	2.01
s17	5.16	15	65.9	17	0.006	0.78
s18	4.29	13.12	52.5	10.66	0.005	0.49
s19	11.11	29.16	45.5	1.9	0.004	1.98
s20	16.3	29.43	60.7	12	0.006	1.2
s21	29.8	34.55	51.2	5.50	0.005	2.1
s22	49	47.98	66.4	8.09	0.005	1.89
s23	12.3	27.9	62.5	13.8	0.02	0.88
s24	5.4	14.5	45.7	8.7	0.02	0.75
s25	24.44	35.16	49.3	5.65	0.006	1.99
s26	8.22	21.19	27	3.92	0.009	1.34
s27	11.23	19	37.8	6.65	0.004	1
s28	15.76	24.19	51.5	10.05	0.007	1.23
s29	17.54	32.12	68.6	16.16	0.008	1.43
s30	26.29	38.9	89	9.11	0.02	1.45
s31	10.13	24.51	48.6	4	0.005	1.29
s32	2.3	12.15	47.5	6.87	0.009	0.70
s33	8.89	19.66	58.7	2.31	0.004	1.56

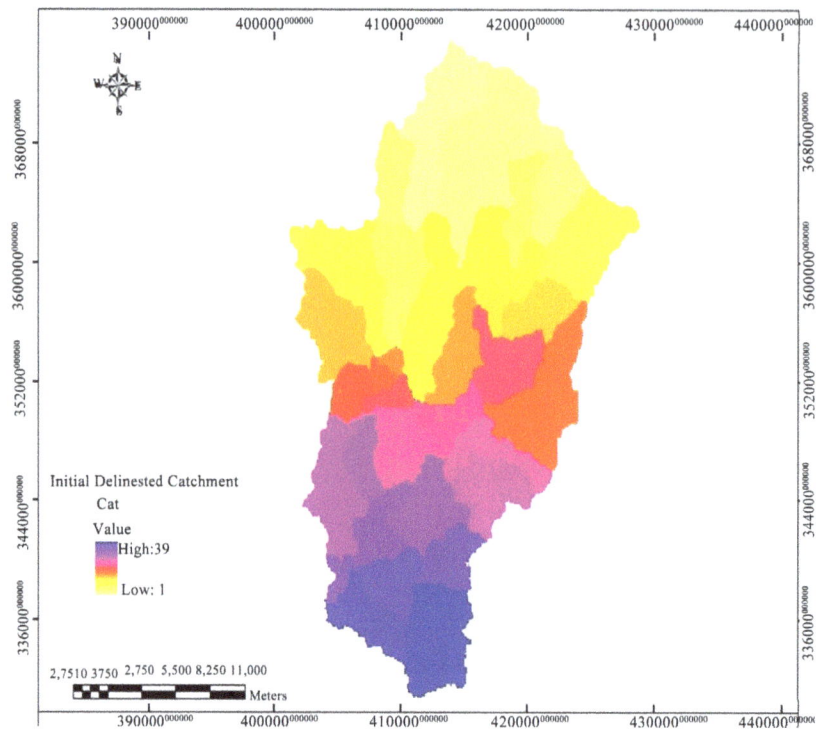

Figure 5. Sub-watershed derived of the Klang watershed.

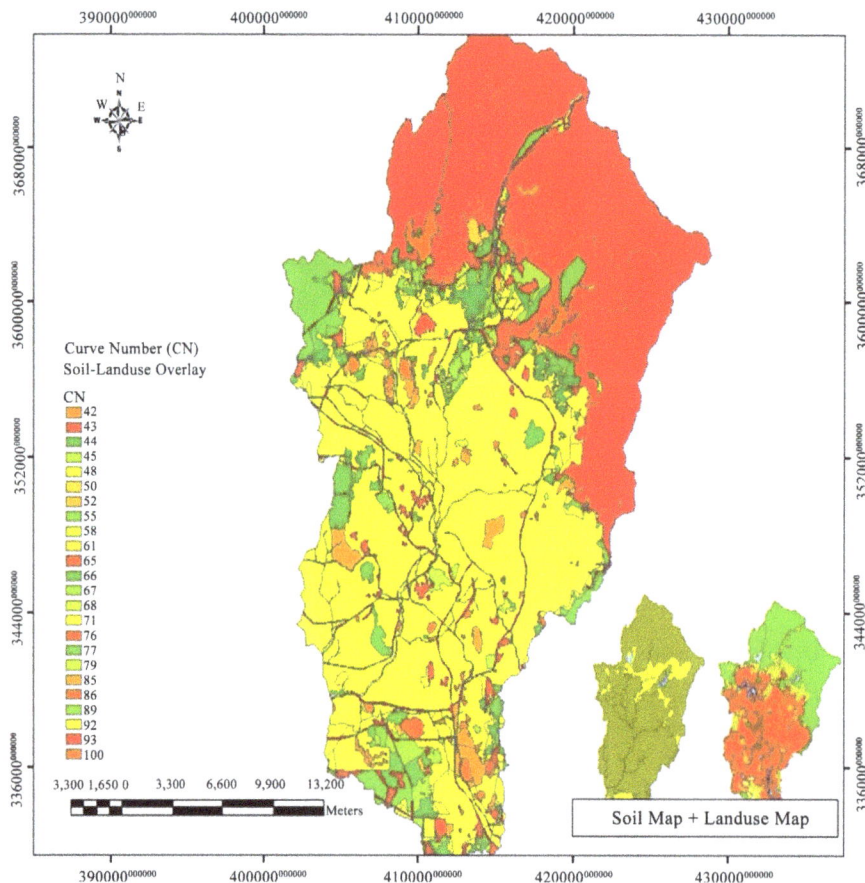

Figure 6. Map of curve number (CN) values for Klang watershed.

Table 4. Curve number of different land use and Hydrologic soil groups (HSGs) in Klang watershed.

Landuse	HSGs	CN	Area (m^2)
Agriculture	A	67	11410612.3
	B	77	27203033.8
	D	87	20830361.9
Forest	A	30	210,681,253
	B	55	26414983.3
	D	77	11109115.4
Pasture	A	30	2752165.6
	B	58	2115919.59
	D	78	1,324,609
Urban	A	48	7368539.92
	B	66	22636343.6
	D	86	304,775,190
Mining areas	B	88	315852.11
	D	93	3786944.33
Newly cleared land	A	39	1401685.1
	B	61	6403586.91
	D	80	771857.73
Swap and water body		100	12,614,850

Table 5. Infiltration parameters in Green-Ampt method for each sub-watershed in Klang area.

Sub-watershed	Hydraulic conductivity (mm/h)	Wetting from suction (mm)	Saturated water content	Initial water content	Impervious (Km^2)	impervious %	CN
1	27.084	64.676	0.42	0.289	0.03	0.07	46
2	29.9	61.3	0.437	0.312	0.16	0.37	43
3	18.103	114.612	0.469	0.241	0.85	1.33	64
4	23.245	86.053	0.439	0.268	0	0.00	45
5	27.626	72.132	0.441	0.299	0	0.00	43
6	25.836	80.438	0.446	0.288	0.49	0.96	51
7	11.021	153.757	0.466	0.202	1	1.39	72
8	2.779	195.299	0.476	0.154	1.75	2.19	80
9	7.714	171.469	0.467	0.183	8.58	11.00	78
10	19.006	116.075	0.45	0.249	2.76	4.31	64
11	3.251	194.358	0.471	0.158	11.8	13.26	89
12	1.163	207.91	0.464	0.147	9.6	10.91	88
13	8.945	166.36	0.462	0.191	6	7.50	80
14	12.569	145.764	0.465	0.211	7.15	9.66	74
15	1	208.8	0.464	0.146	1.65	1.81	91
16	5.005	187.663	0.462	0.169	16.98	21.23	80
17	1	208.8	0.464	0.146	1.67	1.96	85
18	1	208.8	0.464	0.146	1.82	2.09	87
19	1	208.8	0.464	0.146	6.12	6.65	92
20	1	208.8	0.464	0.146	5.65	6.49	87
21	1	208.8	0.464	0.146	14.99	17.43	86
22	1.882	198.9	0.463	0.148	12.25	14.41	85
23	1	208.8	0.464	0.146	4.77	5.36	89
24	1	208.8	0.464	0.146	3.7	4.07	91
25	1	208.8	0.464	0.146	20.33	23.37	87
26	1.76	203	0.469	0.149	5.6	6.59	85
27	1	208.8	0.464	0.146	6.5	7.65	85
28	4.077	185.304	0.485	0.16	2.19	2.61	84
29	7.913	162.656	0.49	0.181	0.9	1.14	79
30	2.665	200.249	0.463	0.156	16.18	18.39	88
31	3.284	196.869	0.463	0.159	7.8	9.18	85
32	1	208.8	0.464	0.146	2.7	3.10	87
33	1.538	205.156	0.466	0.149	3.57	3.84	93

4.3. Generating Green-Ampt Maps

Green-Ampt has essential parameters for flood-runoff modeling. To make Green-Ampt parameters at first all the relevant infiltration values adapted from Rawls and Brakensiek (1983) were assigned into soil texture map in GIS. And then it is attempted to make an average value of the infiltration parameters according to sub-watershed boundary by HEC-Geo-HMS to estimate the loss model maps such as hydraulic conductivity, suction and initial maps and also the percentage of impervious map. **Figure 7** is hydraulic conductivity map as an illustration of Green-Ampt component. **Table 5** presents all the Green-Ampt parameters for each sub-basin.

4.4. Generating Direct Runoff and Peak Discharge

Once all the parameters were setup in HEC-HMS for the both loss models (SCS-CN and Green-Ampt), the models run to obtain the direct runoff and peak discharge for each sub-watershed. **Table 6** displays the output of mod-

els run according to flood event of 6 May 2002.

5. Conclusion

In order to determine the efficiency and suitability of methods used there has been attempted to make a comparison on the results by some correlation coefficients and error indices such as Mean Square Error (RMSE), Mean Absolute Error (MAE), coefficient of determination (R^2), correlation coefficient (r), Nash-Sutcliffe efficiency (NSE) where as RMSE and MAE values of 0 indicate a perfect fit. R^2, r and NSE values of 1 indicate perfect correlation. Model for each methods run and the results are presented in **Table 6**. A comparison is conducted on the results of Green & Ampt to SCS-CN loss methods for estimation of runoff losses (**Table 7**). And also the selection of best method is on the base of considering least difference between the results of simulation to observed events in hydrographs so that it can address which model is suit for runoff-flood simulation in Klang watershed (**Table 8**). The comparison indicates that the

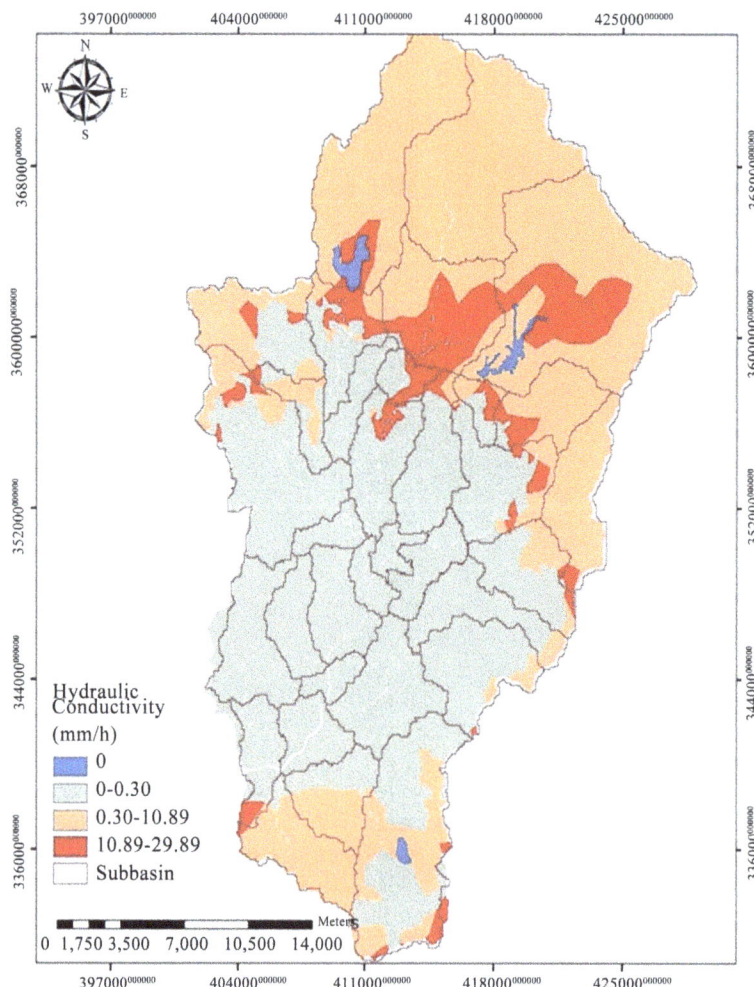

Figure 7. Hydraulic conductivity map of Klang watershed.

Table 6. The comparison of peak discharge and total direct runoff modeled by SCS-CN and Green-Ampt loss methods in HEC-HMS for each sub-watershed in Klang area.

Sub-watershed	Peak discharge (M^3/S) (SCS/CN)	Peak discharge (M3/S) (Green-Ampt)	Total direct runoff (mm) (SCS/CN)	Total direct runoff (mm) (Green-Ampt)
S1	5.7	7.4	1.9	2.36
S2	0.3	0.5	0.09	0.13
S3	5	8.5	1.07	2.71
S4	26	25.9	2.4	5.41
S5	3.5	5.9	3.01	4.23
S6	1.6	1.8	1.23	1.42
S7	10.5	12.8	10.87	11.76
S8	13.2	16.8	15.61	17.73
S9	14.4	14	6.97	6.68
S10	5	5	2.65	2.65
S11	34	33.9	31.03	30.38
S12	72.5	85.5	32.99	37.86
S13	50	45.8	21.36	18.55
S14	45.5	49.4	18.81	18.74
S15	13.5	15.5	19.16	21.37
S16	73.1	78.6	17.91	18.76
S17	30.8	41.9	32.4	41.69
S18	34.5	40.6	43.81	51.27
S19	45.5	48.6	49.06	51.28
S20	85.3	99.2	33.33	42.71
S21	47.9	50.2	27.4	33.14
S22	41.6	52.2	15.7	19.89
S23	57.8	76.2	27.82	34.73
S24	14.6	18.7	24.62	29.64
S25	70.1	84.6	27.41	32.74
S26	32.9	38.3	24.78	28.59
S27	32.8	42.4	19.78	25.33
S28	34.4	40.5	12.44	16.69
S29	21.4	39	9.68	15.3
S30	81.9	91.5	25.8	28.22
S31	59.5	61.2	46.59	45.92
S32	16.2	24.7	32.61	39.56
S33	11.9	12.3	17.29	17.63

Table 7. Evaluation of Green-Ampt and SCS-CN methods for calculating total direct runoff and peak discharge.

Parameters	RMSE	MAE	R^2	r	NSE
Total Direct Runoff (Runoff Depth)	4.15	5.63	0.96	0.98	0.90
Peak Discharge	7.6	7.95	0.97	0.98	0.94

Table 8. The comparison of direct runoff and peak discharge by use of objective functions.

Rainfall Event	Direct Runoff (MM)				Peak Flow (M³/S)			
Date	Green-Ampt Method		SCS_CN Method		Green-Ampt Method		SCS_CN Method	
	Simulated	Observed	Simulated	Observed	Simulated	Observed	Simulated	Observed
06-May-2002	12.31	10.47	11.46	10.47	360.1	361	359.3	361
21-Dec.-2002	9.12	8.42	8.94	8.42	122.6	121.5	121.8	121.5
11-Jun-2002	23.12	25.6	23.1	25.6	447.7	448.9	449.3	448.9

Green-Ampt and SCS-CN loss methods in three events have no significant difference in results of runoff and flood studies in Klang watershed.

REFERENCES

[1] USDA-SCS, "National Engineering Handbook, Section 4-Hydrology," USDA-SCS, Washington DC, 1985.

[2] M. Marsik and P. Waylen, "An Application of the Distributed Hydrologic Model CASC2D to a Tropical Montane Watershed," *Journal of Hydrology*, Vol. 330, No. 3-4, 2006, pp. 481-495.

[3] K. Warrach, M. Stieglitz, H. T. Mengelkamp and E. Raschke, "Advantages of a Topographically Controlled Runoff Simulation in a Soil-Vegetation-Atmosphere Transfer Model," *Journal of Hydrometeorology*, Vol. 3, No. 2, 2002, pp. 131-148.

[4] R. Kumar, C. Chatterjee, R. D. Singh, A. K. Lohani and S. Kumar, "Runoff Estimation for an Ungauged Catchment Using Geomorphological Instantaneous Unit Hydrograph (GIUH) Model," *Hydrological Process*, Vol. 21, No. 14, 2007, pp. 1829-1840.

[5] A. Pandey and A. K. Sahu, "Generation of Curve Number Using Remote Sensing and Geographic Information System," 2002. http://www.GISdevelopment.net

[6] T. R. Nayak and R. K. Jaiswal, "Rainfall-Runoff Modelling Using Satellite Data and GIS for Bebas River in Madhya Pradesh," *Journal of the Institution of Engineers*, Vol. 84, 2003, pp. 47-50.

[7] X. Zhan and M. L. Huang, "ArcCN-Runoff: An ArcGIS Tool for Generating Curve Number and Runoff Maps," *Environmental Modelling and Software*, Vol. 19, No. 10, 2004, pp. 875-879.

[8] M. L. Gandini and E. J. Usunoff, "SCS Curve Number Estimation Using Remote Sensing NDVI in a GIS Environment," *Journal of Environmental Hydrology*, Vol. 12, 2004, p. 16.

[9] G. De Winnaar, G. Jewitt and M. Horan, "GIS-Based Approach for Identifying Potential Runoff Harvesting Sites in the Thukela River Basin, South Africa," *Physics and Chemistry of the Earth*, Vol. 32, 2007, pp. 1058-1067.

[10] C. Michel, A. Vazken and C. Perrin, "Soil Conservation Service Curve Number Method: How to Mend a Wrong Soil Moisture Accounting Procedure," *Water Resources Research*, Vol. 41, No. 2, 2005, pp. 1-6.

[11] L. E. Schneider and R. H. McCuen, "Statistical Guidelines for Curve Number Generation," *Journal of Irrigation and Drainage Engineering*, Vol. 131, No. 3, 2005, pp. 282-290.

[12] S. Mishra, R. Sahu, T. Eldho and M. Jain, "An Improved I_aS Relation Incorporating Antecedent Moisture in SCS-CN Methodology," *Water Resources Management*, Vol. 20, No. 5, 2006, pp. 643-660.

[13] R. K. Sahu, S. K. Mishra, T. I., Eldho and M. K., Jain, "An Advanced Soil Moisture Accounting Procedure for SCS Curve Number Method," *Hydrological Processes*, Vol. 21, No. 21, 2007, pp. 2872-2881.

[14] US Department Agriculture, Soil Conservation Service, "Urban Hydrology for Small Watersheds SCS Technical Release 55," US Government Printing Office, Washington DC, 1986.

[15] S. K. Mishra, M. K. Jain, R. P. Pandey and V. P. Singh, "Catchment Area-Based Evaluation of the AMC-Dependent SCS-CN-Inspired Rainfall-Runoff Models," *Journal of Hydrological Process*, Vol. 19, No. 14, 2005, pp. 2701-2718.

[16] B. P. Wilcox, W. J. Rawls, D. L Brakensiek and J. R. Wight, "Predicting Runoff from Rangeland Catchments: A Comparison of Two Models," *Water Resources Research*, Vol. 26, No. 10, 1990, pp. 2401-2410.

[17] X. C. Zhang, M. A. Nearing and L. M. Risse, "Estimation of Green-Ampt Conductivity Parameters: Part I. Row Crops," *Transactions of the ASAE*, Vol. 38, No. 4, 1995, pp. 1069-1077.

[18] X. C. Zhang, M. A. Nearing and L. M. Risse, "Estimation of Green-Ampt Conductivity Parameters: Part II. Perennial Crops," *Transactions of the ASAE*, Vol. 38, No. 4, 1995, pp. 1079-1087.

[19] M. A. Nearing, B. Y. Liu, L. M. Risse and X. Zhang, "Curve Numbers and Green-Ampt Effective Hydraulic Conductivities," *Journal of the American Water Resources Association*, Vol. 32, No. 1, 1996, pp. 125-136.

[20] S. T. Chu, "Infiltration during Unsteady Rain," *Water Resources Research*, Vol. 14, No. 3, 1970, pp. 461-466.

[21] R. G. Mein and C. L. Larson, "Modeling Infiltration during a Steady Rain," *Water Resources Research*, Vol. 9,

No. 2, 1973, pp. 384-394.

[22] W. J. Rawls and D. L. Brakensiek, "A Procedure to Pre-dict Green and Ampt Infiltration Parameters," *Proceedings of the National Conference on Advances in Infiltration Chicago*, Chicago, 1983, pp. 12-13.

14

Evaluation of a Simple Hydraulic Resistance Model Using Flow Measurements Collected in Vegetated Waterways

Fredrik Huthoff[1,2], **Menno W. Straatsma**[3], **Denie C. M. Augustijn**[1], **Suzanne J. M. H. Hulscher**[1]

[1]Department of Water Engineering & Management, Faculty of Engineering Technology, University of Twente, Enschede, The Netherlands; [2]HKV Consultants, Lelystad, The Netherlands; [3]Department of Earth System Analysis, Faculty of Geosciences, Utrecht University, Utrecht, The Netherlands.

ABSTRACT

A simple idealized model to describe the hydraulic resistance caused by vegetation is compared to results from flow experiments conducted in natural waterways. Two field case studies are considered: fixed-point flow measurements in a Green River (case 1) and vessel-borne flow measurements along a cross-section with floodplains in the river Rhine (case 2). Analysis of the two cases shows that the simple flow model is consistent with measured flow velocities and the present vegetation characteristics, and may be used to predict a realistic Manning resistance coefficient. From flow measurements in the river floodplain (case 2) an estimate was made of the equivalent height of the drag dominated vegetation layer, as based on measured flow characteristics. The resulting height corresponds well with the observed height of vegetation in the floodplain. The expected depth-dependency of the associated Manning resistance coefficient for could not be detected due to lack of data for relatively shallow flows. Furthermore, it was shown that topographical variations in the floodplain may have an important impact on the flow field, which should not be mistaken as roughness effects.

Keywords: Hydraulic Roughness; Vegetation; River Flow; Floodplains; ADCP Measurements

1. Introduction

Various studies have shown that if vegetation penetrates a significant part of the water column then a constant Manning resistance value is no longer adequate to describe the hydraulic resistance for varying flow conditions [1-5]. These studies have shown that the hydraulic resistance due to vegetation (in terms of Manning's n) tends to decrease with increasing water level. No methodology to describe such behavior is generally accepted despite intense research efforts in recent years. This is partly due to the empirical nature of proposed relationships and the difficulties in deriving generally applicable methods due to the complexity of flow around vegetation. In particular, hydraulic resistance parameters determined in the field are unavoidably contaminated with additional external influences, such as geometrical variations of the channel, density currents, sediment interactions or surface waves [6]. Furthermore, collecting flow data in natural vegetated waterways is in itself a tricky task: the presence of vegetation obstructs detailed flow velocity sampling [7]. In particular when vegetation is abundant and has a large impact on the flow field, accurate data sampling is difficult. As a result, studies where flow characteristics are measured in vegetated waterways are relatively scarce [8], they are case-specific and, due to the complexity of the environment, are difficult to interpret (see [9], where flume studies are used as a reference to field measurements).

Most experimental studies on hydraulic resistance of vegetation are conducted in laboratory flumes, where it is possible to minimize hydraulic impacts due to other external influences [10-13]. That way, investigations of vegetative hydraulic resistance allow isolation of the impact of specific vegetation characteristics (such as stem width, height, flexibility). These studies have revealed that flow through vegetation is difficult to describe based on geometrical conditions of the vegetation alone, even if vegetation is described in a simplified way as cylindrical stems. A recurring complication of these hydraulic resistance models is the need for a general representation of energy losses associated with turbulent mixing patterns. Such energy-loss representations may enter the flow models in the form of a turbulent mixing length [14-17]. Flow descriptions that include the mixing length concept only have practical value if the mixing length is directly related to measurable quantities. Such relations have been proposed [16,18,19], but due to lack of theoretical justification their general applicability remains questionable. Other, more general, approaches include k-ε turbulence models, which explicitly describe the transport of turbulent energy [20-22]. However, these methods require considerably larger computational effort, in particular if applied in models on river-reach scales.

Alternatively, a simple vegetation resistance model has been proposed by Huthoff *et al.* [23] that includes only measurable quantities. The model requires knowledge of a vegetation drag coefficient C_D, the average vegetation height k, stem diameter D and a representative spacing between neighboring plants s (see also [24]). Among these parameters, only the drag coefficient cannot be directly obtained from vegetation dimensions, but is to be determined from flow experiments. It appears that the C_D-value is species-specific [25] and is a function of the Reynolds number [26], stem aspect ratio [27], vegetation distribution density [28,29] and foliage and streamlining effects [30,31].

In the current work we investigate whether the simplified model proposed in [23] is consistent with flow measurements in the field, and whether it provides a potential candidate for integration with river-reach flow models. Two case-studies with flow in vegetated waterways are considered (**Figure 1**). First, measured flow velocities and vegetation characteristics in a Green River are evaluated against predictions of the hydraulic resistance model (case 1). It is shown that the model provides an acceptably accurate estimate of the average flow velocity, as based on general vegetation characteristics. Next, flow velocities are measured in a natural floodplain at different discharge magnitudes (case 2). In this case study, no detailed information about the vegetation characteristics was available. Therefore, we evaluate whether the measured dynamic behavior of the hydraulic resistance is consistent with the vegetation resistance model. It is shown that the proposed model describes the vegetation resistance well, but that because of the weak dynamic behavior of the hydraulic resistance, and the lack of data at shallow flows, also a simple wall-roughness model may be used for the considered flow conditions.

2. Case 1: Fixed-Point Measurements

2.1. Study Location

In February 2005, during a high-discharge event in the Rhine river a Green River[1] was deployed to convey surplus river discharge. Due to its relative homogeneous geometrical boundaries, the Green River is suitable for measurements of the hydraulic effect of the present vegetation. The aerial picture in **Figure 2** shows the

study area, including the bridge from where measurements were performed. The flow direction in this picture is from east (right) to west (left). **Figure 3** shows the presence of vegetation on the channel bed of the Green River, as observed during dry conditions. Note that the Green River includes a central sub-channel (indicated in **Figure 2** and also just visible in right panel of **Figure 3**). The sub-channel is always wet but is closed at both ends, and therefore does not convey water.

2.2. Methodology and Results

The bridge that spans the Green River was used to tether a float from, to record the flow velocities below. **Figure 4** shows a picture of the bridge on the 17th of February 2005, the day that data were collected. Three locations

Figure 1. A map of the Netherlands indicating locations of the two case studies (Driel and Bimmen).

Figure 2. Aerial picture of the Green river at Driel. Also, the central (sub-) channel is indicated.

Figure 3. The measurement location at Driel during dry conditions (pictures: M.W. Straatsma).

---

[1]A Green River is a secondary waterway that is only deployed for additional discharge if the discharge in the main river channel reaches a particular critical level.

along the bridge were used for measurements: one in the central sub-channel of the Green River (Location 2) and two locations well-separated to either side of the central sub-channel (Locations 1 and 3). These latter two locations have similar bed coverage characteristics (as seen in **Figure 3**) and are thus expected to be hydraulically equivalent.

An RD Instruments Acoustic Doppler Current Profiler (ADCP) was used to record flow velocity profiles. The ADCP resolves the three-dimensional flow vector at various user-defined depths. Due to the size of the ADCP itself, and the inability of the ADCP to measure flow velocities immediately below the device, flow velocities were measured downwards from 46 cm below the water surface, at depths 10 cm apart. An average ensemble interval of 6 s was used to determine the mean streamwise velocity. Near the bed, large fluctuations in flow velocities were measured (**Figure 5**). The fluctuations are likely due to the presence of vegetation. Vegetation parameters were independently measured after the flood, giving an average height of $k = 37.5$ cm, an average stem width of the vegetation of $D = 3.7$ mm, an average surface density of $m = 51$ stems/m^2, leading to an average spacing between plants of $s = 13.6$ cm.

Figure 5 shows the measured flow velocity profiles for the three locations along the bridge. The graphs also show the profiles of the average streamwise velocities, together with 16- and 84-percentile error boundaries. These boundaries correspond to a 1σ standard deviation if errors were distributed normally. The flow depth h is measured independently by the ADCP, as stated above the graphs in **Figure 5**. Also stated are the depth-averaged flow velocities in the surface layer U_s, representing flow above the vegetation.

The graphs in **Figure 5** show that, between the three measurement locations, the depth-averaged flow velocity in the deepest location (Location 2) is slightly smaller than for the other two locations. This may seem an unexpected result, because for equal bed resistance one would expect larger flow velocities at larger depths.

However, it appears that at Location 2 not the entire water column contributes to discharge capacity. As mentioned before, the sub-channel at Location 2 is closed at both ends, thus partially obstructing flow. Another effect that may cause the flow in the central channel to slow down is the presence of a small bridge (see **Figure 3**). This small bridge was entirely flooded during the high-discharge event in February 2005 (see **Figure 4**), thus obstructing flow near the central sub-channel. Consequently, we consider the flow measurements at Location 1 and Location 3 in **Figure 5** more representative of the hydraulic response due to bed vegetation. In the remainder, the measurements from Location 2 in the central sub-channel are therefore discarded for the analysis of hydraulic resistance due to vegetation.

Finally, the energy slope i was determined by tracking a float from a fixed location on the river bank (for float tracking methodology, see [32]). This resulted in an average value for the energy slope of $i = 9.2 \times 10^{-5}$, having an error of about 20%.

2.3. Comparison to Model Predictions

Here we consider the simple model for the hydraulic resistance of vegetation as proposed by Huthoff *et al.* [23] (also see [24]). In this model, the flow velocity in the vegetation layer (or resistance layer) U_r is described

Figure 4. The measurement location at Driel during high discharge on February 17, 2005 (picture: M.W. Straatsma).

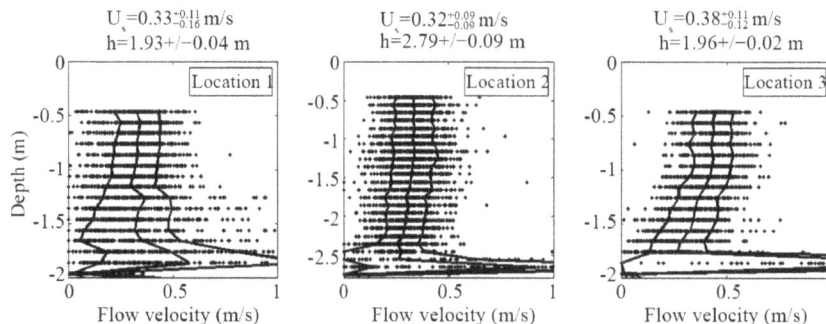

Figure 5. Velocity measurements at three different locations for case 1. The average velocity profile and 16- and 84-percentile boundaries are shown in each of the graphs. Stated above each graph, U_s is the depth-averaged velocity in the surface layer and h the flow depth.

separately from flow in the surface layer U_s. Together, they give an estimate of the average flow velocity over the total depth U_T:

$$U_T = \frac{k}{h}U_r + \frac{h-k}{h}U_s \qquad (1)$$

The depth-averaged flow velocity in the surface layer is represented by the power law

$$U_s = U_{r0}\left(\frac{h-k}{s}\right)^{\frac{2}{3}\left(1-(k/h)^5\right)} \qquad (2)$$

where the power exponent reduces to a constant value of 2/3 if the total flow depth h is at least twice as large as vegetation height k. The representative separation between homogeneously distributed cylindrical vegetation elements vegetation (s), is calculated as [23]:

$$s = \frac{1}{\sqrt{m}} - D \qquad (3)$$

The average flow velocity in between the vegetation (in the vegetation layer) is estimated as

$$U_r = U_{r0}\sqrt{\frac{h}{k}} \qquad (4)$$

In Equations (2) and (4) the characteristic scaling velocity U_{r0} is given by

$$U_{r0} = \sqrt{\frac{2gi}{mDC_D}} \qquad (5)$$

with g the gravitational acceleration, i the (energy) slope, m the number of vegetation elements per unit bed area, D the stem diameter and C_D the dimensionless drag coefficient. Using the earlier stated vegetation parameters in Section 2.2 the average flow velocities can be predicted using Equations (1)-(5). The only unknown parameter is the drag coefficient for which we adopt $C_D = 1.8$ [33].

Figure 6 shows the predicted flow velocities for the resistance layer, the surface layer, the entire flow depth and Manning's n based on a 20% standard deviation in the measured vegetation parameters and a 10% standard deviation in adopted C_D-value. Comparing the measured values for the surface velocity in **Figure 5** ($U_s \approx 0.33$ - 0.38 m/s) with the predicted value in **Figure 6** ($U_s \approx 0.38$ m/s), shows that the applied vegetation resistance model gives quite good results.

2.4. Discussion

For the considered case in the Green River we were only able to measure flow velocities in the surface layer, because the presence of vegetation distorted measurement near the bed (see **Figure 5**). Therefore, we cannot evaluate the predicted depth-averaged flow velocities of the vegetation layer U_r and of the total flow depth U_T (as

given in **Figure 6**) against field measurements. However, using the predicted values for U_T, a Manning resistance parameter n can be calculated, which subsequently allows comparison with values cited in literature for grass-lined channels. Manning's equation is given as:

$$U_T = \frac{R^{2/3}}{n}\sqrt{i} \qquad (6)$$

which has the property that the resistance coefficient n is practically constant for hydraulically rough turbulent flow over a fixed bed [34]. Inserting the measured flow depth (at Location 3, see **Figure 5**) and the predicted values for U_T yields Manning resistance coefficients $n = 0.045 +/- 0.007$ m$^{-1/3}$s. This calculated n-value is consistent with the Manning values cited in literature for grass-lined floodplains. For "high grass" Chow [35] states a general value of $n = 0.035$, with a lower boundary of $n = 0.030$ and an upper boundary of $n = 0.050$ m$^{-1/3}$s. However, as was mentioned in the introduction, n generally does not remain constant for flows over vegetation.

Figure 7 shows how the model Equations (1)-(5) together with Manning's law, Equation (6), lead to a depth-dependent Manning coefficient. It can be seen that at flow depths that are greater than the one considered, the n-value is expected to decrease. However, the projected changes for large flow depths appear to be quite insignificant, and also a constant Manning's n may be used. Shallower flows are expected to yield considerably stronger changes in n, with nearly doubled values for just submerged vegetation. Similar trends have been reported in empirical studies of vegetation resistance [1-5]. In particular, the study of Wilson and Horritt [4] shows that the Manning coefficient becomes practically constant if the flow depth is much larger than the vegetation height.

We conclude that the model by Huthoff *et al.* [23] can be used to give a reliable estimate of the effective roughness under field conditions and that, qualitatively, predicted trends of the roughness model agree with trends cited in the literature.

3. Case 2: Vessel-Borne Flow Measurements

3.1. Study Location

In 1998, a high-discharge event occurred on the Rhine River with a maximum recorded discharge of 9413 m^3/s

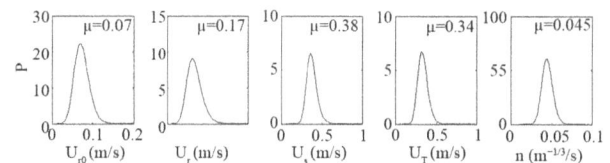

Figure 6. Probability densities of predicted flow velocities (μ = mean value) based on uncertainty in the drag coefficient and variations in measured vegetation characteristics.

(recorded on November 4th at station Lobith). During the high-discharge event, vessel-borne ADCP measurements were carried out along a cross-section of the River [36]. See **Figure 8** for the research vessel and the ADCP device, and **Figures 9** and **10** for the sampled river cross-section. Water flows from the east to the west, which in **Figure 9(b)** is from the lower right to the upper left. The flood-plain on the southern bank is clearly visible in **Figure 9**, as is the embankment that separates it from the main channel. Groynes are present on the northern bank and also further upstream on the southern bank. The flood-plain is covered with grassland.

3.2. Methodology and Results

Flow data were collected along a cross-section at km 863.9 in the Rhine on four consecutive days from November 3 to 6 and then again from November 9 to 11. The discharge peak occurred at November 4 and decreased quite rapidly thereafter. On November 9 the water level had dropped to a level that flow measurements were no longer possible in the floodplain. On each day, flow measurements were collected for multiple transits along the river cross-section (**Figure 10**). The transits outline a narrow cross-sectional strip of the channel, for which the best fit straight-line was defined as the equivalent transect (dotted line in **Figure 10**). In the subsequent analysis, all measurements are mapped onto the equivalent transect.

In **Figure 11**, an example is shown of one of the sampling runs. Flow velocities were measured below the research vessel at depths 25 cm apart with the highest measuring point 1.32 m below the water surface. Lateral sampling separations were on average 4 m, depending on the travelling speed of the vessel. A region of about 60 cm above the channel bed could not be reliably sampled for flow velocities, partly due to presence of obstructing objects (vegetation) or debris. The measurements were performed in *bottom-track mode*, correcting measurements for vessel movement and thus giving flow velocities with respect to the detected fixed bed. However, a bottom layer of sediment may be dragged along with the flow, which the ADCP interprets as a fixed bed and subsequently gives a bias towards underestimation of flow velocities. The measurements used here are corrected for this effect by comparison with independent GPS data of the research vessel. Details of this procedure are described in [36].

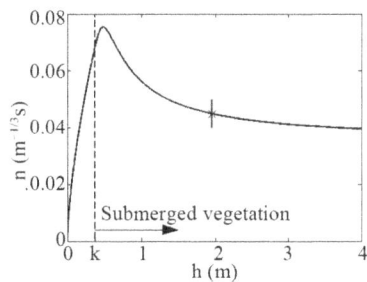

Figure 7. Manning's resistance coefficient *n* as predicted by the model Equations (1)-(5). The "x" symbol corresponds to the average conditions at Location 3 in the Green River.

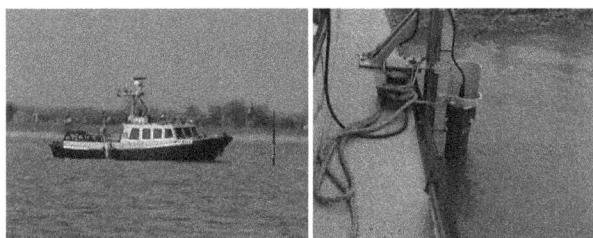

Figure 8. The used research vessel for flow measurements in the Rhine River in November 1998 (left). The picture on the right shows the attached ADCP (pictures: aqua vision).

Figure 9. Aerial pictures of the study area. The white line shows the measurement transect.

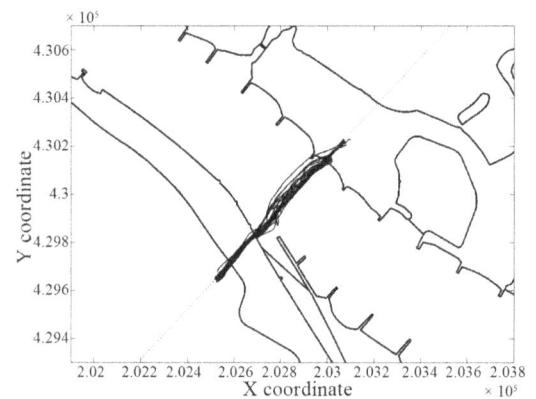

Figure 10. The measurement trajectories near rkm 863.9. The dotted line depicts the equivalent transect.

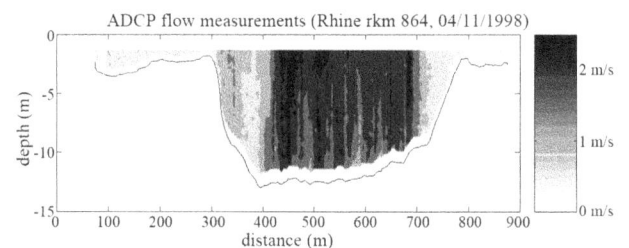

Figure 11. ADCP velocity measurements for one of the measurement runs shown in Figure 10.

Figure 11 shows that lower flow velocities are recorded in the shallower regions of the cross-section. However, a striking feature in the measurement results is that near the deepest part of the channel there appears to be a local minimum in flow velocities (near the lateral coordinate of 380 m). This feature is also observed in the remaining measurement runs (not shown here). The overview of the study area in **Figure 9** may provide an explanation for this observation. Just upstream of the channel cross-section a small channel merges with the Rhine on its southern bank. The local dip in flow velocities is likely due to enhanced mixing in the wake of the confluence zone.

To be able to compare the measured flow velocities in the floodplains to model predictions of vegetation resistance, we depth-averaged the measured flow velocities. Next, we grouped all data collected on the same day, assuming that within this time frame no significant hydraulic changes occurred in the system. **Figure 12** shows the resulting lateral profiles of depth-averaged hydraulic conditions as measured on November 4, 1998. Averaging of data is performed by taking the mean of all measurement points that fall within columns of 10 m wide. The bottom graph in **Figure 12** shows the function $h^{2/3}/U$ set out against the lateral coordinate. Looking at Equation (6), this value represents the quantity $n/\sqrt{i}$. Therefore, if we assume that the downstream slope i is constant across the cross-section, then the bottom graph in **Figure 12** reflects a measure for the effective hydraulic roughness. It thus appears that the hydraulic roughness of the main channel is nearly equal to that of the floodplain ($n/\sqrt{i} \approx 3.5$).

Next, the hydraulic characteristics (depth, flow velocity, effective flow resistance) are spatially-averaged over selected lateral regions, as marked in **Figure 12** by the four vertical grey bars (each having a width of 50 m). Two regions are selected in the floodplain (fp1 and fp2) and also two regions are selected in the main channel: (mc1 and mc2). **Figure 13** shows that the averaged hydraulic conditions in these four regions change during the measurement period. The change in flow depths in the floodplain and in the main channel clearly reflects passing of the peak of the flood wave. After the flood wave has passed, the effective hydraulic roughness in the main channel seemed to have increased. This observation is in line with the study by Wilbers [37], who showed that high-discharge events cause large bed forms on the channel bed, which increases the hydraulic roughness. Note that conditions in the floodplain were only measured during four days around the discharge peak. At lower discharges it was no longer possible to enter the floodplain for flow velocity sampling.

The depth-averaged flow velocities in the main channel could be obtained quite accurately, because the ADCP

was able to sample most of the water column. However, due to the absence of flow measurements near the bed and near the water surface, in the floodplain typically only half of the water column could be sampled (see **Figure 11**). To get a better representative flow velocity in the floodplains we therefore apply a correction procedure. Starting point of the correction procedure is the averaged velocity U_m that is based on the actual sampled data and represents flow in the layer $h_m < z < h - d_m$, (see **Figure 14**). The highest measurement point is 1.32 m below the water surface and the lowest measurement point is approximately 0.6 m above the bed. Measurement depths were 25 cm apart, therefore the depth-averaging range of the ADCP is confined by a height from the bed $h_m = 0.6 - 0.25/2$ m and a depth below the water

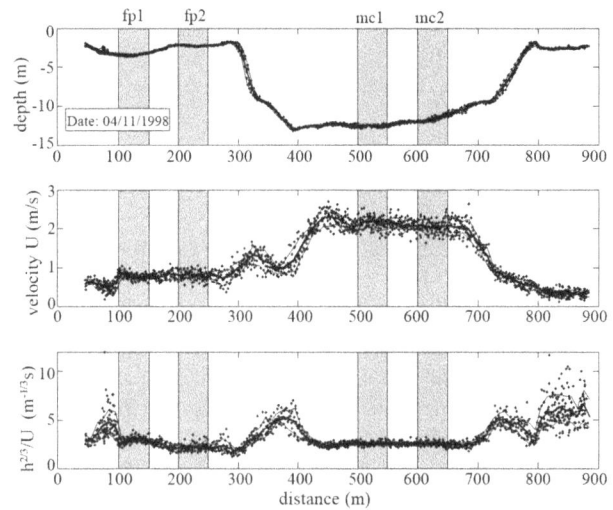

Figure 12. Measured depths (top graph), depth-averaged flow velocities (middle graph) and effective hydraulic resistance (bottom graph).

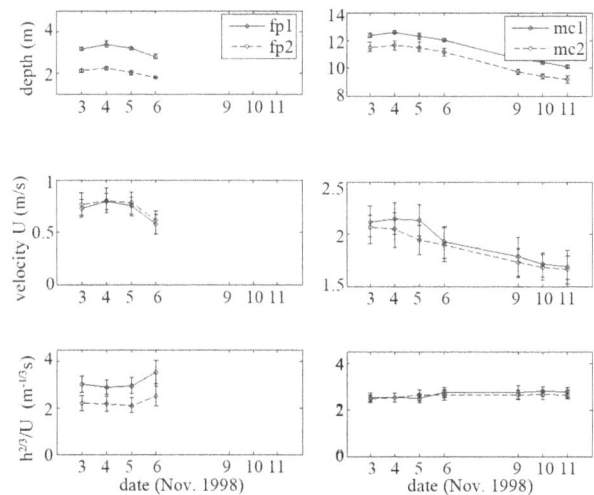

Figure 13. Time-dependency of the hydraulic characteristics within the averaged-regions in the floodplain (fp1 and fp2) and the main channel (mc1 and mc2, see Figure 12).

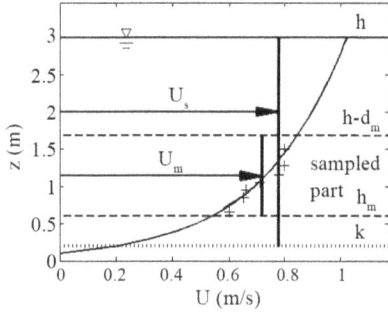

Figure 14. Sketch to illustrate the correction procedure of measured flow velocities.

surface $d_m=1.32 - 0.25/2$ m. The steps in correcting the measured average velocity U_m is as follows:

1) We assume that the flow field above the vegetation follows a vertical logarithmic velocity profile, with a first guess for the corresponding (Nikuradse) roughness height of $k_N = 0.1$ m, which is a reasonable value for grass according to Van Velzen et al. [33].

2) The assumed profile is shifted vertically to fit the sampled flow velocity points (by minimizing least squares).

3) From the fitted profile, the average velocity in the surface layer U_s is determined, assuming an initial value of the vegetation layer height of $k = 0.06$ m.

Figure 14 illustrates the procedure for correcting flow velocities in the surface layer. Effectively, the correction-procedure described above corresponds to multiplying the measured average flow velocity U_m by a factor Φ:

$$U_s = \phi U_m \qquad (7)$$

where

$$\Phi = \frac{h-h_m-d_m}{h-k} \frac{\int_k^h \ln\frac{z}{k_N} dz}{\int_{h_m}^{h-d_m} \ln\frac{z}{k_N} dz} \qquad (8)$$

In **Figure 15(a)** the results after correcting flow velocities are shown. As expected, the corrections are largest for flow measurements in the shallower region fp2.

3.3. Comparison to Model Predictions

We compare the corrected flow velocities in the surface layer (U_s), with values obtained with the simplified flow model proposed by Huthoff et al. [23]. If the depth of the surface layer ($h - k$) is more than three times the vegetation height, then the flow velocity in the surface layer is described by a simple scaling law (see Section 2.3):

$$U_s = U_{r0}\left(\frac{h-k}{s}\right)^{\frac{2}{3}} \qquad (9)$$

This expression can be rearranged to

$$U_s^{3/2} = Ah - B \qquad (10)$$

where $A = (U_{r0})^{3/2}/s$ and $B = Ak$. The bottom graph in **Figure 15** shows the measured and corrected averaged velocities, $(U_m)^{3/2}$ and $(U_s)^{3/2}$ respectively, set out against the flow depth h. For each data set also a best fit straight line is shown. Following Equation (10), the ratio B/A yields an estimate of k, which reflects the height of the drag dominated flow layer (or: the vegetation height). **Table 1** gives an overview of these values.

The results in **Table 1** show that for location fp2 the effective height of the drag-dominated flow layer is approximately $k = 0.4$ m, if using the corrected flow velocities U_s. This value of 0.4 m is higher than expected for the floodplain under consideration (recall that we assumed 0.06 m in the correction procedure), but it is still acceptable for the type of vegetation that can be found in grassed floodplains. In contrast, for location fp1 an effective vegetation height of 1.8 m is found, which is unrealistically high given the nearly uniform vegetation characteristics in the floodplain meadow.

A closer inspection of conditions in the floodplain reveals that the flow depth at location fp1 is about 1.3 m

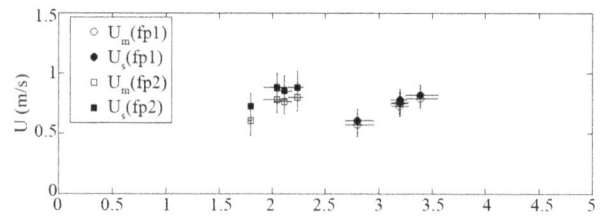

(a) Measured and corrected velocities.

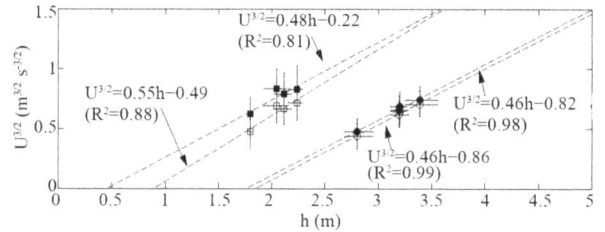

(b) Best fit for $U^{3/2} \sim h - k$.

Figure 15. (a) Depth-averaged velocity U vs flow depth for locations fp1 and fp2 (filled symbols: corrected velocities); (b) Regression relations of $U^{3/2}$ vs depth.

Table 1. Equivalent vegetation heights (k) reproduced from flow measurements and Equation (10).

Location	U	A	B	$k = B/A$	R^2
fp1	U_m	0.46	0.86	1.9 m	0.99
fp1	U_s	0.46	0.82	1.8 m	0.98
fp2	U_m	0.55	0.49	0.9 m	0.88
fp2	U_s	0.48	0.22	0.4 m	0.81

deeper than at location fp2 (**Figure 13**), while the average flow velocities at these locations are nearly the same. It appears that in the deeper part of the floodplain water does not flow freely. **Figure 16** shows the floodplain shortly after a flooding event in 2007, leaving behind a clearly confined pool at location fp1. Therefore, the larger depth at fp1 is a local topographic effect. Consequently, at fp1 we cannot assume (quasi-) uniform flow and therefore we cannot estimate the hydraulic roughness based on the average flow velocity and local flow depth. However, fp2 is not as much affected by local topographical variations, so for this location we can compare measured data to prediction using model Equations (1)-(6). Using the corrected average flow velocities at fp2 (**Figure 15(a)**), a vegetation height of $k = 0.4$ m (**Table 1**) and by adopting a vegetation spacing of $s = 1$ cm for grassed floodplains (see Van Velzen *et al.* [33]), Equation (9) yields an approximate characteristic velocity of $U_{r0} = 0.03$ m/s. Next, if we use a channel slope of $i = 10^{-4}$, which is a representative value for the considered part of the Rhine River [38], model Equations (1)-(6) give the Manning values as shown in **Figure 17**. It can be seen that for shallow flows Manning values can be as high as $n = 0.15$, while for flow depths h > 1.5 m values converge towards a constant value in the range 0.02 - 0.025. Also, the results for an assumed vegetation height of $k = 0.06$ m are shown, indicating that for grass-lined floodplains the vegetation height has only a minor effect on overall flow resistance for relatively deep flows (*i.e.* if h > 2 m). The predicted values in **Figure 17** are slightly lower than those stated in Chow [35], who gives n = 0.030 ± 0.005 for short grass in floodplains.

3.4. Discussion

We studied two cases of flows in natural vegetated waterways to investigate the bed roughness properties and to compare these to predictions from the vegetation roughness model proposed by Huthoff *et al.* [23]. Because of incomplete data sampling we were forced to make assumptions on the general shape of the velocity profiles, in order to obtain representative depth-averaged flow velocities. Despite the uncertainties associated with the obtained flow conditions it was shown that local topographic variations can have important impacts on the flow field, and that these localized effects can easily be misinterpreted as local changes in hydraulic roughness. Therefore, field flow measurements should be treated with extreme care if used to determine hydraulic parameters such as Manning's roughness values.

Several studies have shown that remotely-sensed vegetation characteristics can be obtained with high enough accuracies to use them at as input for roughness parameterizations [32,39,40]. The vegetation roughness model proposed by Huthoff *et al.* [23] is based on such input

Figure 16. A clearly confined pool indicating a local depresssion in the floodplain (picture: F. Huthoff).

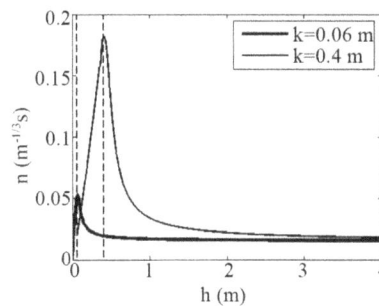

Figure 17. Manning's *n* vs depth according to model Equations (1)-(6) adopting a velocity of $Ur_0 = 0.03$ m/s in the vegetation layer and a vegetation spacing of $s = 1$ cm.

parameters (such as vegetation height and vegetation spacing), which gives opportunities for advanced inundation modeling techniques where measured vegetation characteristics are translated into effective roughness values.

In the current study, data limitations did not allow for an accurate field validation of the considered roughness model. However, it was shown that the model equations give predictions that are in the correct range and also agree with roughness values stated in the literature. For grass-lined floodplains, the model equations suggest that constant Manning values may be appropriate for well-submerged vegetation ($h > 2$ m), which has traditionally been the preferred approach in inundation studies (e.g. [41,42]). In contrast, for situations where the vegetation is just submerged, the model equations predict significant variations in Manning's roughness values. The importance of these roughness variations for evolvement of floodplain inundations should be the topic of more focused research in the future.

4. Conclusion

In the case studies considered here, the waterways with submerged vegetation included predominantly grass species, having average stem heights that are easily an order of magnitude smaller than the flow depths. The vegetative roughness model proposed in Huthoff *et al.* [23] is consistent with measured flow velocities over such vegetation, predicting nearly constant Manning roughness values for well-submerged vegetation. In contrast, model equations predict that constant Manning's roughness values are no longer appropriate when vegetation is just submerged. Future research should focus on the importance of vegetation roughness-changes under such conditions. Further, it was shown that topographical variations of the bed may have major impacts on local flow velocities, underlining that extreme care should be taken when using field data to estimate floodplain roughness coefficients.

5. Acknowledgements

This research is supported by the Technology Foundation STW, applied science division of NWO and the technology programme of the Ministry of Economic Affairs (Netherlands). We gratefully thank Aqua Vision for letting us use their data and Anne Wijbenga (HKV Consultants) for his valuable comments. We appreciate the help from Ivo Thonon and Chris Roosendaal (both Utrecht University), who assisted during the measurement campaign in Driel, and the help from Abe Klaas De Jong (HKV Consultants) and Blanca Pérez Lapeña (University of Twente), who assisted in producing the figures.

REFERENCES

[1] H. L. Cook and F. B. Campbell, "Characteristics of Some Meadow Strip Vegetation," *Agricultural Engineering*, Vol. 20, No. 9, 1939, pp. 345-348.

[2] US Soil Conservation Service, "Handbook of Channel Design for Soil and Water Conservation," United States Department of Agriculture, Annapolis, 1947.

[3] J. E. P. Green and J. E. Garton, "Vegetation Lined Channel Design Procedures," *Transactions of the American Society of Agricultural Engineers*, Vol. 26, No. 2, 1983, pp. 437-439.

[4] C. A. M. E. Wilson and M. S. Horritt, "Measuring the Flow Resistance of Submerged Grass," *Hydrological Processes*, Vol. 16, No. 13, 2002, pp. 2589-2598.

[5] R. Garcia Diaz, "Analysis of Manning Coefficient for Small-Depth Flows on Vegetated Beds," *Hydrological Processes*, Vol. 19, No. 16, 2005, pp. 3221-3233.

[6] A. F. Lightbody and H. M. Nepf, "Prediction of Velocity Profiles and Longitudinal Dispersion in Emergent Salt Marsh Vegetation," *Limnology and Oceanography*, Vol. 51, No. 1, 2006, pp. 218-228.

[7] J. C. Green, "Comparison of Blockage Factors in Modelling the Resistance of Channels Containing Submerged Macrophytes," *River Research and Applications*, Vol. 21, No. 6, 2005, pp. 671-686.

[8] T. Sukhodolova, A. Sukhodolov and C. Engelhardt, "A Study of Turbulent Flow Structure in a Partly Vegetated River Reach," In: G. Carravetta and D. Morte, Eds., *River Flow*, Vol. 1, 2004, pp. 469-478.

[9] J. K. Lee, L. C. Roig, H. L. Jenter and H. M. Visser, "Drag Coefficients for Modeling Flow through Emergent Vegetation in the Florida Everglades," *Ecological Engineering*, Vol. 22, No. 4-5, 2004, pp. 237-248.

[10] J. Järvelä, "Flow Resistance of Flexible and Stiff Vegetation: A Flume Study with Natural Plants," *Journal of Hydrology*, Vol. 269, No. 1-2, 2002, pp. 44-54.

[11] Z. Shi and J. M. R. Hughes, "Laboratory Flume Studies of Microflow Environments of Aquatic Plants," *Hydrological Processes*, Vol. 16, No. 16, 2002, pp. 3279-3289.

[12] C. A. M. E. Wilson, T. Stoesser, P. D. Bates and A. Batemann-Pinzen, "Open Channel Flow through Different Forms of Submerged Flexible Vegetation," *Journal of Hydraulic Engineering*, Vol. 129, No. 11, 2003, pp. 847-853.

[13] A. Armanini, M. Righetti and P. Grisenti, "Direct Measurement of Vegetation Resistance in Prototype Scale," *Journal of Hydraulic Research*, Vol. 43, No. 5, 2005, pp. 481-487.

[14] A. Murota, T. Fukuhara and M. Sato, "Turbulence Structure in Vegetated Open Channel Flow," *Journal of Hydroscience and Hydraulic Engineering*, Vol. 2, No. 1, 1984, pp. 47-61.

[15] T. Tsujimoto, T. Kitamura and T. Okada, "Turbulent Structure of Flow over Rigid Vegetation-Covered Bed in Open Channels," *KHL Progressive Report* 2, Kanazawa University, Kanazawa, 1991.

[16] D. Klopstra, H. J. Barneveld, J. M. van Noortwijk and E. H. van Velzen, "Analytical Model for Hydraulic Roughness of Submerged Vegetation," *The 27th International IAHR Conference*, San Fransisco, 10-15 August 1997, pp. 775-780.

[17] D. Velasco, A. Bateman and V. De Medina, "A New Integrated Hydromechanical Model Applied to Flexible Vegetation in Riverbeds," In: G. Parker and M. Garcia, Eds., *River, Coastal and Estuarine Morphodynamics*, 2005, pp. 217-227.

[18] D. G. Meijer and E. H. van Velzen, "Prototype-Scale Flume Experiments on Hydraulic Roughness of Submerged Vegetation," *The 28th International IAHR Conference*, Graz, 22-27 August 1998, Article ID: D108.

[19] M. G. Khublaryan, A. P. Frolov and V. N. Zyryanov, "Modeling Water Flow in the Presence of Higher Vegetation," *Water Resources*, Vol. 31, No. 6, 2004, pp. 668-674.

[20] Y. Shimizu and T. Tsujimoto, "Numerical Analysis of Turbulent Open-Channel Flow over Vegetation Layer Using a k-ε Turbulence Model," *Journal of Hydroscience and Hydraulic Engineering*, Vol. 11, No. 2, 1994, pp. 57-67.

[21] F. Lopez and M. H. Garcia, "Mean Flow and Turbulence

Structure of Open Channel Flow through Non-Emergent Vegetation," *Journal of Hydraulic Engineering*, Vol. 127, No. 5, 2001, pp. 392-402.

[22] A. Defina and A. C. Bixio, "Mean Flow and Turbulence in Vegetated Open Channel Flow," *Water Resources Research*, Vol. 41, No. 7, 2005, Article ID: W07006.

[23] F. Huthoff, D. C. M. Augustijn and S. J. M. H. Hulscher, "Analytical Solution of the Depth-Averaged Flow Velocity in Case of Submerged Rigid Cylindrical Vegetation," *Water Resources Research*, Vol. 43, 2007, Article ID: W06413.

[24] N. S. Cheng, "Representative Roughness Height of Submerged Vegetation," *Water Resources Research*, Vol. 47, No. 8, 2011, Article ID: W08517.

[25] C. Fischenich and S. Dudley, "Determining Drag Coefficients and Area for Vegetation," *Ecosytem Management & Restoration Research Program*, US Army Corps of Engineers, Washington DC, 2000.

[26] F.-C. Wu, H. W. Shen and Y.-J. Chou, "Variation of Roughness Coefficients for Unsubmerged and Submerged Vegetation," *Journal of Hydraulic Engineering*, Vol. 125, No. 9, 1999, pp. 934-942.

[27] D. Sumner, J. L. Heseltine and O. J. P. Dansereau, "Wake Structure of a Finite Circular Cylinder of Small Aspect Ratio," *Experiments in Fluids*, Vol. 37, No. 5, 2004, pp. 720-730.

[28] H. M. Nepf, "Drag, Turbulence, and Diffusion in Flow through Emergent Vegetation," *Water Resources Research*, Vol. 35, No. 2, 1999, pp. 479-489.

[29] M. R. Raupach, "Drag and Drag Partition on Rough Surfaces," *Boundary-Layer Meteorology*, Vol. 60, No. 4, 1992, pp. 375-395.

[30] M. Fathi-Maghadam and N. Kouwen, "Nonrigid, Non-Submerged, Vegetative Roughness on Floodplains," *Journal of Hydraulic Engineering*, Vol. 123, No. 1, 1997, pp. 51-57.

[31] J. Järvelä, "Determination of Flow Resistance Caused by Non-Submerged Woody Vegetation," *Journal of River Basin Management*, Vol. 2, No. 1, 2004, pp. 61-70.

[32] M. Straatsma, "3D Float Tracking: *In Situ* Floodplain Roughness Estimation," *Hydrological Processes*, Vol. 23, No. 2, 2009, pp. 201-212.

[33] E. H. Van Velzen, P. Jesse, P. Cornelissen and H. Coops, "Stromingsweerstand Vegetatie in Uiterwaarden," Report RIZA, Arnhem, 2003.

[34] B. C. Yen, "Open Channel Flow Resistance," *Journal of Hydraulic Engineering*, Vol. 128, No. 1, 2002, pp. 20-39.

[35] V. T. Chow, "Open-Channel Hydraulics," McGraw-Hill, New York, 1959.

[36] J. M. Eij, "Verwerking van ADCP Stromingsmetingen rond de Pannerdensche Kop tijdens Hoogwater in November 1998," *Technical Report AV DOC* 040211, Aqua Vision, Utrecht, 2004.

[37] A. W. E. Wilbers, "The Development and Hydraulic Roughness of Subaqueous Dunes," Ph.D. Thesis, University of Utrecht, Utrecht, 2004.

[38] P. Y. Julien, G. J. Klaassen, W. B. M. Ten Brinke and A. W. E. Wilbers, "Case Study: Bed Resistance of Rhine River during 1998 Flood," *Journal of Hydraulic Engineering*, Vol. 128, No. 12, 2002, pp. 1042-1050.

[39] D. C. Mason, D. M. Cobby, M. S. Horritt and P. D. Bates, "Floodplain Friction Parameterization in Two-Dimensional River Flood Models Using Vegetation Heights Derived from Airborne Scanning Laser Altimetry," *Hydrological Processes*, Vol. 17, No. 9, 2003, pp. 1711-1732.

[40] M. W. Straatsma and M.J. Baptist, "Floodplain Roughness Parameterization Using Airborne Laser Scanning and Spectral Remote Sensing," *Remote Sensing of Environment*, Vol. 112, No. 3, 2008, pp. 1062-1080.

[41] A. P. Nicholas and C. A. Mitchell, "Numerical Simulation of Overbank Processes in Topographically Complex Floodplain Environments," *Hydrological Processes*, Vol. 17, No. 4, 2003, pp. 727-746.

[42] V. Tayefi, S. N. Lane, R. J. Hardy and D. Yu, "A Comparison of One- and Two-Dimensional Approaches to Modelling Flood Inundation over Complex Upland Floodplains," *Hydrological Processes*, Vol. 21, No. 23, 2007, pp. 3190-3202.

15

Spatial and Temporal Variation of Stable Isotopes in Precipitation across Costa Rica: An Analysis of Historic GNIP Records

Ricardo Sánchez-Murillo[1,2]*, Germain Esquivel-Hernández[1,3], Kristen Welsh[2,4], Erin S. Brooks[5], Jan Boll[2,5], Rosa Alfaro-Solís[1,3], Juan Valdés-González[1,3]

[1]Escuela de Química, Universidad Nacional, Heredia, Costa Rica; [2]Waters of the West-Water Resources Program, University of Idaho, Moscow, USA; [3]Laboratorio de Química de la Atmósfera, Escuela de Química, Universidad Nacional, Heredia, Costa Rica; [4]Division of Research and Development, Centro Agronómico Tropical de Investigación y Enseñanza (CATIE), Turrialba, Costa Rica; [5]Department of Biological and Agricultural Engineering, University of Idaho, Moscow, USA.

ABSTRACT

The location of Costa Rica on the Central American Isthmus creates unique microclimate systems that receive moisture inputs directly from the Caribbean Sea and the Pacific Ocean. In Costa Rica, stable isotope monitoring was conducted by the International Atomic Energy Agency and the World Meteorological Association as part of the worldwide effort entitled Global Network of Isotopes in Precipitation. Sampling campaigns were mainly comprised of monthly-integrated samples during intermittent years from 1990 to 2005. The main goal of this study was to determine spatial and temporal isotopic variations of meteoric waters in Costa Rica using historic records. Samples were grouped in four main regions: Nicoya Peninsula ($\delta^2H = 6.65\delta^{18}O - 0.13$; $r^2 = 0.86$); Pacific Coast ($\delta^2H = 7.60\delta^{18}O + 7.95$; $r^2 = 0.99$); Caribbean Slope ($\delta^2H = 6.97\delta^{18}O + 4.97$; $r^2 = 0.97$); and Central Valley ($\delta^2H = 7.94\delta^{18}O + 10.38$; $r^2 = 0.98$). The water meteoric line for Costa Rica can be defined as $\delta^2H = 7.61\delta^{18}O + 7.40$ ($r^2 = 0.98$). The regression of precipitation amount and annual arithmetic means yields a slope of -1.6‰ $\delta^{18}O$ per 100 mm of rain ($r^2 = 0.57$) which corresponds with a temperature effect of -0.37‰ $\delta^{18}O/°C$. A strong correlation ($r^2 = 0.77$) of -2.0‰ $\delta^{18}O$ per km of elevation was found. Samples within the Nicoya Peninsula and Caribbean lowlands appear to be dominated by evaporation enrichment as shown in *d-excess* interpolation, especially during the dry months, likely resulting from small precipitation amounts. In the inter-mountainous region of the Central Valley and Pacific slope, complex moisture recycling processes may dominate isotopic variations. Generally, isotopic values tend to be more depleted as the rainy season progresses over the year. Air parcel back trajectories indicate that enriched isotopic compositions both in Turrialba and Monteverde are related to central Caribbean parental moisture and low rainfall intensities. Depleted events appear to be related to high rainfall amounts despite the parental origin of the moisture.

Keywords: Stable Isotopes; Costa Rica; Precipitation; GNIP; HYSPLIT Model

1. Introduction

Tropical regions cover approximately 50% of Earth's landmass and are home to three-quarters of the human population. During the last decade, the scientific community, environmental institutions, governments, and communities have increased their awareness of tropical climate variability based on the premise that changes in regional and global circulation trends may lead to inten-

sification (*i.e.*, greater rainfall intensity) or weakening (*i.e.*, prolonged drought) of the hydrological cycle in particular regions [1]. For example, several studies have shown the potential high vulnerability of tropical ecosystems to increasing mean surface temperatures [2-4]. In addition, extreme climate events could be even more severe than those experienced to date [5]. Climate anomalies will increase the vulnerability of certain regions and communities to changes in magnitude, timing, and duration of hydrological responses.

---

*Corresponding author.

Spatial and Temporal Variation of Stable Isotopes in Precipitation across Costa Rica: An Analysis of Historic GNIP Records

125

Understanding precipitation patterns, particularly how they differ spatially and temporally, will provide further information on regional and global hydrologic processes and enable better planning and preparation for future potential changes in climate. A better understanding of the factors that control precipitation patterns will inform stakeholders and environmental agencies so that they can prioritize efforts and resources in ungauged basins where potential future droughts or floods may drastically affect ecological assemblages and socio-economic activities. In Central America, many socio-economic activities, especially agriculture, tourism, and hydropower generation, are dependent on seasonal cycles of precipitation. Therefore, in order to determine the interannual variability in precipitation, it is necessary to describe the regional annual cycle precisely and examine the mechanisms that control it [6].

Across the globe, spatial variation of stable isotopes in the water cycle occurs due to the movement of air masses and fractionation during phase changes, such as condensation and evaporation, creating a directional spatial pattern across regions and continents [7]. Light stable isotope compositions of tropical meteoric waters have proven to be an important indicator of current climate variability [8-16]. In particular, $\delta^{18}O$ values have provided novel insights on El Niño/Southern Oscillation dynamics [12,17-19]. In temperate regions, isotopic variations in precipitation frequently have been correlated with mean surface air temperatures [20-25], whereas, in the tropics, several authors have reported "the amount effect" as the main controlling factor [16,24,26,27]. These consistent processes create relatively strong spatial patterns of isotopic compositions across landscapes [28]. Large-scale regional and global patterns can be depicted through *isoscapes* that illustrate the spatial and temporal variation of isotopes over a geographic region.

Despite novel advances to understand stable isotope precipitation dynamics in the tropics, there is consensus among scientists in the published literature [13,16,27] on the urgent need for long-term sampling networks that may lead to a better spatial and temporal resolution, particularly in mountainous regions such as the Central American Continental Divide (CACD). In the CACD orographic effects, moisture recycling and canopy interception, intense evapotranspiration, and microclimates are relevant factors in isotopic variation. Furthermore, recent studies have demonstrated that variations in the stable isotope composition of precipitation occur even between storm events [29-32], emphasizing the importance of event-based sampling.

Stable isotope research in Costa Rica has been limited to date with no studies conducted on a regional scale examining trends in precipitation. Lachniet and Patterson

(2002) [27] provided a preliminary evaluation of $\delta^{18}O$ and δ^2H in Costa Rican surface waters ($n = 63$ from rivers and lakes). Their study found regional trends in the isotopic composition of surface waters inversely correlated to altitude ranges (-1.4 $\delta^{18}O$ ‰/km) and evaporation enrichment, particularly in the northwestern region. Rhodes *et al.* (2006) studied the seasonal variation of isotopic composition in tropical montane forests of Monteverde's National Park located on the continental divide. Based on seasonal *d*-excess variations, the authors concluded that water evaporated from land is an important component of the water budget to the region during the transitional and dry seasons. Reynolds and Fraile (2009) [33] conducted a groundwater recharge study on one of the main aquifers of the Central Valley of Costa Rica that provides water to approximately 30% of the population. Based upon precipitation, well, and spring records they were able to identify local recharge areas mainly dominated by Pacific-originated storms. Greater understanding of spatial and temporal isotopic variations in tropical regions with similar topographic, vegetation, and climatic attributes can enhance hydrological studies of this nature.

Costa Rica's climate is mainly controlled by four phenomena: northeast trades-winds (*i.e.*, alisios), the shifts of the Intertropical Convergence Zone (ITCZ), cold continental outbreaks, and indirect influence of Caribbean cyclones [34]. These circulation processes produce three distinct seasons. The wet season (May-October) corresponds to the time when the ITCZ travels over Costa Rica, and precipitation is characterized by heavy convective rain storms. The transitional (November-January) and dry (February-April) seasons comprise the months when the ITCZ is located to the south of Costa Rica [35]. During the wet season, the air masses arriving in Costa Rica can be classified as continental winds, reaching Costa Rica's Central Valley from the Pacific Ocean. Most of these air masses move over industrial and urban areas of the country. In the dry season, trade winds bring air masses to Costa Rica from the Caribbean Sea, and they move over non-industrial and less populated areas. Air masses arriving over Costa Rica during the transitional season can be transported by continental or trade-winds, and a weakening in the trade winds reaching Costa Rica is usually observed during the afternoon. Due to the transport pattern of the air masses observed in Costa Rica throughout the year, the isotopic composition of precipitation in Costa Rica is expected to vary depending on its origin from either the Caribbean Sea or the Pacific Ocean and also on the air trajectory across the country (*i.e.* industrial-urban areas versus remote-forested areas). Annual precipitation varies from ~1500 mm in the northwestern region and Nicoya Peninsula to ~7000 mm on the Caribbean side of the Tala-manca range. Temperature seasonality is low. The mean annual temperatures vary from around 27°C on the coastal lowlands, 20°C in the Central Valley, and below 10°C at the summits of the highest mountain range (~4000 m.a.s.l.).

The Global Network for Isotope in Precipitation (GNIP) [36] data set provides isotopic values of precipitation from 46 stations across Costa Rica between 1990 and 2005 (intermittent collection over this period). Most of the reported stations were located within the Central Valley (greatest populated area) and the North Pacific region. Therefore, a considerable gap exists along the Pacific and Caribbean shores, continental divide, and the northern and southern regions. Herein, we present an analysis of the spatial and temporal variation of stable isotopes in meteoric waters in Costa Rica and determine which broad climatic or topographic factors may influence isotopic variations across the region. Results of this analysis provide baseline historical spatial-temporal characterization of meteoric waters in Costa Rica that has been lacking to date.

2. Methods and Materials

2.1. Study Area

The Central Valley is a heavily populated area containing four major cities of Costa Rica located within its boundaries (~60% population) along with significant industrial activity. Weather conditions in the valley are mainly influenced by easterly Caribbean trade-winds during the dry season (February-April) and by frequent westerly continental air masses during the wet season (May-October). About ~50% of GNIP were located within the valley limits. In this region, mean annual precipitation ranges from ~3500 mm in the highlands to ~2000 mm within the foothills (**Figure 2**), and the area experiences 3 - 4 dry months over the summer. This inter-mountainous valley region has an elevation between 1000 and 2500 meters m.a.s.l, with average temperatures that range from 14°C to 18°C. The northern Pacific region of the Nicoya Peninsula receives on average 1500 mm of precipitation and suffers extensive dry periods that can reach up to 6 months. The Caribbean slope is characterized by a short dry season (1 - 2 months); mean annual precipitation in this region ranges from 4500 to 6000 mm.

2.2. Historic GNIP Records Analysis

We analyzed stable isotope records ($n = 679$) of 46 monitoring stations across Costa Rica obtained from the GNIP data base (**Table 1**, **Figure 1**). Station altitude ranges from 0 - 3000 m.a.s.l with a mean altitude of 930

m.a.s.l. Sampling frequency was typically monthly except for one station (Monteverde) where event-based sampling was conducted by Rhodes *et al.* (2006). The entire data set is comprised of three relative short-term sampling campaigns: 1990-1992, 1997-1998, and 2002-2005. Weighted $\delta^{18}O$ and δ^2H means were calculated for 42 stations (87%). Linear regression analysis of elevation and arithmetic $\delta^{18}O$ values was conducted considering 18 stations which are unlikely to be influenced by evaporation enrichment: La Selva, Turrialba, Santa María, San Pablo, Santa Lucía, Birrí, Monteverde, Zurquí, Heinz Hoffman, Monte de la Cruz, Pacayas, Fraijanes, Vara Blanca, La Giralda, Sacramento, Paso Llano, Poás, and Irazú. Likewise, linear regression analysis of precipitation amount and $\delta^{18}O$ values was determined using data from 11 stations which are representative of all four regions: Turrialba, Puriscal, Fraijanes, Orotina, Santa Barbara, Rancho Redondo, Puntarenas, Hacienda Tempisque, and Sacramento.

Spatial analysis was conducted using the Atlas of Costa Rica 2008 database [37]. The natural neighbor procedure in ArcGIS 9.3 was applied to calculate deuterium excess, $\delta^{18}O$, and δ^2H interpolations. High station densities within the Central Valley and Nicoya Peninsula may overemphasize the isotopic signature of these regions. Data limitations constrained this and additional analyses. First, the majority of the stations (61%) generally reported isotopic values over a 1 year period with monthly values ranging from $n = 4$ to 28 with a mean of $n = 15$. Second, most of the reported stations were located within the Central Valley (the most populated area) and the North Pacific region (**Figure 1**). Therefore, a considerable gap exists along the Pacific and Caribbean shores, continental divide, and the northern and southern regions (**Figure 1**). Third, there was an omission of temperature values which prevented calculation of $\delta^{18}O$ and temperature correlation.

2.3. HYSPLIT Model Simulations

To complement the historic GNIP station data analysis, air parcel back trajectories were calculated for particular locations using the hybrid single particle Lagrangian integrated trajectory model (HYSPLIT) [38]. Isotopic values in precipitation from two distinct study locations, Monteverde (continental divide) and Turrialba (Caribbean Slope), were analyzed. Both a depleted and enriched isotopic event during the 2011 and 2012 water years were chosen from these two locations. Single (24 hours) five days back trajectories were calculated using the vertical velocity model calculation method. Trajectory ensembles were done using the GDAS1 meteorological database. Back trajectories were calculated starting at 2300 UTC. Rainfall rate, downward radiation,

Spatial and Temporal Variation of Stable Isotopes in Precipitation across Costa Rica: An Analysis of Historic GNIP Records

127

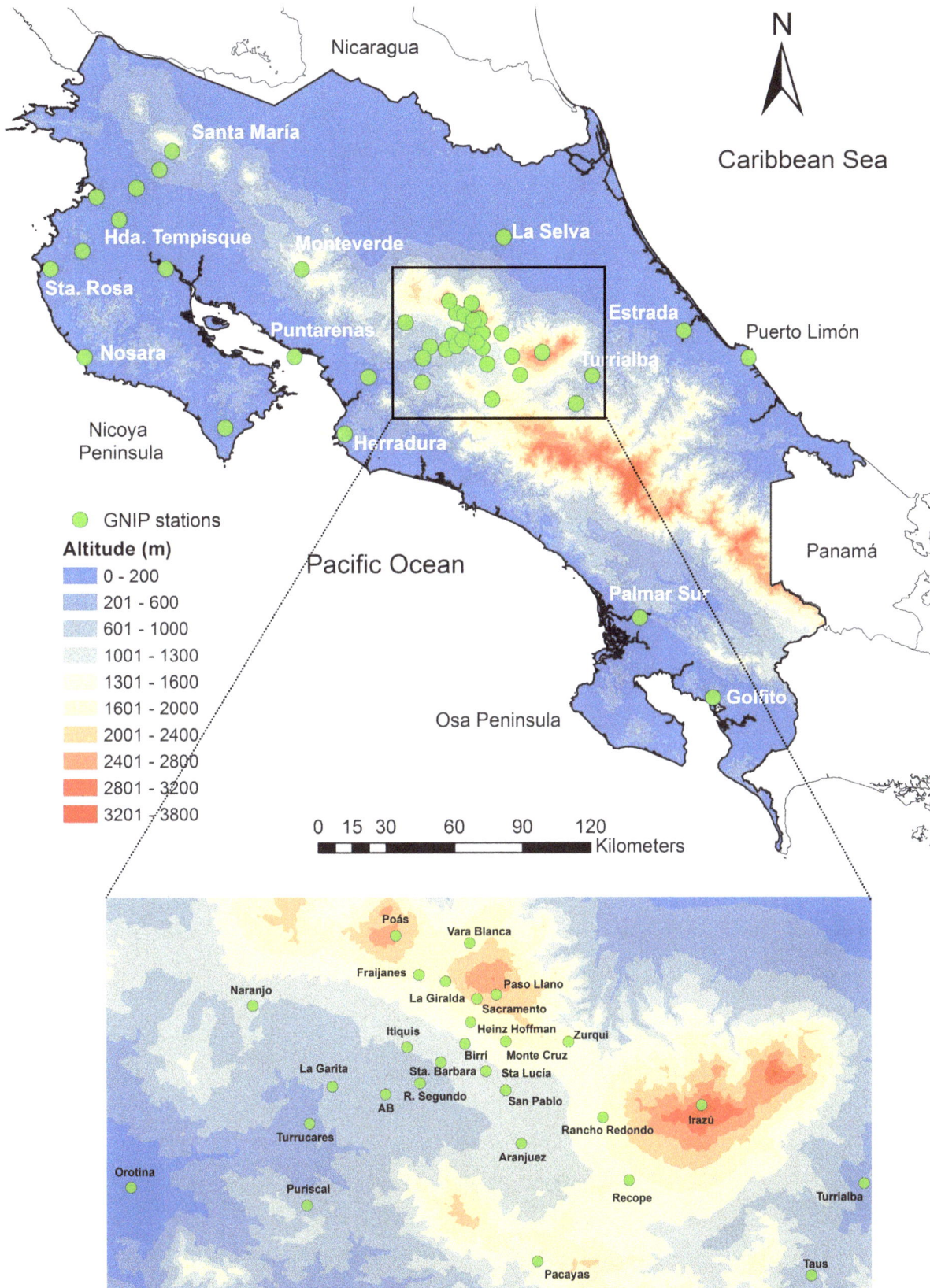

Fgure 1. (Top) Digital elevation map of Costa Rica and location of historic GNIP stations; (bottom) Inset showing Central Valley stations. DEM source: Atlas of Costa Rica 2008 (ITEC); 100 m resolution.

Table 1. Summary of GNIP stations operated in Costa Rica from 1990 to 2005.

Station	Latitude (dec.deg)	Longitude (dec.deg)	Elevation (m)	Climate	Sampling period	n	LMWL	r^2	Total precipitation (mm)	Arithmetic means			Weighted means		
										$\delta^{18}O$	δH	d	$\delta^{18}O$	δH	d
Estrada	10.0667	−83.2833	6	Af	10/90-03/92	21	$\delta H = 7.59\delta^{18}O + 7.40$	0.94	1055	−3.5	−19.3	8.8	−3.2	−19.7	5.5
La Selva	10.4314	−84.0029	45	Aw	01/02-01/04	23	$\delta H = 8.48\delta^{18}O + 14.03$	0.99	4304	−1.8	−17.3	12.3	−3.8	−20.0	10.4
Pto.Limón	9.9620	−83.0251	0	Af	6/03-05/04	12	$\delta H = 8.51\delta^{18}O + 10.74$	0.99	4162	−3.8	−21.2	8.8	−3.9	−20.6	10.2
Taus	9.7833	−83.7167	900	Cbf	01/90-12/90	8	$\delta H = 7.97\delta^{18}O + 11.10$	0.97	4256	−7.1	−45.5	11.3	−7.2	−45.8	11.5
Turrialba	9.8913	−83.6529	604	Aw	01/04-02/02	24	$\delta H = 8.62\delta^{18}O + 16.53$	0.99	3177	−5.0	−26.4	13.4	−5.2	−30.2	11.1
AB	9.9958	−84.2319	860	Aw	12/03-10/04	12	$\delta H = 7.91\delta^{18}O + 10.11$	0.99	2017	−6.7	−42.8	10.7	−7.3	−49.1	9.5
Aranjuez	9.9380	−84.0696	1191	Aw	02/02-01/04	19	$\delta H = 8.16\delta^{18}O + 12.11$	0.99	1910	−8.3	−55.6	10.8	−9.3	−64.5	9.8
Birrí	10.0550	−84.1370	1318	Aw	12/03-10/04	12	$\delta H = 8.40\delta^{18}O + 14.20$	0.99	2916	−6.6	−41.2	11.6	−7.3	−49.0	9.4
Fraijanes	10.1359	−84.1917	1737	Aw	01/02-01/12	24	$\delta H = 8.47\delta^{18}O + 16.58$	0.99	3506	−6.8	−40.9	13.4	−7.8	−51.4	10.7
Heinz Hoffman	10.0807	−84.1299	1650	Aw	12/03-10/04	12	$\delta H = 8.16\delta^{18}O + 12.45$	0.99	2940	−6.6	−41.1	11.4	−7.1	−46.9	9.8
Irazú	9.9833	−83.8500	3000	Cbf	01/90-12/90	8	$\delta H = 7.85\delta^{18}O + 8.91$	0.98	N.A.	−10.1	−70.4	10.4	N.A.	N.A.	N.A.
Itiquis	10.0511	−84.2061	1110	Aw	10/03-9/04	11	$\delta H = 7.93\delta^{18}O + 10.71$	0.99	3026	−7.1	−45.8	11.2	−8.4	−56.7	10.3
La Garita	10.0052	−84.2957	760	Aw	10/03-9/04	12	$\delta H = 7.97\delta^{18}O + 10.99$	0.99	2349	−6.5	−40.8	11.2	−7.1	−47.6	9.1
La Giralda	10.1282	−84.1600	2014	Cfa	01/02-12/03	20	$\delta H = 8.49\delta^{18}O + 16.71$	0.99	2708	−6.8	−41.3	13.3	−7.3	−47.9	10.6
Monte de la Cruz	10.0577	−84.0879	1700	Aw	10/03-09/04	13	$\delta H = 8.24\delta^{18}O + 12.80$	0.99	2189	−7.3	−47.1	11.0	−9.2	−62.9	11.1
Naranjo	10.1001	−84.3917	1051	Aw	01/02-12/03	15	$\delta H = 7.98\delta^{18}O + 9.80$	0.99	2064	−7.6	−51.0	10.0	−11.1	−76.8	12.2
Pacayas	9.8000	−84.0500	1735	Cwb	01/90-12/90	8	$\delta H = 8.37\delta^{18}O + 13.72$	0.99	1942	−7.9	−52.4	10.8	−7.9	−51.9	10.9
Paso Llano	10.1128	−84.0991	2397	Cfa	12/03-10/04	12	$\delta H = 8.40\delta^{18}O + 16.27$	0.99	4131	−6.9	−41.8	13.5	−6.6	−41.6	11.1
Poás	10.1818	−84.2193	2500	Cbf	12/97-06/99	14	$\delta H = 8.12\delta^{18}O + 14.52$	0.99	N.A.	−7.7	−48.3	13.6	N.A.	N.A.	N.A.
Puriscal	9.8656	−84.3267	836	Aw	01/02-01/04	22	$\delta H = 8.03\delta^{18}O + 12.21$	0.99	2694	−6.8	−43.4	11.0	−8.3	−55.9	10.5
Rancho Redondo	9.9686	−83.9713	1662	Aw	02/02-01/04	21	$\delta H = 8.20\delta^{18}O + 13.20$	0.98	2435	−8.2	−53.7	11.5	−9.9	−68.7	10.7
Recope	9.8949	−83.9388	1563	Aw	02/02-01/04	19	$\delta H = 8.44\delta^{18}O + 15.45$	0.99	1179	−8.4	−55.3	11.8	−10.7	−73.3	12.0
Río Segundo	10.0089	−84.1908	980	Aw	12/03-10/04	11	$\delta H = 8.59\delta^{18}O + 16.11$	0.99	2092	−7.0	−43.7	13.6	−8.1	−54.5	10.3
Sacramento	10.1081	-84.1221	2260	Cbf	01/90-12/90 01/03−12/04	21	$\delta H = 8.39\delta^{18}O + 13.62$	0.99	2577.5	−7.4	−47.3	11.5	−8.6	−57.7	11.5
San Pablo	10.0008	−84.0882	1215	Aw	12/03-10/04	12	$\delta H = 8.30\delta^{18}O + 13.19$	0.99	2051	−6.6	−41.4	11.2	−7.6	−51.4	9.4
Santa Barbara	10.0334	−84.1660	1102	Aw	02/02-01/04	21	$\delta H = 8.19\delta^{18}O + 12.01$	0.99	2522	−7.5	−49.4	10.3	−9.3	−63.6	10.8
Santa Lucía	10.0230	−84.1118	1251	Aw	01/02-12/03	18	$\delta H = 8.19\delta^{18}O + 12.01$	0.99	1793	−6.4	−40.1	10.8	−10.8	−74.8	12.0
Turrucares	9.9617	−84.3237	642	Aw	12/03-10/04	11	$\delta H = 7.93\delta^{18}O + 9.66$	0.99	1945	−6.5	−41.7	10.1	−7.6	−51.2	9.5
Vara Blanca	10.1732	−84.1310	1845	Aw	12/03-10/04	12	$\delta H = 8.40\delta^{18}O + 14.84$	0.99	4579	−6.0	−36.0	12.4	−5.2	−31.5	8.2
Zurquí	10.0573	−84.0124	1548	Aw	03/02-05/04	28	$\delta H = 8.57\delta^{18}O + 17.53$	0.99	5074	−5.6	−30.1	14.4	−6.0	−30.4	17.4
Capulin	10.6228	−85.4611	125	Aw	11/97-05/98	8	$\delta H = 7.79\delta^{18}O + 6.82$	0.99	N.A.	−8.4	−59.0	8.6	N.A.	N.A.	N.A.
Cartagena	10.3800	−85.6750	63	Aw	01/91-12/91	7	$\delta H = 7.83\delta^{18}O + 1.95$	0.97	1010	−5.7	−42.5	2.9	−6.3	−47.1	3.5

Continued

Cóbano	9.6900	−85.1100	160	Aw	01/90-12/90	8	$\delta^2H = 8.42\delta^{18}O + 13.57$	0.98	2400	−6.5	−41.4	10.8	−7.1	−46.3	10.5
Est. Ecológica	10.6944	−85.3706	340	Aw	11/97-05/98	7	$\delta^2H = 8.03\delta^{18}O + 8.77$	0.95	N.A.	−8.2	−57.4	8.5	N.A.	N.A.	N.A.
Hacienda Tempisque	10.5000	−85.5300	22	Aw	05/90-07/90	13	$\delta^2H = 7.81\delta^{18}O + 5.52$	0.92	897	−5.2	−35.1	6.5	−6.2	−42.3	7.2
Monte Galan	10.5900	−85.6200	60	Aw	01/90-12/90	8	$\delta^2H = 6.06\delta^{18}O − 2.69$	0.91	1185	−5.4	−35.4	7.8	−6.0	−37.4	10.6
Nosara	9.9667	−85.6667	15	Aw	05/90-06/91	14	$\delta^2H = 8.43\delta^{18}O + 10.89$	0.92	1591	−6.3	−42.5	8.2	−6.3	−43.2	7.6
Puerto Humo	10.3100	−85.3450	10	Aw	01/90-12/90	9	$\delta^2H = 8.60\delta^{18}O + 14.50$	0.99	1374	−6.6	−42.5	10.5	−6.5	−40.8	10.8
Santa María	10.7667	−85.3200	825	Am	01/90-12/92 01/97−12/98	26	$\delta^2H = 8.04\delta^{18}O + 9.67$	0.99	1901	−5.5	−34.8	9.5	−5.9	−38.7	8.6
Santa Rosa	10.3100	−85.8000	25	Aw	01/90-12/90	4	$\delta^2H = 8.24\delta^{18}O + 13.30$	0.99	992	−9.2	−62.1	11.1	−8.6	−57.5	11.5
Golfito	8.6400	−83.1667	15	Af	01/90-12/90	7	$\delta^2H = 8.08\delta^{18}O + 11.60$	0.99	2633	−7.0	−44.8	11.0	−7.0	−44.7	11.0
Herradura	9.6667	−84.6333	3	Am	01/90-12/90	8	$\delta^2H = 8.18\delta^{18}O + 12.32$	0.99	3082	−6.9	−44.5	11.1	−7.3	−47.2	11.0
Orotina	9.8870	−84.5387	168	Aw	03/02-01/04	19	$\delta^2H = 7.93\delta^{18}O + 9.70$	0.99	2256	−6.3	−40.1	10.1	−8.5	−56.1	11.7
Palmar Sur	8.9500	−83.4600	16	Am	01/90-12/90	8	$\delta^2H = 8.50\delta^{18}O + 14.17$	0.99	3293	−7.6	−50.1	10.4	−7.9	−52.9	10.0
Puntarenas	9.9667	−84.8333	3	Aw	01/93-12/93 01/03-12/04	15	$\delta^2H = 8.10\delta^{18}O + 12.00$	0.99	1443.35	−7.2	−46.7	11.3	−8.9	−59.7	11.6
Monteverde	10.3067	−84.8047	1460	Am	06/03-03/05	42	$\delta^2H = 8.60\delta^{18}O + 14.28$	0.99	N.A.	−6.1	−37.9	10.7	N.A.	N.A.	N.A.
Mean										−6.7	−43.2	10.8	−7.4	−49.0	10.3
Max										−1.8	−17.3	14.4	−3.2	−19.7	17.4
Min										−10.1	−70.4	2.9	−11.1	−76.8	3.5
S.D.										1.4	10.3	2.0	1.8	13.7	2.1

Köppen-Geiger code Aw = tropical wet and dry; Am = tropical trade-wind littoral; Af = tropical rainforest; Cbf = tropical highland; Cfa = humid; Cwb = tropical highland. N.A. = not available.

relative humidity, and elevation of the mixing layer depth were also computed.

2.4. Stable Isotope Analysis

Precipitation was collected on event-basis. Monteverde sampling was carried out during July, August, and September of 2011. Turrialba's sampling period used in this study was from September to December, 2012. The collector for individual events consisted of a 10.16 cm plastic funnel coupled with a metal filter mesh (*i.e.*, to prevent external contamination). The funnel was connected to 4 L high density polyethylene (HDPE) container. An approximately 3 cm mineral oil layer was added to prevent fractionation according with standard sampling protocols. Later, mineral oil was separated using a 250 - 500 mL separatory funnel. Samples were stored upside down in a HDPE bottles with conic and polyseal inserts and parafilm seals until analysis. Stable isotope analyses were conducted at the Chemistry School of the National University (Heredia, Costa Rica) and the Idaho Stable Isotope Laboratory (Moscow, Idaho) using a Cavity Ring Down

Spectroscopy (CRDS) water isotope analyzer L2120-I (Picarro, CA) and L1120-i (Picarro, CA), respectively. Ratios of $\delta^{18}O/\delta^{16}O$ and δ^2H/δ^1H are expressed in delta units (‰, parts per mil) relative to Vienna Standard Mean Ocean Water (V-SMOW). The current analytical analyzers precision is 0.1‰ $\delta^{18}O/\delta^{16}O$ and 0.5‰ δ^2H/δ^1H.

3. Results and Discussion

3.1. Precipitation

Precipitation variability in Costa Rica reflects regional and local complex dynamics. **Figure 2** shows mean annual precipitation throughout Costa Rica and precipitation amounts recorded during isotopic sampling campaigns. GNIP precipitation records are consistent with the Atlas of Costa Rica reported values and ranged from 5074 mm (Zurquí) to 897 mm (Hacienda Tempisque) with a mean of 2479 mm (**Table 1, Figure 2**). The lowlands of the Nicoya Peninsula is the region with the lowest precipitation regime (<1500 mm) while the Talamanca Range and northeastern Caribbean lowlands received the greatest amount, between 4500 - 8000 mm

Spatial and Temporal Variation of Stable Isotopes in Precipitation across Costa Rica: An Analysis of Historic GNIP Records

129

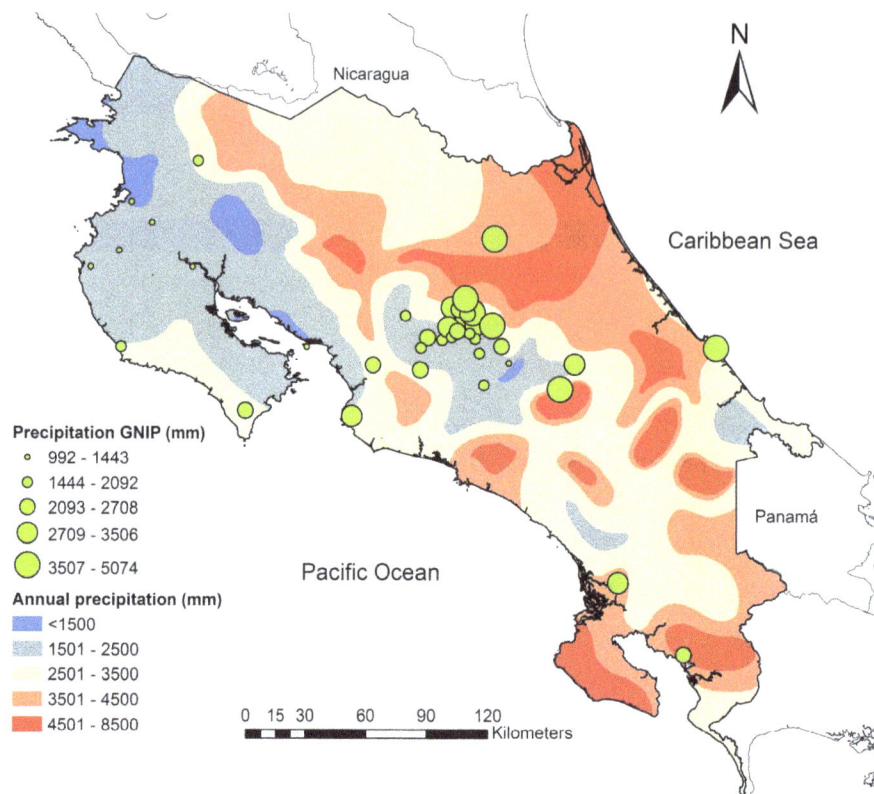

Figure 2. Mean annual precipitation throughout Costa Rica with GNIP stations showing precipitation recorded during sampling periods.

annually. These precipitation records provide a reliable basis for further stable isotope spatial analysis.

3.2. $\delta^{18}O$ and δ^2H Isotopic Variations

Stable isotope signatures in precipitation across Costa Rica were variable during the sampling campaigns. $\delta^{18}O$ composition ranged from −1.8‰ to −10.1‰ with an arithmetic mean of −6.7 ± 1.4 (‰) (**Figure 3**, **Table 1**). Weighted $\delta^{18}O$ values range from −3.2‰ to −11.1‰ with a mean of −7.4 ± 1.8 (‰) (**Table 1**). Deuterium composition varied from −17.3‰ to −70.4‰ with an arithmetic mean of −43.2 ± 10.3 (‰) (**Figure 3**, **Table 1**). Weighted δ^2H values ranged from −19.7‰ to −76.8‰ with a mean of −49.0 ± 13.7 (‰) (**Table 1**). **Figure 4** shows the regional water meteoric lines for each of the four main regions for the GNIP Stations (Nicoya Peninsula, Pacific Coast, Caribbean Slope, and Central Valley). The best fit straight line to the data for the Nicoya Peninsula yields an arithmetic water line equation of $\delta^2H = 7.84\delta^{18}O + 7.60$ (**Figure 4**), whereas the weighted meteoric water line yields a distinct best fit straight line equation of $\delta^2H = 6.65\delta^{18}O − 0.13$ (**Figure 4**), likely resulting from small precipitation amounts. For the Pacific Coast, the best fit straight line to precipitation-weighted annual means is $\delta^2H = 7.60\delta^{18}O + 7.95$ (**Figure 4**), which is consistent with the Global Meteoric Water Line (GMWL) of $\delta^2H = 8\delta^{18}O + 10$ [39]. Interestingly, Pacific Coast stations are close to equilibrium conditions rather than exhibiting evaporation enrichment. The Caribbean Slope precipitation-weighted water line $\delta^2H = 6.98\delta^{18}O + 4.97$ (**Figure 4**) reflects orographic distillation of Caribbean-sourced moisture. Overall, the best fit line to the precipitation-weighted data for all regions combined yields a water line of $\delta^2H = 7.61\delta^{18}O + 7.40$ ($r^2 = 0.98$) which corresponds closely to the GMWL (**Figure 5**). Enriched values above the annual mean of −6.7‰ $\delta^{18}O$ were often observed during January to March. The greater variation occurred in May and June when the ITCZ starts its migration over Costa Rica. Towards the middle of the wet season (August, September, and October) $\delta^{18}O$ variation appears to decrease due to increase in precipitation amounts. The intensification of northeast trade winds in December produced a decrease in precipitation, and consequently, enrichment in isotopic values (**Figure 5**).

3.3. The Altitude Effect

The variability in isotopic composition of precipitation with elevation is not consistent across the GNIP stations indicating a complex system. **Figure 6** shows three lon-

Spatial and Temporal Variation of Stable Isotopes in Precipitation across Costa Rica: An Analysis of Historic
GNIP Records

131

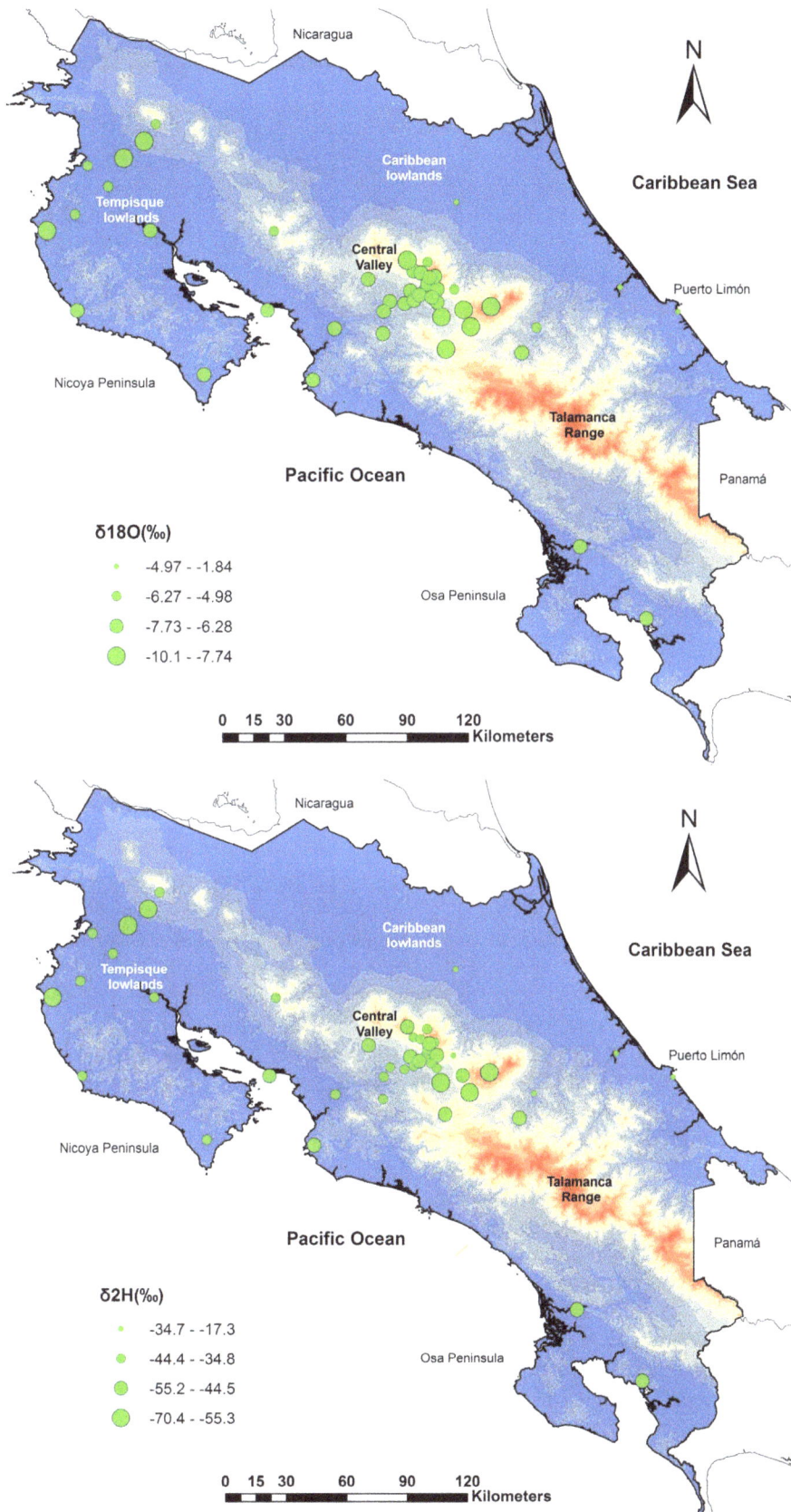

Figure 3. Spatial variations of $\delta^{18}O$ (top) and δ^2H (bottom) arithmetic means.

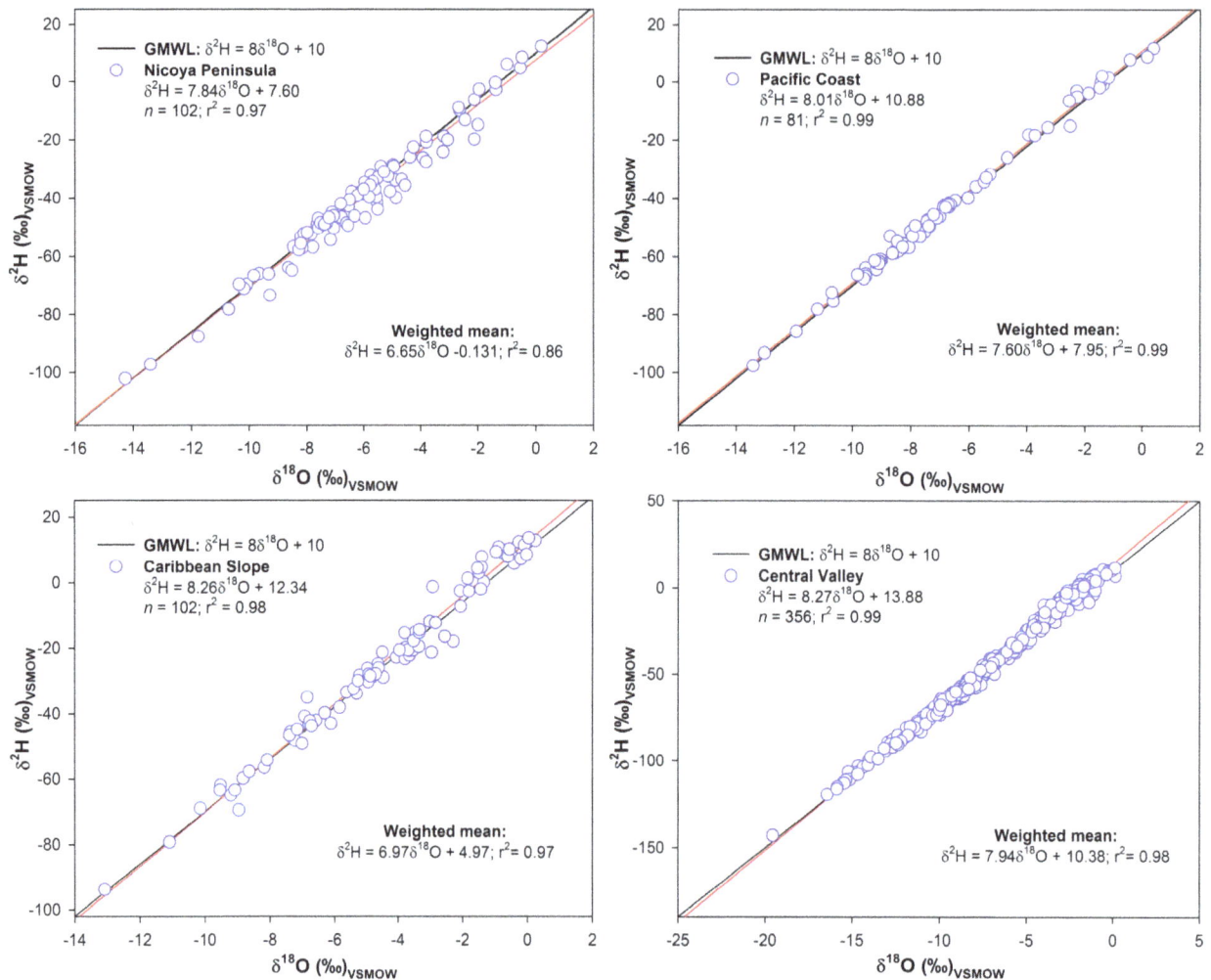

Figure 4. Meteoric water lines for four major geographic regions of Costa Rica.

gitudinal transects. For the Caribbean slope, there is an evident orographic distillation effect from Puerto Limón (0 m.a.s.l.) to Taus (900 m.a.s.l.) of $-3.5‰$ $\delta^{18}O$ /km. The ~75 km Pacific slope transect from Herradura (3 m.a.s.l.) to Paso Llano (2397 m.a.s.l.) represents a more complex system. For instance, a rainout effect is observed up to the Santa Bárbara station (1102 m.a.s.l.) (**Figure 6**). However, at the two higher elevation stations (Birrí and Heinz Hoffman) isotopic compositions are more enriched, suggesting another moisture source. A similar phenomena is also observed along the Nicoya Peninsula transect of ~66 km from the coast to the Rincón de la Vieja volcano (**Figure 6**).

Considering the 18 stations mainly from the Central Valley cluster, the correlation of elevation with arithmetic mean annual $\delta^{18}O$ is moderately strong ($r^2 = 0.77$). The best fit straight line has a slope of $-0.2‰$ $\delta^{18}O$ per 100 m, which is close to the range of 0.1‰ and 0.36‰ reported by Leibungdgut *et al.* (2009) [40]. Since the altitude effect is the combined result of the temperature

effect and moisture depletion by adiabatic cooling, a temperature effect can be calculated by knowing the atmospheric lapse rate. Lachniet and Paterson (2002) [27] reported a lapse rate of $-5.4°C$/km for the San José area. Therefore, an altitude effect of $-0.2‰$/km corresponds to $-0.37‰/°C$, in agreement with temperature effects reported in the pioneer work done by Dansgaard [23].

3.4. The Amount Effect and d-Excess Isoscape

One important factor controlling isotopic variations in the tropics is the "amount effect" [23,24,41]. This effect has been attributed to three main factors according to Scholl *et al.* (2009) [42]: 1) the isotopic composition of the condensation in a cloud decreases as cooling and "rainout" occurs; 2) smaller raindrops equilibrate to a larger degree with the water vapor and temperature conditions below the cloud; and 3) small raindrops evaporate more than larger raindrops on their way to the land

Spatial and Temporal Variation of Stable Isotopes in Precipitation across Costa Rica: An Analysis of Historic
GNIP Records

133

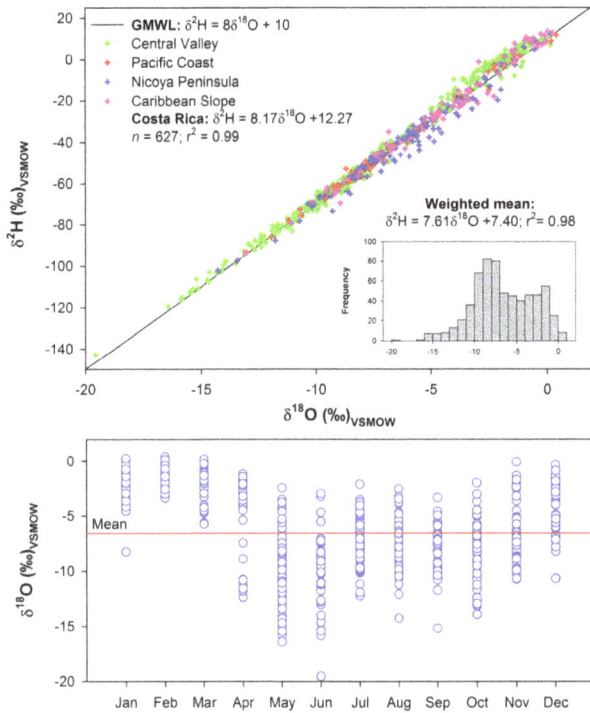

Figure 5. (Top) Meteoric water line for Costa Rica. Contributions from all four regions are color coded. Inset: distribution of $\delta^{18}O$ values; (Bottom) Temporal variation of $\delta^{18}O$ values.

surface. The slope of the best fit straight line to precipitation amount and $\delta^{18}O$ from 11 stations, which are representative of the four main regions described previously, was found to be −1.6‰ per 100 mm decrease in rainfall ($r^2 = 0.57$). This effect is clearly present during the times of lowest precipitation (January to April).

The "amount effect" was also observed in the *d-excess isoscape* presented in **Figure 7**. D-excess composition ranged from 2.9‰ to 14.4‰ with an arithmetic mean of 10.8 ± 2.0 (‰) (**Table 1**, **Figure 7**). Weighted *d-excess* values range from 3.5‰ to 17.4‰ with a mean of 10.3 ± 2.1 (‰) (**Table 1**, **Figure 7**). Low *d-excess* (~3‰) is predominantly present within the Nicoya Peninsula and Caribbean lowlands and may also be a common phenomena across the northern lowlands and Nicaragua trough. High *d-excess* (>17‰) was found in the dense forested areas east of the Central Valley (Zurquí and Irazú mountain range) and the Pacific slope of the Monteverde tropical cloud forest (**Figure 7**), which may indicate successive evaporation and precipitation cycles. Based on *d-excess* values, Rhodes *et al.* (2006) [43] concluded that moisture flux from evaporation is an important input to the region during the transitional and dry seasons when trade winds from the Caribbean slope dominate. Therefore, it appears that intense moisture recycling processes govern isotopic variations within the

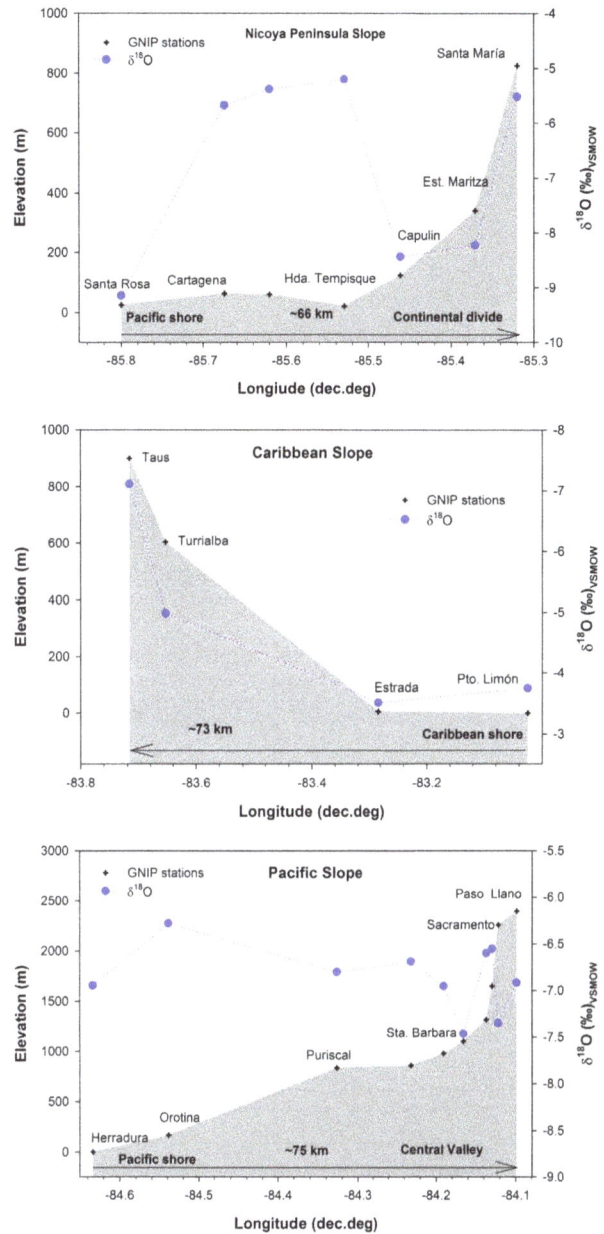

Figure 6. Longitudinal elevation and isotopic transects of Nicoya Peninsula, Pacific and Caribbean slopes.

inter-mountainous regions of Costa Rica.

3.5. Air Mass Back Trajectories and Precipitation Events

Figure 8 shows daily variation in isotopic composition in precipitation at Monteverde and Turrialba. For Monteverde, a remarkable depletion in $\delta^{18}O$ was observed on 7/18/2011 and an enriched event was sampled on 7/26/2011 (**Figure 8**). For Turrialba, sampling coincided with the wet to dry season transition, but sampling events for the purposes of the air mass trajectories were selected

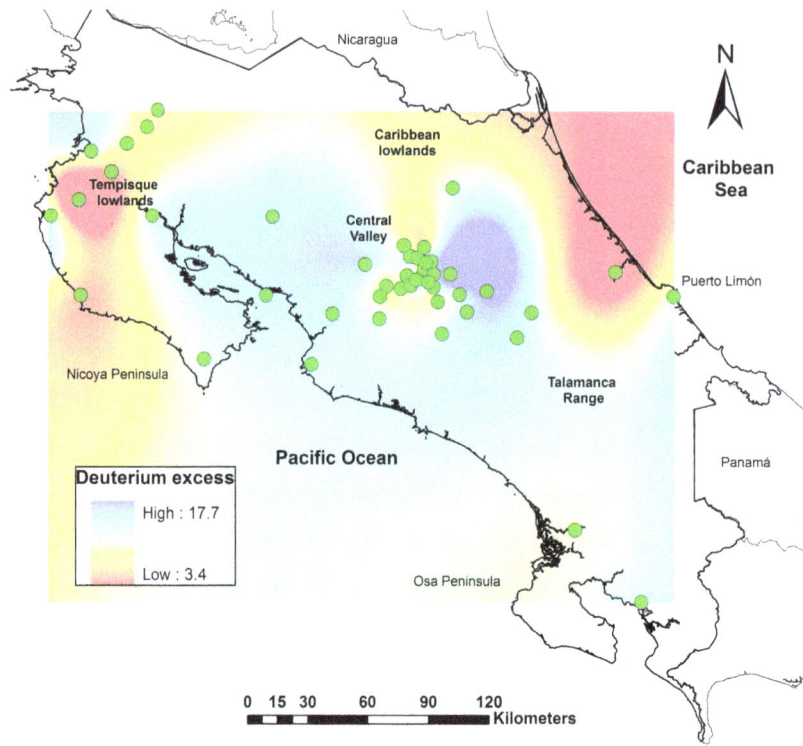

Figure 7. Preliminary *d-excess* isoscape for Costa Rica.

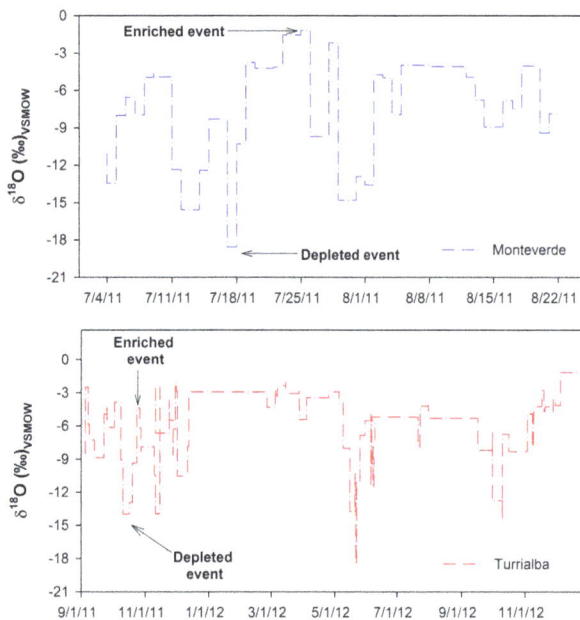

Figure 8. Daily $\delta^{18}O$ variations in precipitation at Monteverde and Turrialba stations.

during the wet and transitional seasons (September-December, 2011). Considerably depleted and enriched $\delta^{18}O$ values were recorded on 10/17/2011 and 10/27/2011, respectively (**Figure 8**). **Figure 9** shows air mass back trajectories of five days prior to the depleted or enriched

precipitation event coupled with rainfall intensities. HYSPLIT simulated trajectories indicate that enriched isotopic compositions both in Turrialba and Monteverde correspond to central Caribbean-sourced moisture and small rainfall intensities (**Figures 9(a)** and **(b)**). Depleted events appear to correspond to high rainfall amounts despite the parental origin of the moisture. In the case of Turrialba, the depleted $\delta^{18}O$ event was related to continental air masses arriving over Costa Rica from the southwest, producing convective precipitation as they travel along and over the Pacific Coast. Air masses intrusions from the Pacific Ocean are typically observed during the wet season and are less frequent when the wet to dry season transition begins in November. The depleted precipitation sampled in Monteverde was related to greater rainfall rates and parental moisture from the southeastern Caribbean basin.

4. Conclusions

Stable isotope signatures in precipitation and water vapor are useful indicators of climate variability and provide critical information regarding regional and global water processes. In temperate regions, isotopic sampling has been conducted for decades, providing novel insights on water cycle dynamics. However, isotopic records from the humid tropics, such as Central America, have been quite limited to date. Long-term $\delta^{18}O$ and δ^2H records

Spatial and Temporal Variation of Stable Isotopes in Precipitation across Costa Rica: An Analysis of Historic GNIP Records

135

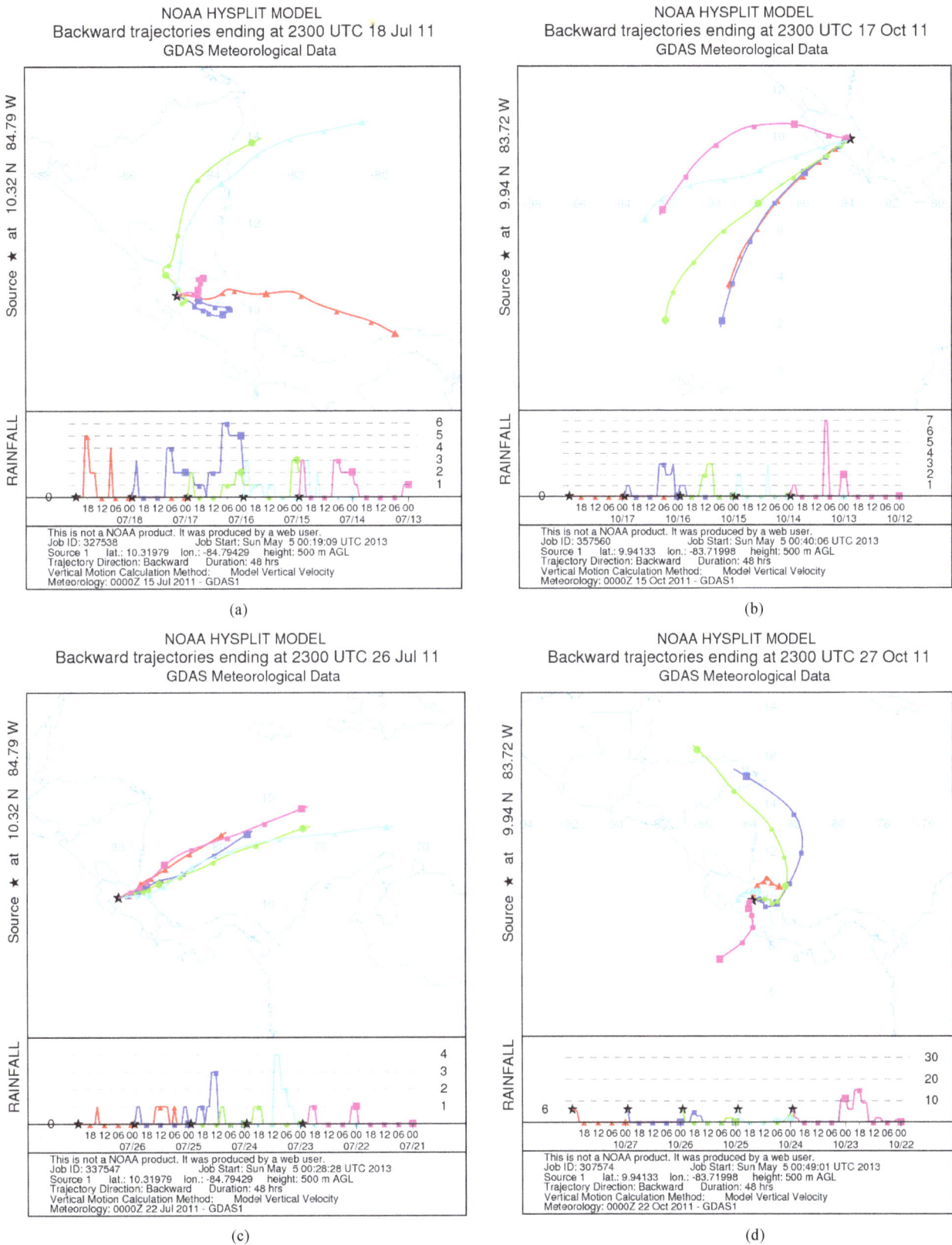

Figure 9. HYSPLIT model simulations including rainfall intensity in mm/hr. *Depleted events*: (a) Monteverde 7/18/11 and (b) Turrialba 10/17/11. *Enriched events*: (c) Monteverde 7/26/11 and (d) Turrialba 10/27/11. Five-day air mass back trajectories are color coded.

offer a unique opportunity to study precipitation and ground and surface water processes at a relatively low cost compared to continuous streamflow or climate monitoring, which is often not feasible in tropical developing countries due to high costs and limited resources. Based on historic GNIP records, we conducted a spatial and temporal characterization of stable isotopes in four main geographical regions of Costa Rica: Nicoya Peninsula, Pacific Coast, Caribbean slope, and Central Valley. Despite data limitations several correlations were found. The regression analysis of precipitation amounts with annual arimethic means in samples from all four regions yields a slope of $-1.6‰$ $\delta^{18}O$ per 100 mm of rain ($r^2 = 0.57$), which corresponds with a temperature effect of $-0.37‰$ $\delta^{18}O/°C$. A strong correlation ($r^2 = 0.77$) of $-2.0‰$ $\delta^{18}O$ per km of elevation was also found in these data.

D-excess has been identified as a potential indicator of recycled moisture. Therefore, meteoric *d-excess* isoscapes coupled with isotopic monitoring in surface waters may provide quantitative estimates of the contribution of recycled moisture, which would greatly enhance our understanding of water budgets in Costa Rica. *D-excess* interpolation from 46 GNIP stations suggests that evaporation enrichment often occurs over the Nicoya Peninsula, especially during periods of lowest precipitation (January-April). The Nicoya Peninsula, which receives less than 1500 mm annually, has been recently identified [44] as a hydrologically vulnerable region due to the increase of water demand and susceptibility to experience prolonged drought periods. Greater *d-excess* values were observed in the inter-mountainous region of the Central Valley and the Pacific slope. It is still unknown if moisture recycling is occurring due to evaporation within dense canopy covers (e.g., Zurquí-Irazú or Monteverde areas) or if instead moisture is available from multiple sources such as flood plains, dams or reservoirs, or soil moisture. However, this moisture input may play a significant role in late afternoon and overnight convective storms along the Pacific slope.

The illustrative example using the HYSPLIT model indicates that enriched isotopic compositions both in Turrialba and Monteverde are related to central Caribbean parental moisture and small rainfall intensities. Depleted events appear to be related to high rainfall amounts despite the parental moisture origin. Further event-based sampling in strategic locations across the country coupled with air mass trajectory analysis is needed to evaluate the effects of parental moisture origin and isotope signatures. Overall, we believe that this preliminary stable isotope analysis across Costa Rica may provide a useful baseline for future isotopic studies in the region as well as enhance our understanding of moisture recycling

processes within tropical inter-mountainous systems. Further research should evaluate spatial and temporal isotopic variations in surface and groundwater systems to ultimately determine critical recharge zones across the country.

5. Acknowledgements

We thank the International Atomic Energy Agency for granting access to historic GNIP records. A National Science Foundation-IGERT scholarship (Grant No. 0903479) US Borlaug Fellowship for Food Security supported the isotopic fieldwork in Turrialba, Costa Rica. Funding provided by the Chemistry School at the National University of Costa Rica and the National University Council of Costa Rica was essential to acquire a laser water isotope analyzer. The authors are also grateful for access to the field sites and logistical support by the Tropical Science Center and the Monteverde Cloud Forest Reserve.

REFERENCES

[1] P. Kabat, R. E. Schulze, M. E. Hellmuth and J. A. Veraart, "Coping with Impacts of Climate Variability and Climate Change in Water Management: A Scoping Paper," DWC-Report No. DWCSSO-01 International Secretariat of the Dialogue on Water and Climate, Wageningen, 2003.

[2] R. K. Colwell, G. Brehm, C. L. Cardelús, A. C. Gilman and J. T. Longino, "Global Warming, Elevational Range Shifts, and Lowland Biotic Attrition in the Wet Tropics," *Science*, Vol. 322, No. 5899, 2008, pp. 258-261.

[3] S. R. Loarie, P. P. Duffy, H. Hamilton, G. P. Asner, C. B. Field and D. D. Ackerly, "The Velocity of Climate Change," *Nature*, Vol. 462, No. 7276, 2009, pp. 1052-1055.

[4] J. J. Tewksbury, R. B. Huey and C. A. Deutsch, "Putting the Heat on Tropical Animals," *Science*, Vol. 320, No. 5881, 2008, pp. 1296-1297.

[5] Y. P. Zhou, K. M. Xu, Y. C. Sud and A. K. Betts, "Recent Trends of the Tropical Hydrological Cycle Inferred from Global Precipitation Climatology Project and International Satellite Cloud Climatology Project Data," *Journal of Geophysical Research*, Vol. 116, No. D6, 2011, Article ID: D09101.

[6] V. Magaña, J. A. Amador and S. Medina, "The Midsummer Drought over Mexico and Central America," *Journal of Climate*, Vol. 12, No. 6, 1999, pp. 1577-1588.

[7] G. Bowen, "Statistical and Geostatistical Mapping of Precipitation Water Isotope Ratios," In: J. B. West, *et al.*, Eds., *Isoscapes: Understanding Movement, Pattern, and*

Process on Earth through Isotope Mapping, Springer Science, New York, 2010, pp. 139-160.

[8]   L. J. Araguás-Araguás, K. O. Froehlich and K. Rozanski, "Stable Isotope Composition of Precipitation over Southeast Asia," *Journal of Geophysical Research*, Vol. 103, No. D22, 1998, pp. 28721-28742.

[9]   K. M. Cobb, J. F. Adkins, J. W. Partin and B. Clark, "Regional-Scale Climate Influences on Temporal Variation of Rainwater and Cave Dripwater Oxygen Isotopes in Northern Borneo," *Earth and Planetary Science Letters*, Vol. 263, No. 3-4, 2007, pp. 207-220.

[10] Y. Ishizaki, K. Yoshimura, S. Kanae, M. Kimoto, N. Kurita and T. Oki, "Interannual Variability of $H_2^{18}O$ in Precipitation over the Asian Monsoon Region," *Journal of Geophysical Research*, Vol. 117, No. D16, 2012, Article ID: D16308.

[11] K. R. Johnson and B. L. Ingram, "Spatial and Temporal Variability in the Stable Isotope Systematics of Modern Precipitation in China: Implications for Paleoclimate Reconstructions," *Earth and Planetary Science Letters*, Vol. 220, No. 3-4, 2004, pp. 365-377.

[12] M. Lachniet, "Sea Surface Temperature Control on the Stable Isotopic Composition of Rainfall in Panama," *Geophysical Research Letters*, Vol. 36, No. 3, 2009, Article ID: L03701.

[13] M. Lachniet and W. Patterson, "Oxygen Isotope Values of Precipitation and Surface Waters in Northern Central America (Belize and Guatemala) Are Dominated by Temperature and Amount Effects," *Earth and Planetary Science Letters*, Vol. 284, No. 3, 2009, pp. 435-446.

[14] M. Vuille, R. S. Bradley and F. Keimig, "Interannual Climate Variability in the Central Andes and Its Relation to Tropical Pacific and Atlantic Forcing," *Water Resources Research*, Vol. 105, No. D10, 2000, pp. 12447-12460.

[15] M. Vuille, R. S. Bradley and F. Keimig, "Climatic Variability in the Andes of Ecuador and Its Relation to Tropical Pacific and Atlantic Sea Surface Temperature Anomalies," *Journal of Climate*, Vol. 13, No. 14, 2000, pp. 2520-2535.

[16] M. Vuille, R. S. Bradley, R. Healy, M. Werner, D. R. Hardy, L. G. Thompson and F. Keimig, "Modeling d18O in Precipitation over the Tropical Americas 2: Simulation of the Stable Isotope Signal in Andean Ice Cores," *Journal of Geophysical Research*, Vol. 108, No. D6, 2003, p. 4175.

[17] K. Ichiyanagi and M. D. Yamanaka, "Interannual Variation of Stable Isotopes in Precipitation at Bangkok in Response to El Niño Southern Oscillation," *Hydrological Processes*, Vol. 19, No. 17, 2005, pp. 3413-3423.

[18] H. O. Panarello and C. Dapeña, "Large Scale Meteorological Phenomena, ENSO and ITCZ, Define the Paraná River Isotope Composition," *Journal of Hydrology*, Vol. 365, No. 1-2, 2009, pp. 105-112.

[19] M. Vuille and M. Werner, "Stable Isotopes in Precipitation Recording South American Summer Monsoon and ENSO Variability: Observations and Model Results," *Climate Dynamics*, Vol. 25, No. 4, 2005, pp. 401-413.

[20] P. K. Aggarwal, O. A. Alduchov, K. O. Froehlich, L. J. Araguas-Araguas, N. C. Sturchio and N. Kurita, "Stable Isotopes in Global Precipitation: A Unified Interpretation Based on Atmospheric Moisture Residence Time," *Geophysical Research Letters*, Vol. 39, No. 11, 2012, Article ID: L11705.

[21] L. J. Araguás-Araguás, K. O. Froehlich and K. Rozanski, "Deuterium and Oxygen-18 Isotope Composition of Precipitation and Atmospheric Moisture," *Hydrological Processes*, Vol. 14, No. 8, 2000, pp. 1341-1355.

[22] G. Bowen, "Spatial Analysis of the Intra-Annual Variation of Precipitation Isotope Ratios and Its Climatological Corollaries," *Journal of Geophysical Research*, Vol. 113, No. D5, 2008, Article ID: D05113.

[23] W. Dansgaard, "Stable Isotopes in Precipitation," *Tellus*, Vol. 16, No. 4, 1964, pp. 436-468.

[24] K. Rozanski, L. J. Araguas-Araguas and R. Gonantini, "Isotopic Patterns in Modern Global Precipitation," In: P. K. Swart, K. C. Lohmann, J. McKenzie and S. Savin, Eds., *Climate Change in Continental Isotopic Records*, American Geophysical Union, Washington DC, 1993, pp. 1-36.

[25] L. I. Wassenaar, P. Athanasopoulos and M. J. Hendry, "Isotope Hydrology of Precipitation, Surface and Ground Waters in the Okanagan Valley, British Columbia, Canada," *Hydrology*, Vol. 411, No. 1, 2011, pp. 37-48.

[26] G. Bowen and J. Revenaugh, "Interpolating the Isotopic Composition of Modern Meteoric Precipitation," *Water Resources Research*, Vol. 39, No. 10, 2003, p. 1299.

[27] M. Lachniet and W. Patterson, "Stable Isotope Values of Costa Rican Surface Waters," *Journal of Hydrology*, Vol. 260, No. 1-4, 2002, pp. 135-150.

[28] G. J. Bowen, J. B. West and J. Hoogewerff, "Isoscapes: Isotope Mapping and Its Applications," *Geochemical Exploration*, Vol. 102, No. 3, 2009, pp. 5-7.

[29] V. Barras and I. Simmonds, "Observation and Modeling of Stable Water Isotopes as Diagnostics of Rainfall Dynamics over Southeastern Australia," *Journal of Geophysical Research*, Vol. 114, No. D23, 2009, pp. 1-17.

[30] H. Celle-Jeanton, R. Gonfiantini, Y. Travi and B. Sol, "Oxygen-18 Variations of Rainwater during Precipitation: Application of a Rayleigh Model to Selected Rainfalls in Southern France," *Journal of Hydrology*, Vol. 289, No. 1-4, 2004, pp. 165-177.

[31] T. B. Coplen, P. J. Neiman, A. B. White, J. M. Landwehr, F. M. Ralph and M. D. Dettinger, "Extreme Changes in Stable Hydrogen Isotopes and Precipitation Characteristics in a Landfalling Pacific Storm," *Geophysical Research Letters*, Vol. 35, No. 21, 2008, Article ID: L21808.

[32] N. C. Munksgaard, C. M. Wurster, A. Bass and M. I. Bird, "Extreme Short-Term Stable Isotope Variability Revealed by Continuous Rainwater Analysis," *Hydrological Processes*, Vol. 26, No. 23, 2012, pp. 3630-3634.

[33] J. Reynolds-Vargas and J. Fraile, "Use of Stable Isotopes in Precipitation to Determine Recharge Areas within the Barva Aquifer, Costa Rica," Latinoamerican Stable Isotopes Studies, International Atomic Energy Agency, Viena, 2006, pp. 83-96.

[34] P. Waylen, C. N. Caviedes and M. F. Quesada, "Interannual Variability of Monthly Precipitation in Costa Rica," *Journal of Climate*, Vol. 9, No. 10, 1996, pp. 2506-2613.

[35] A. J. Guswa, A. Rhodes and S. E. Newell, "Importance of Orographic Precipitation to the Water Resources of Monteverde, Costa Rica," *Advances in Water Resources*, Vol. 30, No. 10, 2007, pp. 2098-2112.

[36] GNIP, "Global Network of Isotopes in Precipitation," 2012.
http://wwwnaweb.iaea.org/napc/ih/IHS_resources_gnip.html

[37] Instituto Tecnológico de Costa Rica (ITEC), "Atlas de Costa Rica. Escuela de Ingeniería Forestal. Cartago, Costa Rica," Digital Atlas, ITEC, Cartago, Costa Rica, 2008.

[38] R. R. Draxler and G. D. Rolph, "HYSPLIT-Hybrid Single Particle Lagrangian Integrated Trajectory Model," 2013.
http://ready.arl.noaa.gov/HYSPLIT.php

[39] H. Craig, "Isotopic Variations in Meteoric Waters," *Science*, Vol. 133, No. 3465, 1961, pp. 1702-1703.

[40] C. Leibundgut, P. Maloszewski and C. Kulls, "Tracers in hydrology," Wiley-Blackwell, Oxford, 2009.

[41] R. Gonfiantini, M. A. Roche, J. C. Olivry, J. C. Fontes and G. M. Zuppi, "The Altitude Effect on the Isotopic Composition of Tropical Rains," *Chemical Geology*, Vol. 181, No. 1, 2001, pp. 147-167.

[42] M. A. Scholl, J. B. Shanley, J. P. Zegarra and T. B. Coplen, "The Stable Isotope Amount Effect: New Insights from NEXRAD Echo Tops, Luquillo Mountains, Puerto Rico," *Water Resources Research*, Vol. 45, No. 12, 2009, Article ID: W12407.

[43] A. L. Rhodes, A. J. Guswa and S. E. Newell, "Seasonal Variation in the Stable Isotopic Composition of Precipitation in the Tropical Montane Forests of Monteverde, Costa Rica," *Water Resources Research*, Vol. 42, 2006, Article ID: W11402.

[44] MINAET-IMN, "Mejoramiento de las Capacidades Nacionales para la Evaluación de la Vulnerabilidad y Adaptación del Sistema híDrico al Cambio climÁtico en Costa Rica, Como Mecanismo para Disminuir el Riesgo al Cambio climÁtico y Aumentar el Índice de Desarrollo Humano," In: J. A. Retana, Eds., *Ministerio de Ambiente, Energía y Telecomunicaciones*, Instituto Meteorológico Nacional San José, Costa Rica, 2012, 46 p.

16

Remote Monitoring of Vegetation Managed for Dust Control on the Dry Owens Lakebed, California

David P. Groeneveld, David D. Barz

HydroBio Advanced Remote Sensing, Santa Fe, USA.

ABSTRACT

A monitoring program was developed to assess the cover of saltgrass managed for dust control on the saline dry Owens Lake. Although the original intent was to manage the vegetation as total cover that included green and senesced leaf and stem material, aged leaves that make up a large proportion of total cover were not differentiable spectrally from the background salt and lakebed. Hence, greenness-based indices were explored for detection of plant recruitment. Since all plant cover begins as green and growing, greenness indices provide a measure of all future cover whether living or senesced. The criteria for judging compliance were changed so that spatially variable vegetation cover measured as a milestone will need to be met in the future. A derivative of NDVI, $NDVI_x$, calculated using scene statistics, proved highly accurate, to about 0.001 of this index and with an average signal to noise ratio of 64. This high level of accuracy allowed detection of small changes in vegetation growth and vigor. Performance according to the benchmark-as-par standard was determined through combined use of cumulative distribution functions and derivative maps.

Keywords: Dust Control; Remote Sensing; Monitoring; Managed Vegetation; NDVI; Owens Lake; California

1. Introduction

Prior to dust control efforts that began during the first decade of this millennium, the Owens Dry Lake, located in Eastern Central California (**Figure 1**) was the single largest source of anthropogenic PM_{10} in the western hemisphere (GBUAPCD 2003). PM_{10} is particulate matter less than 10 μm in aerodynamic diameter that has been implicated in reducing respiratory health [1]. This dust was caused by hydrologic changes due to diversion of water from the 280 km² lake.

The Owens River that supplies water to the Owens Lake was diverted nearly entirely to the City of Los Angeles beginning in 1913. Prior to diversion, Owens Dry Lake was filled with water but was declining due to irrigation diversion. The lake desiccated completely by 1927 after complete diversion to Los Angeles [2].

Owens Lake has been the terminus for Owens River during the past several thousand years, concentrating the salts received from regional runoff [3,4], and hence, was highly saline before desiccation. Salts stranded in the lakebed are a causal factor for dust releases from the surface due to temperature-controlled salt crystal pre-cipitation that destroys soil cohesion and permits winds of only 7.5 m/s (15 kts) to ablate the dusty surface [5]. Large scale dust releases occur during the winter, generally starting about October 1 and lasting until the end of June, a period recognized as the dust season by the Great Basin Unified Air Pollution Control District (District) whose responsibility it is to monitor and enforce regional air quality.

According to the State Implementation Plan (SIP) required by California state law, SB270, Owens Lake dust control is mandated for the Los Angeles Department of Water and Power (Department) that exports the water [6]. The District is the authority for ensuring compliance and has identified three dust control measures that provide some form of surface cover: gravel, vegetation, or shallow flooding. The Department has opted to apply shallow flooding nearly exclusively, while developing approximately 912 ha (2239 ac) of saltgrass (*Distichlis spicata* L.), known as the Managed Vegetation Area (MVA) (**Figure 1**). Saltgrass forms an effective dust control measure on the MVA because its canopy creates aerodynamic roughness that reduces or eliminates erosive wind energy. The vegetation's rhizomatous root system also holds the soil in

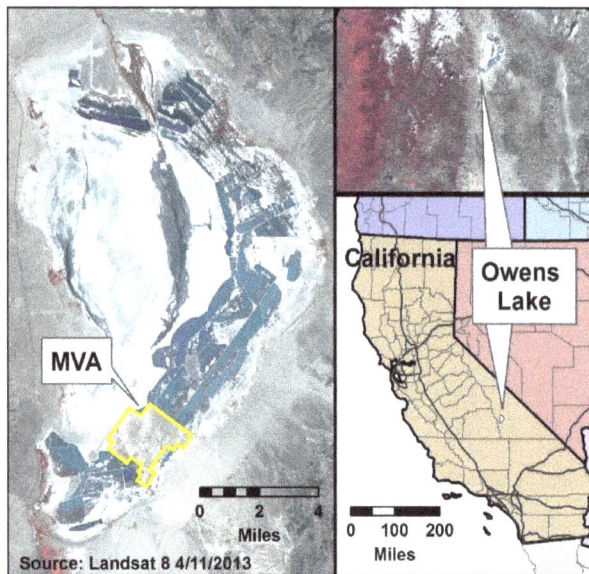

Figure 1. Location map for the Owens Lake and the MVA.

Figure 2. Dust storm from a strong northerly wind across the location of the MVA nine years before construction. The width of the dust plume is about 2 km.

place.

This paper describes development and application of an operational remote sensing program to evaluate MVA vegetation. A second purpose is to provide guidance for those interested in generation of highly accurate vegetation indices appropriate for evaluating vegetation performance through time. A third purpose is to provide a description of the MVA program and the complexities for monitoring in case a similar program is ever contemplated at another location.

1.1. Description of the Managed Vegetation Area

The location of the MVA was formerly one of the most productive PM10 sources on the lakebed primarily because mesoscale topographic influences force winds generated during winter frontal passage into a strong north-south trend, often impinging upon this region of the lakebed (**Figure 2**). The MVA soils consist of clay to sand sized particles with the sand having been moved and concentrated by aeolian processes [7].

The Owens Lake environment is highly arid with annual average precipitation less than nine centimeters per year [8]. Supplementary irrigation is necessary to supply the planted vegetation and to leach salts from the highly saline lakebed. The site was prepared by constructing a series of north-south trending rows separated by furrows spaced 1.5 m (5 ft) apart. A single drip irrigation tube was placed along the center of each row to provide leaching to move salts downward from each row crown. Preparation of the MVA site required over 5000 km of drip irrigation tubing to serve the row tops, alone.

For planting the saltgrass (*Distichlis spicata*, L.), accessions were collected from around the lake, cultivated

separately, and then planted as plugs. Spaced every 0.3 m, this required over 17 million separate saltgrass plugs, each planted by hand. The saltgrass cover is irrigated with a blend of brine drained from the site and fresh water diverted from the Los Angeles Aqueduct that is acidified with dilute sulfuric acid to alleviate salt precipitation within the drip irrigation emitters and tubing. The plantings were made in blocks of 16.2 hectares (40 acres) separated by roadways (**Figure 3**). A system of French drains was established that run either parallel, or at 45 and 90 degree angles to the alignment of the planted rows. Associated pumps and piping operate this infrastructure.

Remote sensing is required to evaluate the vegetation across the large area of the MVA. Vegetation cover varies spatially and so must be assessed spatially because some areas are problematic for establishing and growing saltgrass, while other locations may foster heavy vegetation cover (**Figure 3**). Accurate monitoring also provides advance indications of trends in growth or decline, thus allowing proactive changes in management to avoid loss of cover that could eventually compromise air quality. Because of the scale of the MVA, earth observation satellite data (EOS; e.g., Landsat TM, Quickbird, SPOT, etc.) are required to make the evaluation.

1.2. Criterion for Judging MVA Compliance

The 2003 SIP required that each 0.405 ha (1.0 acre) of the MVA contain at least 50% cover by vegetation for dust control [6]. The MVA was planted in 2002 and during the next five years it became apparent that vegetation growth on portions lagged severely from impaired drainage and concentration of salts due to evapotranspiration of the brine irrigation (**Figure 3**). These constraints prevented meeting the SIP criterion for 50% total cover on all areas.

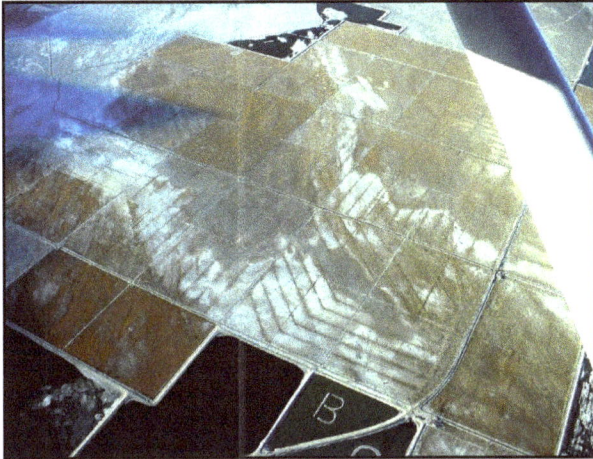

Figure 3. Oblique wintertime air photograph of the MVA showing vegetated blocks and roadways. Extensive light-colored patterns have poor drainage, shallow groundwater, and are salt affected. This pattern is crossed with darker vegetated lines (some at 45° angles) that overlie drains.

No significant areas of blowing dust occurred within the MVA during the initial five years and so the vegetation cover criteria were adjusted at the discretion of the District's Air Pollution Control Officer. The MVA vegetation cover, though poorly established in some areas, provided adequate protection from wind erosion and so a new criterion for vegetation cover was adopted to achieve existing or better cover on all areas. The new criterion recognized the stabilizing effect of the established cover while allowing low or non-emissive areas of poor cover to be accepted. This yardstick must be applied spatially, since concentration of low cover in one portion of the MVA could well lead to emissions. This low cover was set equal to the spatially-discrete peak annual cover that was measured on October 4, 2007 using a TM5 image. This spatial limit to be equaled or exceeded will be called "par", meaning a minimum condition to be met. Application of par requires determination whether peak green growth during a subsequent year attained parity with the 2007 baseline.

The replacement criterion for achieving par vegetation growth was an important change that enabled accurate remote sensing of the MVA vegetation.

1.3. Spectral Properties and Detection of Vegetation Cover

As the MVA saltgrass grew and matured, different remote sensing methods were required to estimate cover. All initial methods used regression of remotely-sensed indices against ground truth cover that was estimated by point frequency frame [9]. The point frequency frame provides an estimate of ground cover through tallies of the first contact of vegetation by the tips of pins that are manually passed vertically downward, while the observer carefully judges the first contact of each point with leaves, stems or flowers of the vegetation. The pins are mounted in a frame and have sharpened tips for exact judgment of contacts. Sharpening the pins is important because the pin tip must be regarded as dimensionless. The total number of contacts divided by the total number of pins is an accurate estimate of fractional plant cover. Because it is labor intensive, point frequency measurements are not appropriate for monitoring at the scale of the MVA, but only for calibration of remotely-sensed data.

When the saltgrass plugs were establishing during the first season, the canopy consisted solely of green plant material and this enabled use of the normalized difference vegetation index, NDVI, that was calibrated to ground-truth supplied by point frame measurements of total cover made at discrete points across the site. During the second season and for the next two years, the ground cover matured and senesced each year, adding a thatch of dead material that contributed cover toward the goal of 50% cover that was required by the SIP. When senesced vegetation constituted the majority of cover, NDVI could no longer be used to evaluate total cover because the majority of the canopy consisted of non-green material that produced little NDVI response. This was solved by adoption of an index formed by the ratio of near infrared (NIR)/Green (Landsat TM and Quickbird bands 4/2) that was initially relatively accurate for evaluating total senesced cover (regression $r^2 = 94\%$). This index was applied in early winter after the saltgrass canopy had completely senesced but before loss of leaves through wind action or flattening of the canopy due to rare winter snow loads.

After 2005, senesced vegetation indices were no longer accurate due to a progression in coloration for aging canopy cover: 1) first as green, living material; 2) as yellowing leaves during tissue dieback; 3) as a golden orange during the first and second seasons of senescence; and finally 4) as a silver color due to weathering. In a spectrometry study using a FieldSpec Pro™, the data were converted into Landsat ETM7+ equivalent spectra using published band sensitivities. As shown in **Figure 4**, the last stage, silver, was found to be indistinguishable from bare substrate because ratios of the various bands could not be used to develop an index that would distinguish plant material from the background soil and saltcrust. Hence, total cover containing silver cannot be detected using EOS remote sensing. Locations were observed on the MVA to have silver material that occupied up to 90% of the total canopy cover.

Changing the initial SIP goal from 50% total cover to a spatially-discrete par approach enabled a new and accurate monitoring method that evaluated green living cover. Back-comparison of green cover from the previous growing season provides a simple assessment for whether the

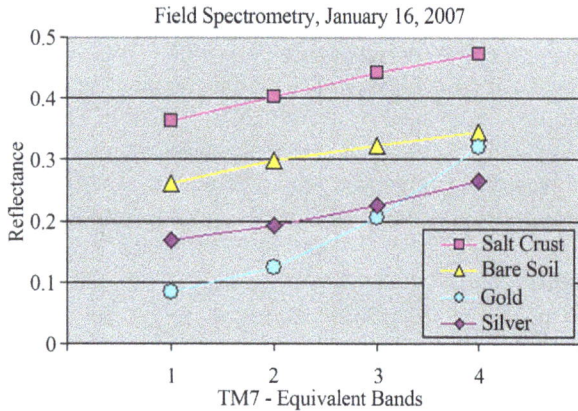

Field Spectrometry, January 16, 2007

Figure 4. Aged and senesced saltgrass attains a silver color with the same ratiometry as in all bands and cannot be distinguished from bare soil or salt crust.

measured value attained par, or not. The changed criterion enabled the operational remotely-sensed monitoring system described in this paper.

Green, living cover can be measured with self-calibrated indices that have been shown to be highly accurate and useful for analysis of plant ecohydrology [10-12]. Such indices have potential for high levels of accuracy on the MVA because the vegetation is a monoculture of one species, saltgrass. During the growing season active living tissues are green, and therefore detectable with such indices.

1.4. Maintaining the Integrity of the Specifications

High levels of accuracy and measurement robustness are desired features for a competent vegetation monitoring program. Initially, the 2003 SIP required use of point frame measurements for linear regression calibration of indices using EOS data [6]. The accuracy and robustness of the point frame for measuring green living cover is important because of the spectral limitations for detection of total cover. Groundtruth using the point frame method, plus regression, were evaluated against the use of an EOS vegetation index calibrated using only scene statistics. For this comparison, a study was undertaken to understand the limitations of the point frame method.

To assess potential error, point frequency data were collected using two separate methods, each employed by the Department and the District during June 13-15, 2007. The measurements were made on a range of 14 carefully chosen sites with homogeneous cover and canopy colors. The timing of the measurement, early in the growing season, enabled separate point frame evaluation of all colors of the saltgrass canopies.

The District gathered point frame data manually using a metal frame that held a rack of 14 pins. The frame was deployed at 28 locations selected to evenly sample within

a 5 m radius circle surrounding a central marking stake. The Department used a photographic method, whereby eight nadir-look photos were located according to a set scheme that was deployed around the center stake. These photos were taken to the office and displayed with a grid overlay for judging the grid contacts with vegetation—a direct analog of the manual method. The color of the contacted plant material was recorded for both methods.

A comparison of total vegetation cover for point frame and digital point frame methods showed excellent agreement ($r^2 = 0.979$) (**Figure 5**). These differences were minor compared to the distribution of the two data sets when evaluating constituent canopy colors of green, gold and silver. While the percent silver and gold cover were moderately well correlated (r^2 values of 0.877 and 0.817, respectively) green cover measurements were rather weakly correlated (r^2 of 0.662; **Figure 5**). Given the high correlation for total cover, the lack of agreement between the two data sets was due, mostly, to the subjectivity in deciding canopy color. Large errors introduced by subjective color judgment during point frequency data acquisition necessarily reduces accuracy of the resulting classification of vegetation cover using EOS data.

Rather than making subjective dichotomous decisions for what is green in canopies containing gradations, such gradations are well captured by EOS data, alone. Thus, using point frame data to calibrate images was bypassed in favor of calibration of EOS data using scene statistics.

2. Methods Chosen for Processing EOS Data

An EOS index, $NDVI_{offset}$ first described by Baugh and Groeneveld [10] was selected as the basis for monitoring vegetation on the managed vegetation area. $NDVI_{offset}$ is derived from normalized difference vegetation index (NDVI) by subtracting an offset value determined from

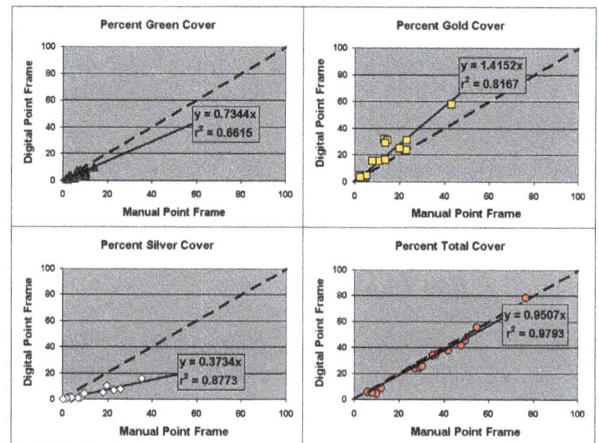

Figure 5. Comparison of percent cover green, gold and silver material measured at 14 sites by manual and digital point frame.

scene statistics. NDVI$_{offset}$ was shown to yield results superior to NDVI and nearly all other indices in prediction of a known linear relationship for promotion of vegetation growth by antecedent precipitation [6]. The generation of NDVI$_{offset}$ corrects for the effects of attenuation and scattering from atmospheric aerosols and for soil background coloration in a single operation.

Peak season green vegetation growth is the target for analysis of MVA performance since this is the greatest expression of plant cover recruitment that maintains surface stability for dust control. For the persistent leaves of saltgrass plants receiving repeated irrigation through the growing season, peak expression has occurred most commonly in early fall. **Table 1** provides the dates and the EOS platforms that were used, all generated by the Landsat program.

A late season image may not always capture the peak of the green growth expression. Therefore, in addition to evaluating late season images, to confirm that the peak conditions have been recorded and that the image is not unduly affected by atmospheric aerosols, it is advisable to evaluate images through the entire growing season for confirmation. The vegetation captured in an annual series of images should increment through the season in a rational fashion, growing or declining smoothly through the season, rather than showing high-low fluctuations. Cumulative distribution functions (CDFs) displaying the cumulative pixel counts on the y axis and the NDVI derivative on the x, were used to compare the imagery through the growing season. In all cases, the annual image series yielded gradually changing vegetation cover through the season, taken to indicate that processing was correct and that the peak season image was acceptable.

2.1. EOS Imagery

Although any of the EOS platforms that supply red and near infrared bands could be used for a program to detect the green cover at the MVA, Landsat TM data were used for development because they were made available free of charge through http://glovis.usgs.gov/.

Six years of imagery were chosen for the comparison (**Table 1**). Landsat TM5 images were selected to represent 2007 through 2011 and in 2012, this EOS platform ceased operation. Landsat ETM7+ data were selected for 2012, however, these data are flawed due to stripes of missing data. The missing data stripes caused an average loss of coverage of only about 5.2% of the MVA, and so,

do not greatly affecting the quality of the comparison. The MVA is located near the center of the overall ETM7+ scene in a location where the missing data constitute only small areas with mixes of high and low vegetation cover—omissions were judged to not induce bias. For comparison of CDFs, the pixel counts for the 2012 scenes were adjusted to be equivalent to other years.

2.2. Image Processing

Calculation of top of atmosphere reflectance for Landsat data products followed methods outlined in the Landsat 7 Data User's Handbook [13]. Following image processing, all images were geocorrected to a high resolution Quickbird panchromatic base image (0.6 m pixels) that was previously corrected to a net of carefully surveyed ground targets.

Following geocorrection, a 0.405 ha grid file was applied to mask out roads and edges from consideration in the statistical examination (**Figure 6**). Masking removed any potential gridcell that could possibly contain edge pixel contamination. Inclusion of edge pixels was judged to be a significant problem since unvegetated pixels among images would cause grid cell values to fluctuate according to the variable amount of unvegetated roadway that could be covered. Grid cells that could contain edge pixels were eliminated if a portion of at least one roadway or boundary pixel fell within a pixel hypotenuse (42.4 m).

2.3. Calculating NDVI and NDVI$_{offset}$

NDVI, the first step in creation of the derivative indices

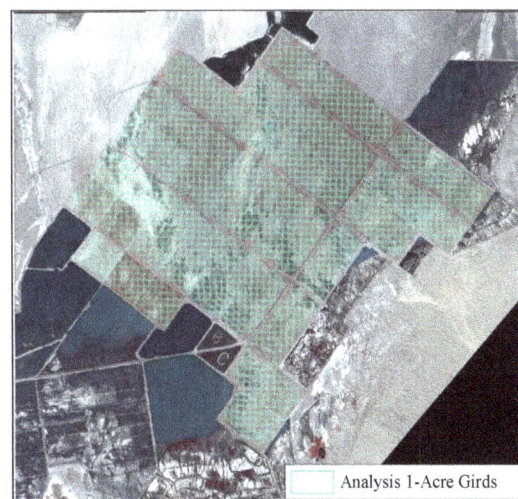

Figure 6. Grid cells (0.405 ha) that are used twice in the data processing—first to clip edge pixels and roadways and after statistical analysis and further processing, to sort the data into the grids for comparison of performance over time.

Table 1. Images chosen for analysis of trend. All images were Landsat TM5 except 2012 that were Landsat ETM7.

2007	2008	2009	2010	2011	2012
Jun-13	Jul-13	Jul-13	Jul-13	Jun-13	Jul-13
Jul-13	Jul-13	Aug-13	Aug-13	Jul-13	Jul-13
Sep-13	Aug-13	Sep-13	Sep-13	Aug-13	Aug-13
Oct-13	Oct-13	Oct-13	Oct-13	Sep-13	Sep-13
				Sep-13	Oct-13

that are used for change assessment, was calculated from red and NIR TM data according to Equation (1), appropriate for the ith pixel:

$$NDVI_i = (Red_i - NIR_i)/(Red_i + NIR_i) \qquad (1)$$

$NDVI_{offset}$ is a simple shift of the NDVI distribution leftward so that zero vegetation cover approximates the soil background. To calculate $NDVI_{offset}$ the raw NDVI data are displayed as a CDF such as that shown in **Figure 7**. The amount of the shift is determined from linear regression calculations on the lowest, near-linear limb of the NDVI CDF the x-intercept of the regression line, $NDVI_0$ and is termed $NDVI_0$, approximating the soil background NDVI value [10]. For the ith pixel:

$$NDVI_{i,\,offset} = NDVI_i - NDVI_0 \qquad (2)$$

The regression interval chosen for calculating $NDVI_0$ is best established after study of a number of images. The lower CDF region contains decreasing expression of the green canopy over a background of non-green mixed cover of soil and senesced vegetation canopies. To use Landsat TM data, it is necessary to examine the CDF to check that the interval does not contain waviness as is visible in **Figure 8**, due to low bit precision (Landsat TM5 data are 8 bit). If so, the interval must be expanded to bridge across the wavy region to emulate the result that would occur were the CDF smooth. The CDF interval chosen for fitting the regression relationships in **Figure 8** was from 500 to 1000 pixels, constituting an interval of between six and 13 percent of the approximate total 7900-pixel distribution (software derived pixel counts may vary).

The regression of pixel intervals chosen on the CDF is solved for y = 0, and this yields an x intercept, $NDVI_0$, that is then subtracted from all NDVI values to shift the NDVI curve leftward (**Figure 9**). This method provides an unbiased estimate of a point of near-zero vegetation,

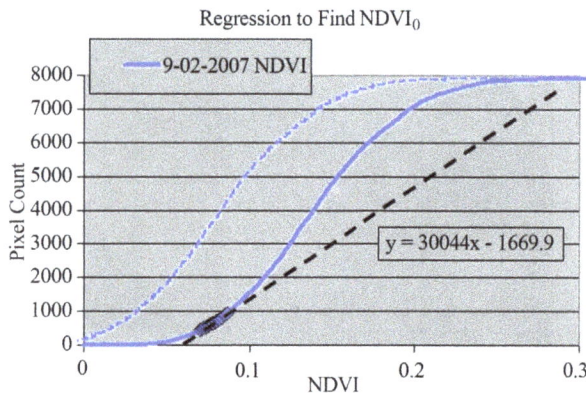

Figure 8. Four NDVI CDFs from 2007 with points chosen for regression. The regression equations were solved for y = 0 to find the x-intercepts of the curves ($NDVI_0$). Waviness in the region of 0.10 to 0.13 NDVI for the October image is a factor of bit precision.

$NDVI_0$, and removes most of the effects of atmospheric scatter and attenuation and is effectively a statistically-derived estimate of the value for bare soil. Equation (2) collects the CDF curves together so that they increment correctly with increasing green expression as the season progresses. As the canopy becomes more and more green, the curves push rightward, In **Figure 9** the rightward push of the increasing greenness culminates with the October image that generally captured the season peak.

The calculation of $NDVI_{offset}$ provides results that are reproducible and consistent, confirming that the calculation of $NDVI_{offset}$ corrects the vegetation signal relationally. $NDVI_{offset}$ is only a first order estimate of bare soil that places the images in correct relation. A second step, indexing to $NDVI_x$, was necessary to further calibrate the images for comparison to the October 4, 2007 benchmark. As shown in **Figure 9**, $NDVI_x$ solves a problem of truncation that occurs at the lowest levels of plant cover.

2.4. Calculating $NDVI_X$

In the middle graph of **Figure 9**, $NDVI_{offset}$ truncates the vegetation distribution values below zero that still may be slightly vegetated. If zero $NDVI_{offset}$ is logically used for the threshold for evaluating MVA vegetation performance, there will be many pixels that are excluded and whose progress cannot be tracked. Since the analysis of vegetation performance for the MVA is made for each grid cell, pixels assigned a zero value also skew grid cell averages for spatial assessments and induce greater variability in the grid cell values that have poor cover. To overcome this problem an additional step was devised, indexing, that corrects all CDF's to be relative to a benchmark image. For the MVA, this benchmark image is the October 4, 2007 to which all other images are compared.

A standard bare ground NDVI value for the 2007

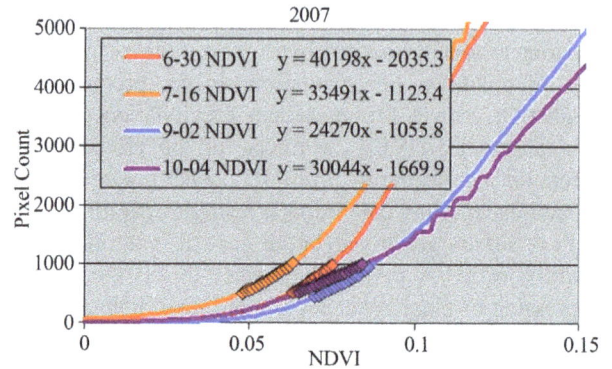

Figure 7. $NDVI_0$, the x-intercept for the regression of the selected points, is the solution y = 0. $NDVI_0$ is subtracted from all NDVI values to shift the distribution to the left, yielding $NDVI_{offset}$ (dotted blue).

benchmark image was established by extraction of raw NDVI values from each of the late season images listed in **Table 1**. Two locations in the MVA, both devoid of vegetation, were chosen for this extraction. A ground view of a location used to measure the reflectance of a bare surface that was used is shown in **Figure 10**.

The value chosen for the standard, $NDVI_{bare\ ground}$, was 0.0357, the NDVI value of bare soil measured on the benchmark 2007 EOS image (**Table 2**). This value was chosen because it was consistent with three other values that were extracted from the seven late season images; the average of the four being a very close value, 0.0355. Three divergent values in **Table 2** were inconsistent and much lower than the four consistent values. These inconsistent values can be explained by NDVI depression from highly variable surface wetting, often occurring for these

Table 2. NDVI bare ground NDVI extracted for each peak cover year. Bolded values have close numerical agreement. Statistics of bolded data are statistically tight while values that deviate from these four have lower magnitude and much greater variability. Coefficient of variation is group standard deviation divided by group average. Images were TM5 except for 2012 that was ETM7+.

Year	Image Date	$NDVI_0$	Bare Ground
2006	10/17/2006	0.0540	0.0260
2007	**10/4/2007**	0.0435	→ **0.0357**
2008	**10/6/2008**	0.0609	**0.0335**
2009	10/9/2009	0.0633	0.0252
2010	10/12/2010	0.0606	0.0127
2011	**9/29/2011**	0.0628	**0.0369**
2012	**10/25/2012**	0.0690	**0.0360**
		Average	0.0294
Bare Ground Overall		St. Dev.	0.0088
		Coeff. of Variation	29.9%
		Bolded Average	0.0355
Bare Ground		Bolded St.Dev.	0.0014
		Coeff. of Variation	4.1%

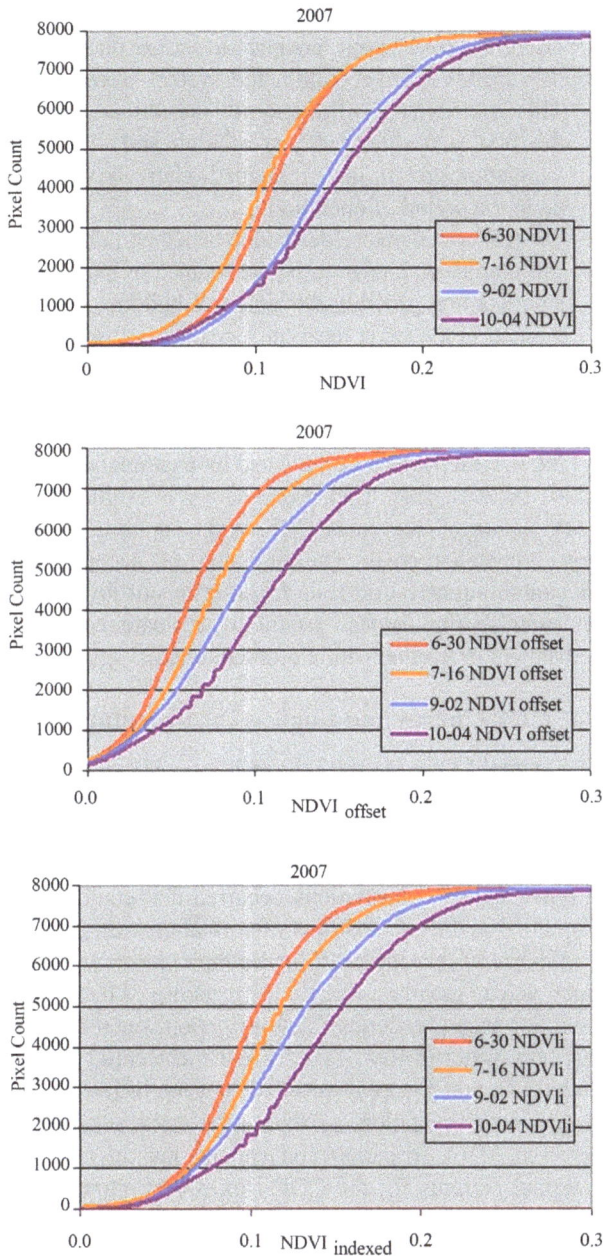

Figure 9. Three graphs of CDFs of NDVI that were generated for 2007, the benchmark year. Raw NDVI calculated from reflectance according to Equation 2 is shown in the top graph. Because of atmospheric effects and background soils, the curves do not increment in the expected order for MVA green cover that increased through summer to reach the annual peak in October. In the middle graph, $NDVI_{offset}$ places the NDVI CDFs into the expected, incremental order peaking in October. Note that the $NDVI_{offset}$ distribution is a first approximation that cuts the tails of these curves. In the lowermost graph, CDF curves are indexed to the 2007 $NDVI_{offset}$ value and the 2007 estimated NDVI value for bare ground to yield $NDVI_x$. The way to read these images is a rightward progression of the curves which denote increased green vegetation cover, while leftward progression denotes a decrease.

Figure 10. A January, 2007 view of a vegetation-free location used in calculation of $NDVI_x$. The bare ground value was obtained two months prior to this photo. Salt enrichment of the surface was due to irrigation and poor drainage.

bare soil targets due to irrigation and poor drainage. Wetting on some of the divergent images in **Table 2** was confirmed using Landsat Band 5, a surrogate measurement for surface wetting that is in use to judge compliance for shallow flood dust control of the Owens Lakebed [14].

Calculation of $NDVI_{offset}$ for each pixel of each image collects the CDFs of all images together relationally as shown in **Figure 9(middle)**. Calculation of $NDVI_x$ then indexes all values to the trusted bare soil value, 0.0357, measured on October 4, 2010 (Equation (3)) that corrects the CDFs of the images by re-shifting all images to the right by the same amount (**Figure 9(bottom)**). This removes truncation resulting from the $NDVI_{offset}$ calculation. For the ith pixel and the jth image:

$$NDVI_{xj} = \left[NDVI_{ij} - NDVI_{0j} \right] + NDVI_{bareground.} \quad (3)$$

The bracketed portion of Equation (3) is the formula for $NDVI_{offset}$. $NDVI_x$ that allows any image from any year to be compared to any other image.

2.5. Resampling to Grid Cells

After calculation of $NDVI_x$ for each image, the last procedure before making comparisons with the benchmark image was to resample the images to conform to the 0.405 ha grid cells. This operation was performed using the cubic convolution algorithm that was modified by an extra step in order to calculate a precise spatial average for the grid cell. Cubic convolution averages all centroid points that are included within a polygon, in this case a grid cell. Each acre contains 4.5 pixels whose centroids may be variously sampled through inclusion of 4, 6 or 9 pixels depending upon how rows and columns of pixel centroids fall into or are excluded from individual grid cells. This yields estimates of $NDVI_x$ that are systematically skewed inducing additional spatial uncertainty into the estimates, especially when comparing across years. This scaling-induced sampling problem was overcome by resampling each pixel into a 10×10 grid of 3-m pixels within each 30m Landsat pixel. This $100\times$ finer grid was then resampled into 0.405 ha cells for a more precise and spatially-correct grid cell average.

3. Results

The data processing workflow for MVA vegetation performance has multiple steps and is relatively complex, so the first evaluation of the results was to confirm the validity of the estimates through the following steps:

1. Determine that the CDF curves incremented logically within each year, lacking any jumps from high to low to high or vice versa. Such jumps indicate unacceptable levels of uncertainty because vegetation in the target environment has not been observed to pulse in

that manner.

2. Using the cover maps, compare to ensure that changes of vegetation cover within the system were logical. This is a spatial confirmation of the observations for the first step since patterns of high and low MVA vegetation growth have remained relatively unchanged in the period of operation.

3. On the cover maps, determine whether poorly performing areas, known to be constrained by poor drainage, shallow groundwater and salt buildup, have remained in the same location. Locations of poor vegetation growth are a more sensitive spatial indication of all factors that cause uncertainty, including geocorrection.

The first step was accomplished by examination of the CDFs for each year. The second step was evaluated as a combination of examining CDFs and examining maps of green vegetation cover. The third step was accomplished by examining acre grid cells for positions of low values, an exercise that is also useful to evaluate cover improvement or decline within problem areas.

3.1. CDF Curves and Logical Incrementation

The annual CDFs for 2007 through 2012 showed that all monthly values incremented in a rational manner, with the lowest values of $NDVI_x$ occurring early in the season generally increasing to a peak in the summer or fall (**Figure 11**). No fluctuations occurred that could not be explained as the ebb and flow of normal growth.

Aspects of management, for example the timing of irrigation, can greatly influence the response of the vegetation. For example, during 2009, irrigation was reduced to test how water-thrifty MVA management could be. Reduction in water supply resulted in very little response for growing vegetation, indicating the direct relationship between MVA irrigation and canopy health. Different seasonal patterns for the CDF curves of **Figure 11** are likely the responses to irrigation amounts, quality and timing.

The 2012 data shown in **Figure 11** were generated with Landsat ETM7+ images that contained missing data stripes that were masked out of the calculations. The total pixel counts of the CDFs for 2012 lacked an average of 5.2% of the potential pixels within the MVA due to missing data stripes and so were scaled upwards to enable direct comparison with the Landsat TM5. The total correct number of pixels was about 7900. Pixel counts may vary according to the number of centroids of pixels included or excluded within the set 0.405 ha grid file of **Figure 6**.

3.2. Maps of Cover for 2007 through 2012

Figure 12 presents acre grid cell maps of the end-of-

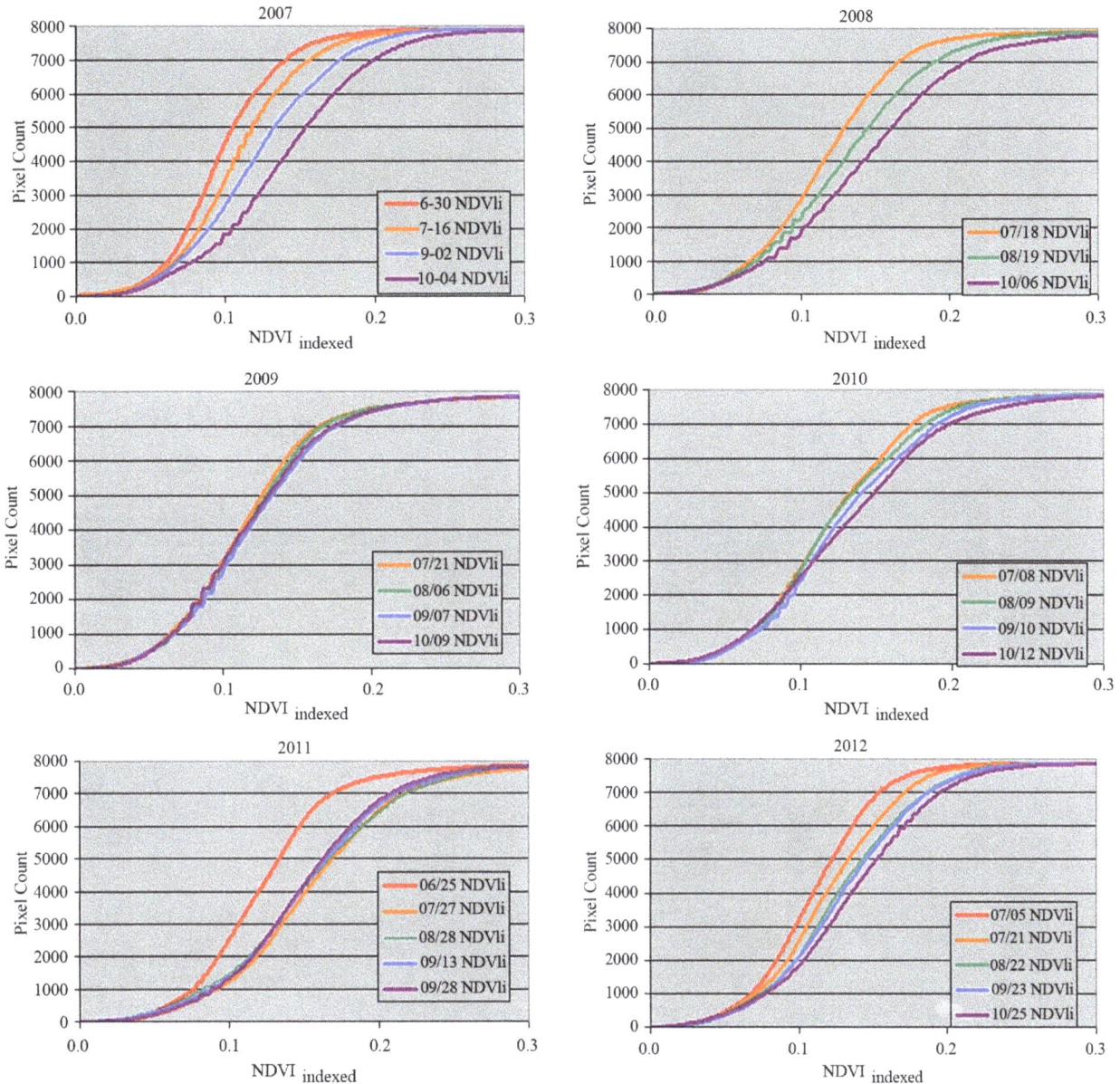

Figure 11. CDFs for 2007, the benchmark year and five years that follow. All images increment reasonably.

season vegetation cover across the MVA for the six years, 2007 through 2012. Examination of these images shows that the spatial vegetation performance follows the relative vegetation performance interpreted through examination of CDFs (**Figures 9** through **12**). Overall, the best performance occurred in 2011 and the poorest occurred in 2009.

Areas of very poor cover exist in the same locations year after year (**Figure 12**). These locations were identified as being caused by poor drainage leading to shallow groundwater that led to near surface salt buildup through capillarity. Given the acre grid cell discretization, these areas can be tracked by each individual acre to plan amelioration to foster more cover. However, since the "this

or better" criterion is tied to 2007 and these areas lacked cover then, nothing different in management would be required.

The consistent spatial distribution of high and low vegetation performance confirms that the data are logical in a spatial sense, fulfilling the second step in the evaluation of the $NDVI_x$ processing.

3.3. Low Vegetation Cover, 2007 through 2012

The third step in the evaluation of the $NDVI_x$ processing methods was used to evaluate the distribution of the poorly performing vegetation cover. These are the areas with the highest potential to give rise to windborne dust within the MVA.

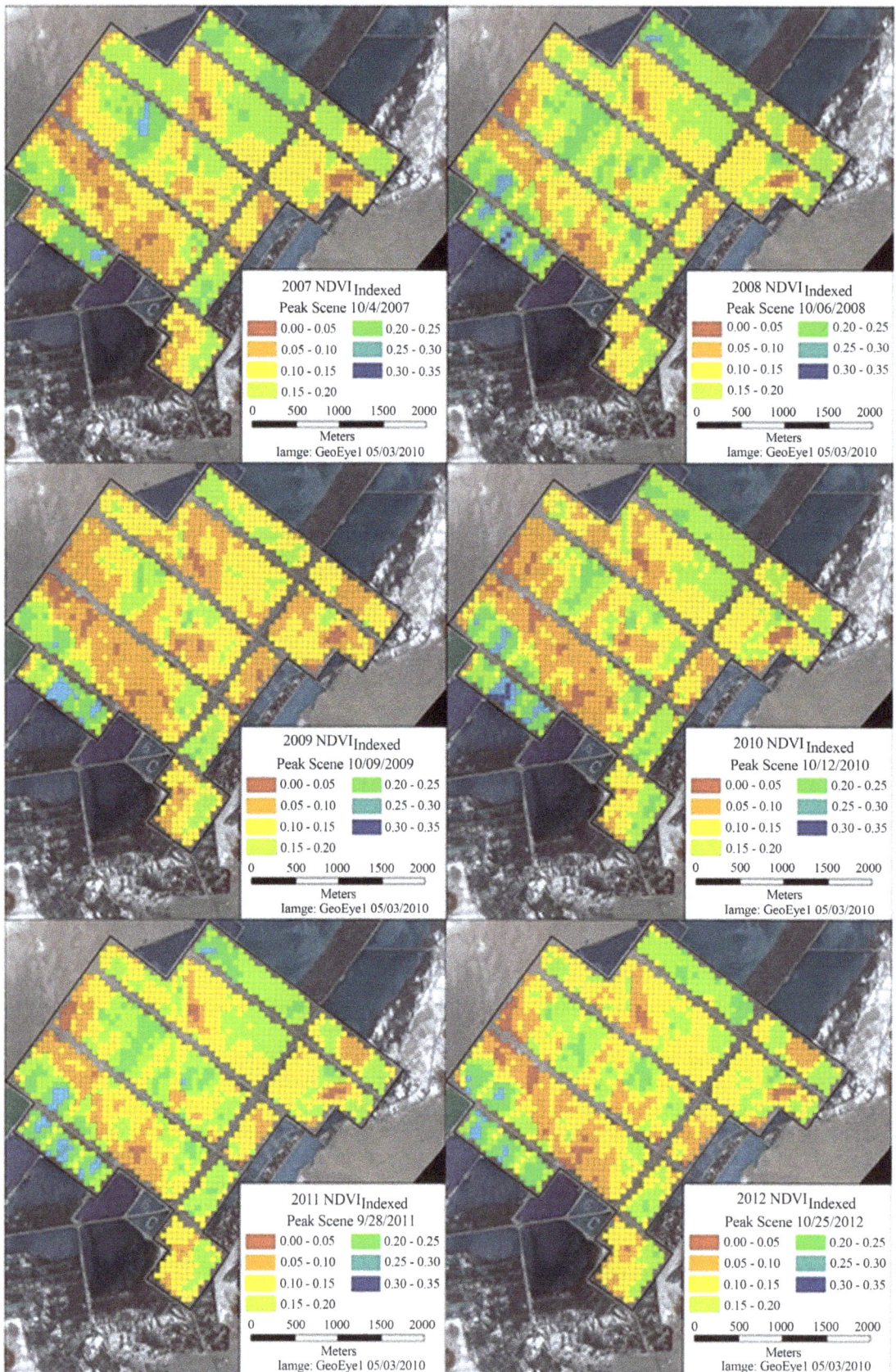

Figure 12. Mapped NDVI$_x$ on the MVA.

Patterns of low cover are due to drainage, shallow groundwater and salinity in static locations. Hence, examining patterns of low cover enables cross checking to determine the spatial accuracy of the $NDVI_x$ analysis. The maps displayed in **Figure 13** show acre grid cells with cover less than 0.10 $NDVI_x$ displayed as transparent red. Locations with low 2007 vegetation cover are displayed as transparent blue and locations for 2007 and each comparison year are displayed as purple.

An examination of the areas in **Figure 13** indicates

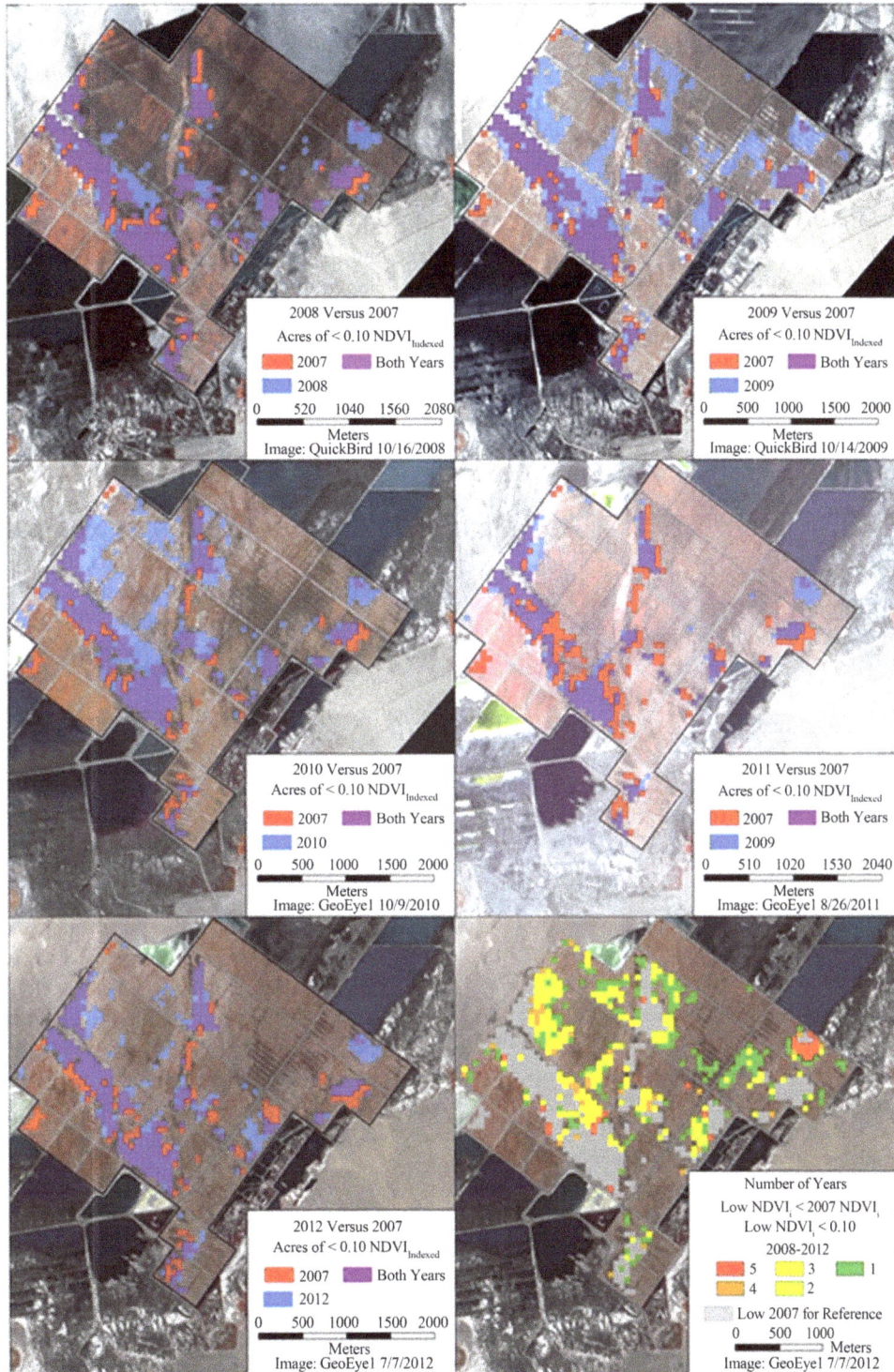

Figure 13. Locations of green growth less than $NDVI_x$ of 0.10 for each year, 2008 to 2012, compared to 2007. The map in the lower right corner is a composite of the five annual comparisons.

that areas of low cover have not changed significantly through the five-year period of this analysis, as expected. This provides spatial confirmation that processing the data to NDVI$_x$ provides accurate and dependable comparisons.

The acre grids of **Figure 13** and especially the threshold level of 0.10 NDVI$_x$, provide relatively coarse comparison relative to the CDFs of **Figure 11** that enable much better discrimination at the important low end of NDVI$_x$ distribution. Red grid cells indicate improvement since 2007, while blue indicates poorer conditions. The spatial expression of improvement in the 2011 and 2012 maps shows that the improving (light gray) gridcells are found on the periphery of areas with low cover in 2007.

Both 2009 and 2010 contain much larger areas of acre grids that are expressed as blue, having less NDVI$_x$ response than in the 2007 benchmark year. The patterning of the blue colored acres, indicating decline from 2007 conditions, are in areas that were repeatedly below this benchmark year. Traces of these patterns are visible in all years.

The map at the lower right hand corner of **Figure 13** is a composite image that shows the number of years that each acre experienced NDVI$_x$ less than 2007 benchmark conditions. There are two general foci of concern for management. An area of concern is located on the far east of the MVA experienced all five years with cover less than 2007. The complex light green to orange gridcells adjacent to the northwest border of the MVA and other areas surrounding the poorest growth zones should be watched for future changes.

3.4. Low Magnitude Vegetation Performance Judged by CDFs

CDFs, alone, provide a means to track relative performance of the MVA, and especially to detect whether vegetation has performed to par. **Figure 14** presents the same CDFs as in **Figure 11** with the lower end of the distribution scaled to facilitate interpretation. Again, it is the lower end of the distribution of the vegetation cover that is important for air quality.

From the CDFs of **Figure 14** it is apparent that there is a great deal of variability among years. Also evident is that the 2007 peak expression of green cover has not always been attained in subsequent years. Of the five comparisons shown, only one exceeded 2007 growth (2011), two were equivalent to 2007 (2008 and 2012), and two failed to reach the 2007 green cover levels (2009 and 2010). Even though there is a range of growth that occurred each year, the lower portions of the six CDFs for 2007 through 2012 illustrate how tight the data are at the lowermost end of the distribution. All curves originnated from the same location near the origin and all in-

cremented comparably to the peak-of-season growth in benchmark 2007 as is expected through the conversion of NDVI$_{offset}$ to NDVI$_x$.

The variability demonstrated in **Figure 14** for attaining peak 2007 green cover levels illustrates that the three factors that can influence MVA saltgrass growth—amount, quality and timing of irrigation—need to be carefully managed in order to achieve growth at par with the 2007 benchmark. Otherwise, there is a 40% chance of failing to meet this goal (two in five years). The three irrigation-related factors mentioned above are the only factors that influence vegetation performance on the MVA since no pests or disease have been identified for the saltgrass, there is no grazing pressure, and the drainage and irrigation infrastructure have remained relatively unchanged through the five year period.

In **Figure 14**, it is apparent that the low MVA performance in 2010 may have been constrained by even lower performance in 2009. Canopy growth in 2010 may have still been impacted by the decreases that occurred the previous year, for example runners of saltgrass growth that failed and needed to regrow from a base that survived irrigation shortfalls. Another hypothesis is that the slow plant growth may have been due to irrigation practices that were extended into 2010. Combining irrigation records with these data offers the opportunity to identify MVA irrigation practices that should be avoided, and those that should be continued. The five years of vegetation growth offer the means to adjust irrigation quantity, quality and timing to achieve the green cover that existed in 2007 while conserving against an over supply that would waste water. Water conservation is an important goal in the Owens Lake second only to dust control.

Using the CDFs for evaluating compliance, reaching the 2007 par levels of green growth, is a simple and accurate way to evaluate the performance over the entire MVA. We know from **Figures 12** and **13** that the patterns of high and low vegetation performance did not change markedly through the seven years and so, the CDFs can be expected to represent nearly the same grid cells, relationally, throughout the history of the MVA. In other words, at each given level of grid cell count, the CDFs compare about the same grid cells throughout the seven years. In these comparisons, 2012 appears to be optimal.

3.5. NDVI$_X$ Precision

CDF curves for 2009 were chosen as a means to evaluate the precision of the NDVI$_x$ processing. Because 2009 recorded very poor growth due to low water supply, the lowermost limb of the curves are measuring vegetation growth that essentially remains static through the grow-

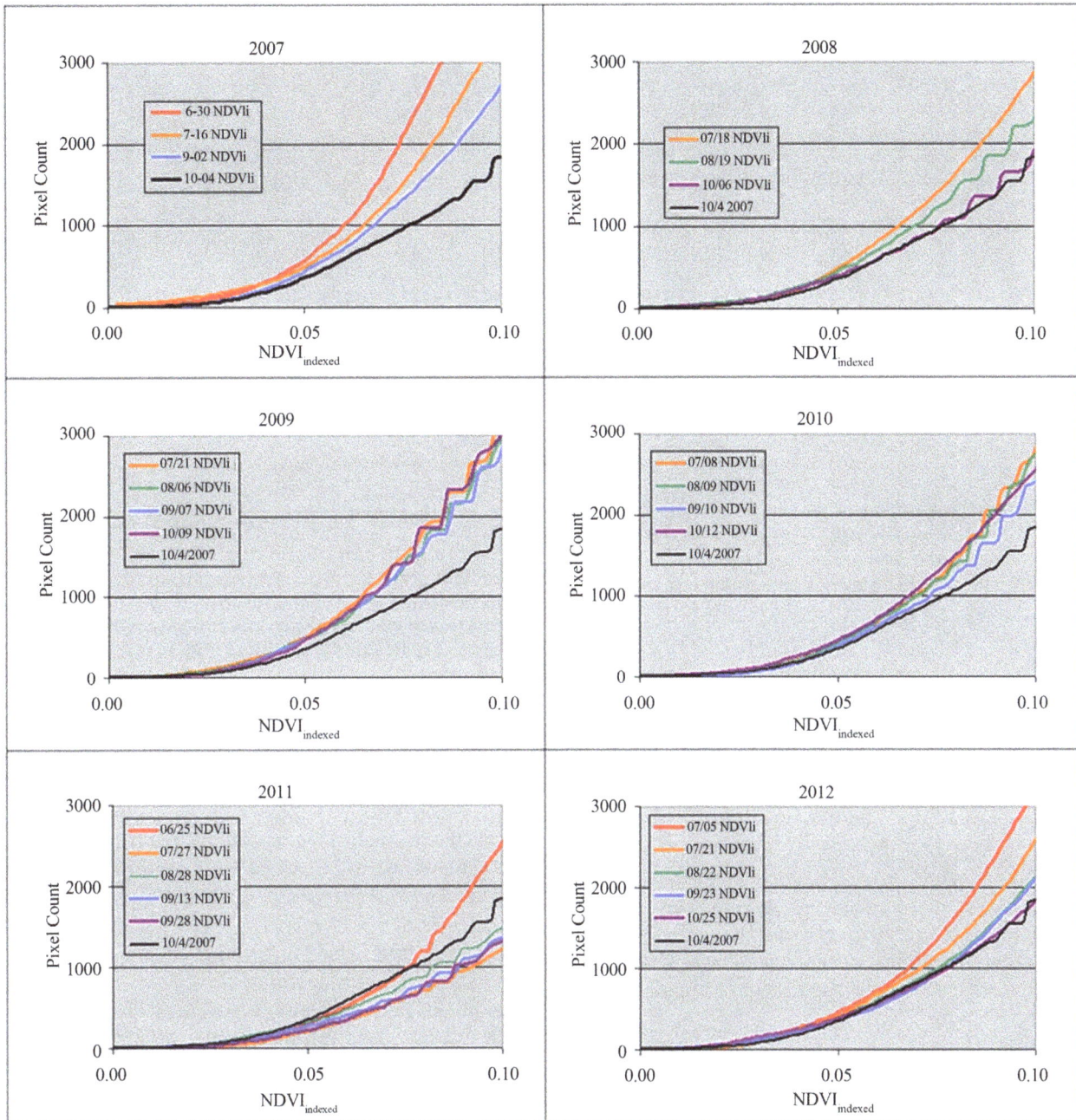

Figure 14. The lower portion of CDF curves for 2009, a year with stringent irrigation supply and quasi static growth response. Values for each curve were extracted in the locations indicated by the arrows.

ing season. Any apparent fluctuations in values can, therefore, be used to make a minimal estimate of precision—minimal, insofar as any differences may actually be the result of plant growth rather than statistical uncertainty. The graph shown in **Figure 15** presents the same data for 2009 as shown in **Figures 12** and **13** but with the lower limb enhanced to examine the variability of the curves. Four lines of equal pixel counts are indicated for 300, 400, 500, and 600 pixels in the region where the curves approach linearity. By assuming that these data

represent no change in cover but simply the fluctuations resulting from uncertainty as the limit of precision for EOS data and $NDVI_x$ calculation is approached. An assessment of the variability at each of these four pixel count values provides an estimation of precision.

Table 3 calculates signal to noise ratio (SNR) for the data extracted at four pixel counts indicated in **Figure 15**. The measure of signal to noise is taken to be the mean as a representation of the signal, divided by the standard deviation, representing noise [15]. These data show that

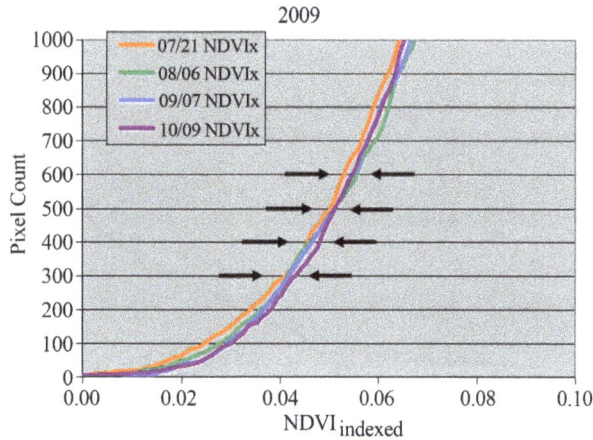

Figure 15. All CDFs displayed for the NDVI$_x$ region below 0.10. The black line in all graphs is the peak-of-season CDF from October 4, 2007.

the calculation of NDVI$_x$ has high precision with the average SNR of 64.2, but consistently at a level above 50. Again, since some of the differences in the CDF curves of 2009 (**Figure 15**) may be actual fluctuations, these values represent a minimal estimate of the SNR.

3.6. NDVI$_X$, MVA Compliance, and Their Relationship to Air Quality

Deviating below the benchmark line of the 2007 peak season CDF has documentable consequences for air quality. **Figure 16** presents an image that mates the 2010 dust camera results with low vegetation cover (0.075 NDVI$_x$ and below) detected in 2010. The dust camera program, operated by the District to identify blowing areas, has permanent digital video cameras mounted around the lake to identify blowing areas. The dust camera output transforms camera coordinate system into world coordinate system based upon a program developed by Corripio [16].

The threshold of 0.075 NDVI$_x$ was chosen through examination of the lower region of the CDFs shown in **Figure 14** that experienced declines in NDVI$_x$ during 2010 from 2007 benchmark conditions. There is a moderately strong relationship between low vegetation vigor and the incidence of blowing dust. Although there are discrepancies in the locations shown by the dust camera mapping, it should be recognized that such dust source identification methods are only relatively accurate—actual field confirmation of evidence for dust release should also be gathered. The point made here is that documented areas of low cover have good potential to release fugitive dust.

The accuracy of the dust camera program suffers in some locations by lacking multiple views of the same dust source locations. For example, the red areas of high incidence for blowing dust in **Figure 16** are closer to the

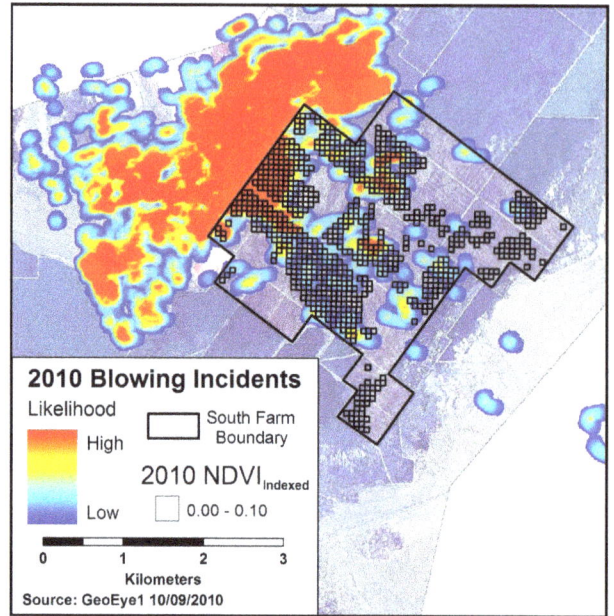

Figure 16. 2010 dust camera data for likelihood of dust release that are color coded for comparison to 0.405 ha grid cells identified to have low cover in 2010—north is to the top. Areas downwind (southeast) of highly active areas may be undercounted due to screening from blowing dust during windstorms.

camera position located to the northwest. Heavy releases of dust in the foreground of the dust camera images may, therefore, occlude active sources in the background during a windstorm.

There are mapped regions of dust release in 2010 that do not coincide with mapped areas of low vegetation in **Figure 16**. This may indicate that dust can be released from areas of thick vegetation cover. Such spot releases may result from wind vortices that entrain dust and salt crystals deposited on leaves of the vegetation. Salt crystal buildup on saltgrass leaves occurs naturally from excretion through glands on their upper surfaces. Hence, although vegetation is effective in reducing fugitive dust, it may contribute dust during high winds depending on the degree of buildup of dust and salt crystals in the canopy. Such mapped areas may generally have low likelihood for emitting dust but may do so under special circumstances.

Dust emission from the MVA may contain a large component of re-entrained particles. This is borne out by the worst areas for dust release being the locations along the northwest edge of the MVA that received significant particle deposition from northerly winds during 2010. **Figure 16** shows areas of high activity dust release as red. The largest of these areas arose from wind entrainment to the northwest of the MVA that translocated up to a meter depth of soil onto the MVA where it partially buried vegetation and continued to be an active dust source during the 2010-2011 dust season.

3.7. Proposed Monitoring Program for Vegetation Compliance

At the time of this paper, the monitoring program is undergoing review for a potential upgrade to accommodate the results from this research. A logical step for monitoring using $NDVI_x$ would have the following steps:

1. Processing as described in this paper to arrive at $NDVI_x$ within the set acre grid cells and after removal of pixels potentially influenced by edge affects.
2. Comparison by subtraction from the October 10, 2007 acre-wise $NDVI_x$ to determine whether each acre grid cell has increased or decreased.
3. Extraction of acre grid $NDVI_x$ and display as CDF for comparison to 2007 conditions
4. Display $NDVI_x$ for the year in question and $NDVI_x$ for 2007 (as in **Figure 11**). A curve to the left of the 2007 curve falls short of par.
5. Display the $NDVI_x$ difference map (as in **Figure 12**) to check for changes that may form patterns of concern.
6. Repeat steps 4 and 5 for $NDVI_x$ of less than 0.1 (as in **Figures 13** and **14**). Such locations of low cover are more critical for air quality.

4. Conclusions

$NDVI_x$ is proposed as the standard for assessing vegetation on the MVA because it is a highly accurate metric of vegetation performance that was shown to be superior to other methods tested in this work and in other comparisons. Because $NDVI_x$ is an index of green vegetation cover, it must be applied during the growing season. To identify and measure peak green vegetation cover, multiple images obtained during the growing season can be evaluated in order to catch the true peak, especially if this occurs at a different time than have typically occurred at the MVA—at season end. Because satellite data are archived, this reconstruction can take place after the growing season has ended. Verification that CDF curves increment logically provides a quality assurance step for the data processing.

Assessment of MVA vegetation performance using $NDVI_x$ requires the following steps for each image to be compared:

1. Select and acquire appropriate EOS imagery
2. Geocorrect imagery by image to image registration with a standard base image
3. Display the image in 432 RGB to examine for haze or other visible aerosols
4. Convert the image to reflectance
5. Calculate NDVI
6. Display NDVI as a CDF

7. Calculate $NDVI_0$ for the image using a set regression interval
8. Subtract $NDVI_0$ from all NDVI values to yield $NDVI_{offset}$
9. Add 0.0357, the benchmark image's bare soil value, to index the images
10. Screen out pixels that may contain non-vegetated surfaces of roadways and areas outside of the MVA boundary.
11. Resample each pixel using cubic convolution to a higher resolutions—a 10×10 resampling was performed for this paper
12. Resample the 100x pixels into 0.45 ha grid cells using cubic convolution

The results from these measurements indicated poor vegetation response in 2009 that carried over into 2010. MVA performance in both years failed to equal or better the benchmark year of 2007. Carryover of lower cover vegetation into 2010 may have played a role in 2010 dust releases documented to have been generated in the MVA. Comparison of irrigation records from good and poor vegetation growth years should be performed to assure that the vegetation cover in all future years remains at or above the 2007 levels.

5. Acknowledgements

The authors thank the Great Basin Unified Air Pollution Control District and staff for supporting this work, especially Ted Schade, Air Pollution Control Officer and Grace McCarley-Holder, Geologist.

REFERENCES

[1] C. A. Pope, D. W. Dockery, J. D. Spengler and M. E. Raizenne, "Respiratory Health and PM10 Pollution: A Daily Time Series Analysis," *The American Review of Respiratory Disease*, Vol. 144, No. 3, 1987, pp. 668-674.

[2] W. L. Kahrl, "Water and Power," University of California Press, Berkeley, 1982.

[3] H. S. Gale, "Salines in the Owens, Searles and Panamint Basins, Southeastern California," US Geological Survey Bulletin 580, 1915, pp. 251-323.

[4] A. S. Jayko and S. N. Bacon, "Late Quaternary MIS 6-8 Shoreline Features of Pluvial Owens Lake, Owens Valley, Eastern California," In: M. C. Reheis, R. Hershler and D. M. Miller, Eds., *Late Cenozoic Drainage History of the Southwestern Great Basin and Lower Colorado River Region: Geologic and Biotic Perspectives*, GSA Special Papers 439, 2008, pp. 185-206.

[5] P. Saint-Amand, C. Gaines and D. Saint-Amand, "Owens Lake, an Ionic Soap Opera Staged on a Natric Playa. Centennial Field Guide," *Cordilleran Section of the Geol. Soc.Am.*, Vol. 1, 1987, pp. 145-150.

[6] Great Basin Unitified Air Pollution Control District, "2003 Owens Valley PM_{10} Planning Area Demonstration

of Attainment State Implementation Plan," 2013. http://www.gbuapcd.org/Air%20Quality%20Plans/2008S IPfinal/2008%20SIP%20-%20FINAL.pdf

[7]   Soil and Water West Incorporated, "Owens Lake Bed Soil Survey," 2013. ftp://gbuapcd.org/HydroReports/Soil%20Survey%202000 .pdf

[8]   California Irrigation Management Information System, "Monitoring Data from Owens Lake," 2013. http://wwwcimis.water.ca.gov/cimis/welcome.jsp

[9]   D. W. Goodall, "Some Considerations in the Use of Point Quadrats for the Analysis of Vegetation," *Australian Journal of Biological Sciences*, Vol. 5, No. 1, 1952, pp. 1-41.

[10]   W. M. Baugh and D. P. Groeneveld, "Broadband Vegetation Index Performance Evaluated for a Low-Cover Environment," *International Journal of Remote Sensing*, Vol. 27, No. 21, 2006, pp. 4715-4730.

[11]   D. P. Groeneveld and W. M. Baugh, "Correcting Satellite Data to Detect Vegetation Signal for Eco-Hydrologic Analyses," *Journal of Hydrology*, Vol. 344, No. 1, 2007, pp. 135-145.

[12]   D. P. Groeneveld, "Remotely-Sensed Groundwater Evapotranspiration from Alkali Scrub Affected by Declining Water Tables," *Journal of Hydrology*, Vol. 358, No. 2-3, 2008, pp. 294-303.

[13]   R. Irish, "Landsat 7 Science Data Users Handbook," 2013. http://landsathandbook.gsfc.nasa.gov/data_prod/prog_sect 11_3.html

[14]   D. P. Groeneveld, R. P. Watson, D. D. Barz, J. B. Silverman and W. M. Baugh, "Assessment of Two Methods to Monitor Wetness to Control Dust Emissions, Owens Lake, California," *International Journal of Remote Sensing*, Vol. 31, No. 11, 2010, pp. 3019-3035.

[15]   D. C. Montgomery, "Design and Analysis of Experiments," 4th Edition, Wiley, New York, 1997.

[16]   J. P. Corripio, "Snow Surface Albedo Estimation Using Terrestrial Photography," *International Journal of Remote Sensing*, Vol. 25, No. 24, 2004, pp. 5705-5729.

17

Variation of Hyporheic Temperature Profiles in a Low Gradient Third-Order Agricultural Stream—A Statistical Approach

Vanessa Beach, Eric W. Peterson

Department of Geography-Geology, Illinois State University, Normal, USA.

ABSTRACT

Sediment size governs advection, controlling the hydraulic conductivity of the stratum, and conduction, influencing the amount of surface area in contact between the sediment particles. To understand the role of sediment particle size on thermal profiles within the hyporheic zone, a statistical approach, involving general summary statistics and time series cross-correlation, was employed. Data were collected along two riffles: Site 1: gravel (d_{50} = 3.9 mm) and Site 2: sand (d_{50} = 0.94 mm).Temperature probe grids collected 15-minute temperature data at 30, 60, 90, and 140 cm below the streambed surface over a 6-month period. Surface water and air temperature were recorded. Diel temperature signal penetration depth was limited to the upper 30 cm of the streambed and was driven by advection. Surface seasonal trends were detected at greater depths, indicating that thermal pulses are transmitted initially by advection and by conduction to areas deeper in the hyporheic zone. Site 1 showed a high degree of thermal heterogeneity via a localized downwelling zone within a gaining stream environment. Site 2 exhibited a vertically and horizontally homogenized thermal environment attributed to an increased amount of sand sediments that limited advection and significant groundwater discharge that mediated the effects of downwelling surface water.

Keywords: Hyporheic Zone; Temperature; Time Series Analysis; Cross-Correlation

1. Introduction

Temperature is a basic parameter that controls physical, ecological, and biogeochemical activities in aquatic systems [1-3]. Water temperature studies have had significant impacts on our knowledge of hydrogeology. Evaluation of streambed temperature profiles has been used to quantify groundwater/stream interactions [4], delineate flow paths in the hyporheic zone (HZ) [5], and assist in the evaluation of factors that generate change within thermal profiles [6]. The thermal regime of the HZ controls organic matter decomposition, fish egg incubation, and invertebrate diapauses [7,8].

The HZ is the area below the stream channel where surface and groundwater mix [5,9]. HZ temperatures are controlled by the mixing of groundwater and surface water, reflecting the rates of infiltrating surface water and upwelling groundwater [10], disregarding geothermal influences, and surface water temperatures show both diel and seasonal fluctuations [11-13]. Differences between surface water temperature and subsurface temperature are a function of diel temperature cycles [14]. Dogwiler and Wicks [15] show that with increasing depth and/or distance from infiltration sites, the diel and seasonal fluctuations of surface water become attenuated and lagged. These patterns can be a valuable tool in defining HZ depth and extent [13,16-18]. However, delineations of HZ extent are not constant through time, as shown by Fraser and Williams [19], whose results suggest that the extent of the HZ varies seasonally as well as with event-based fluctuations [20].

While HZ temperatures are dominantly controlled by advection, conduction can also play a significant role [21, 22]. The influence of both advection and conduction on hyporheic water temperatures suggests that sediment particle size can impact the effectiveness of both by 1) partially defining the hydraulic conductivity of the stratum, and effectively constraining advection; and 2) controlling the amount of surface in contact among the sediment particles, thereby limiting conduction. Vaux [23] and Cooper [24] suggest that larger objects in or on the streambed surface respectively alter the flow paths of hyporheic and stream waters. With respect to finer sediments, Ringler and Hall [25] showed that the largest gradients between stream and hyporheic water temperatures occur at heavily silted sites, where slow flows persist. Additionally, variations in hydraulic conductivity of the streambed may result in uneven discharge and flow geometry [26].

This study focuses on variations in temperature profiles of the HZ at two sites: a gravel dominated HZ and a sand dominated HZ, with the hope of furthering existing knowledge of water temperature in the environment and providing another tool for characterizing HZs. The use of time-series analysis allows the identification of data trends otherwise concealed. A similar statistical based approach taken by Malard *et al.* [6] successfully assessed temperature patterns within a glacial floodplain system. Specific interests lie in transmission of both seasonal and diel surface temperature signals into the subsurface, the comparison of lateral and longitudinal temperature profiles, and the possibility of quantitatively delineating the HZ using temperature data.

2. Study Site

Field investigations focused on two sites along a stretch of the Little Kickapoo Creek (LKC) running through the Illinois State University Randolph Well Field (**Figure 1**),

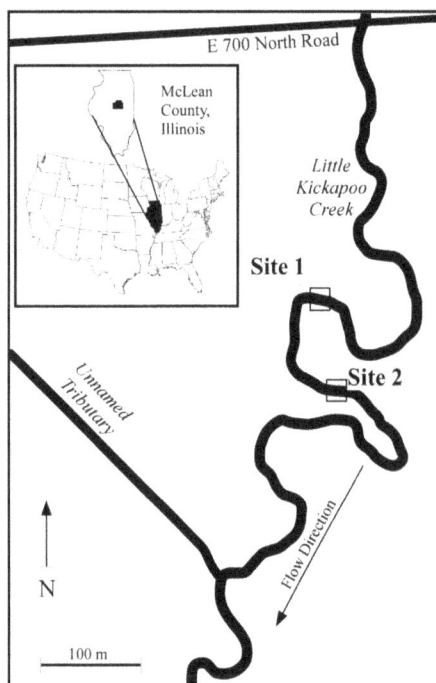

Figure 1. Location of the two study sites within little kickapoo creek inset shows the location within the USA.

located in McLean County, central Illinois, USA. Central Illinois has a temperate climate, with cold, snowy winters and hot, wet summers. Mean annual air temperature for the period from 1950 to 2002 was 11.2°C [12].

The site has been described by Peterson and Sickbert [12] and presented here are the relevant data.Originating in an urban area approximately 11 km north of the study site, LKC is a low gradient third-order perennial stream that meanders (sinuosity of 1.8) through Wisconsinan glacial plains. Regionally LKC is a gaining stream, with a gradient of 0.002. Locally, the meander containing the two study sites has a gradient of 0.003. The reach under investigation is unmodified and meanders through an approximately 300 m wide alluvial valley. Land bordering the stream is predominantly used for agriculture.

Three geologic units comprise the alluvial valley through which LKC meanders: the Wedron Formation, the Henry Formation, and the Cahokia Formation (listed from oldest to youngest). Being a clay-rich low-permeability till, the Wedron Formation acts as a lower confining unit to the Henry Formation. Within the outwash valley, the Henry Formation functions as an aquifer due to its poorly sorted gravels and sands, having an average hydraulic conductivity of 10 m/day and an average thickness of 5 - 7 m. Above the Henry Formation lies the Cahokia Formation, consisting of fine-grained sand and mud, with a thickness of up to 2 m. The LKC channel is inset into the Cahokia Formation, cutting into the top of the Henry Formation. LKC streambed sediments are composed of mostly Henry Formation materials, consisting primarily of gravel and coarse sand with interstitial silt. Surface sediments vary with distance along the channel.

Both sites are located in riffle sections of the stream channel. Site 1 is the further upstream site, featuring predominantly gravel, greater than 2 mm, while Site 2 lies further downstream with predominantly sand size sediments (0.0625 mm to 2 mm).Particles larger than 2 mm comprise 61%, sand accounts for 36% and silts and clays are 3% of the material at Site 1. Overall the median particles size (d50) at Site 1 is 3.9 mm. At Site 2, sand size particles are dominant, comprising 62% of the material. Gravel account for 36% and 2% are silts and clays, resulting in a median particles size (d50) of 0.94 mm at Site 2. Based upon grain size, the hydraulic conductivity (K) at Site 1 is 4.60 cm/s and at Site 2 is 0.02 cm/s [12].

3. Methodology

3.1. Temperature Measurements

Identical temperature probe grids were set up along riffles at two LKC sites. Each grid consisted of five vertical logger nests (referred to as wells) creating both lateral and longitudinal profile lines across the channel. The two profile lines intersected roughly in the stream's thalweg, where one nest provided data for both profiles (**Figure 2(a)**). Within each 6.35 cm PVC well, temperature loggers were positioned at depths of 30 cm, 60 cm, 90 cm, and 140 cm (**Figure 2(b)**). To partition off the different depths and to reduce vertical mixing, foam sealant was used and the wells were capped. At each depth, two 12.7 mm diameter holes drilled into the walls at each depth provided connection to the matrix. Two additional temperature loggers recorded surface water temperatures. HOBO® StoyAwayTidbiT Temperature Loggers with an accuracy of ±0.2°C and a resolution of 0.16°C at 20°C were used in this work. All loggers were programmed to record temperatures at 15-minute intervals. Data collection started on the June 30, 2007 and ended on the December 10, 2007, when all loggers were removed from the substrate. A complementing study examined the amount of scour and fill at each location during the study period. No scour and fill event greater than 10 cm occurred during the period of monitoring.

Additional data collection included stream stage and air temperature. The stream stage was recorded at a permanent stilling-well located 20 m upstream of Site 1. Air temperature was obtained from a weather station 220 m away. Both stream stage and air temperature were recorded on a 15-minute interval.

During the data collection period, several unforeseen problems were encountered. Temperature loggers located at 1A-90 cm, 1E-90 cm, 2B-90 cm, and 2D-90 cm failed completely. Furthermore, due to extensive beaver dam construction upstream of both sites, stream flow intermittently became unmeasureable from approximately August 2, 2007 to October 1, 2007, resulting in low flow conditions at both Site 1 and Site 2. The temperature effects of this can be seen in **Figure 3**. Initially, Site 2 surface stream temperatures closely mimic Site 1 surface stream temperatures. However, near the beginning of August, Site 2 surface stream temperatures show an increase in diel amplitude, approximating the variability of daily air temperatures. Additionally, surface stream temperatures at Site 2 are warmer than at Site 1, beginning near October 1, 2007. This temperature difference is likely due to a greater insolation at Site 2 once trees begin to lose their foliage.

3.2. Statistical Methods

For all statistical calculations, 15-minute (n = 15711) or hourly (n = 3904) temperature values from June 30, 2007 to December 10, 2007 were used. Statistical analyses were conducted using SPSS version 16.0 [27].

Using 15-minute data, box plots were created for both the summer (June 21, 2007 to September 23, 2007) and autumn (September 24, 2007 to December 22, 2007) seasons (defined by the use of equinoxes and solstices, which coincided with the temperature reversal), although data for both periods are incomplete. Summer collection started late on June 30, 2007 while autumn collection ended early on December 10, 2007 due to a stream log-

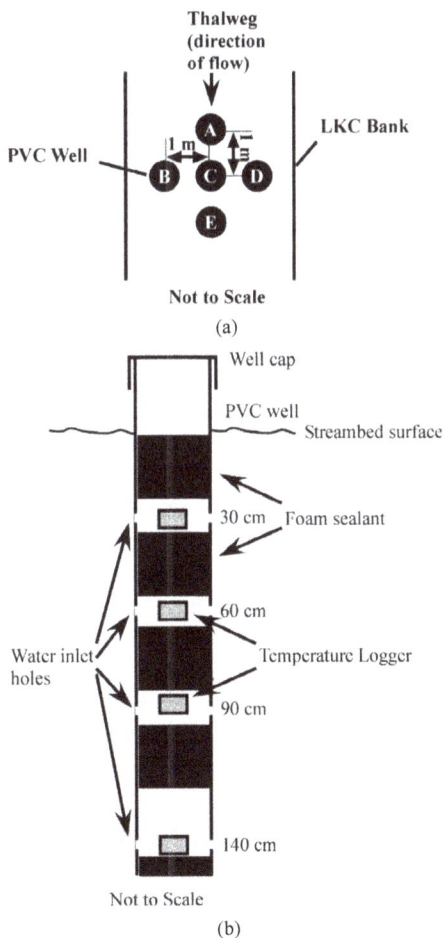

Figure 2. (a) Birds-eye view of well setup in the stream channel; (b) Detailed view of individual well design.

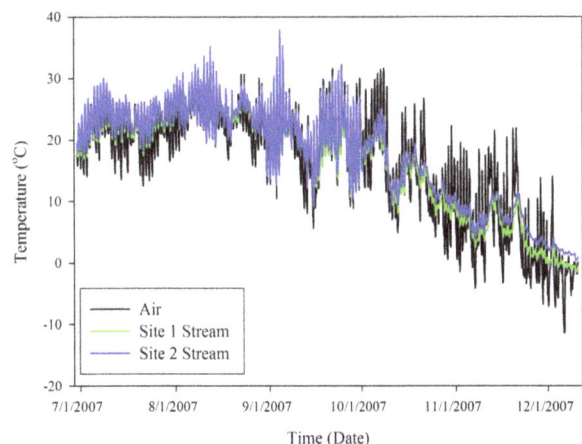

Figure 3. Air and stream (Site 1 and Site 2) temperature 15-minute incrementing time series for entire data collection period.

ger failure. The temperature reversal (an isothermal period during which air temperatures change from warm to cool) occurring in early autumn, requires the separation into seasons for unbiased summary statistics, and though neither season is fully complete, the separation into seasons gives a more illustrative overview of temperatures than a grouped approach.

Time series cross-correlation, as described by Mangin [20], was used to understand the relationships between streambed temperatures within each site, as well as between sites in more detail. Time series cross-correlation measures the relationship between two quantitative time series, i.e. surface water temperature compared to hyporheic water temperature. The observations of two series are correlated as various lags and leads, where the relationship is expressed by a cross-correlation coefficient (r) equal to a value between -1 and 1, with values closest to 1 indicating the strongest relationship between the time series.

The cross correlation coefficient (r) was obtained using the formula proposed by Jenkins and Watts [28] and as used by Malard et al. [6] to analyze streambed time series temperature data:

$$r_{+k} = \frac{C_{x,y(k)}}{S_x S_y}$$

With

$$C_{x,y(k)} = n^{-1} \sum_{i=1}^{n-k} \left(x_i - \overline{x} \right)\left(y_{i+k} - \overline{y} \right) \quad (1)$$

$$r_{-k} = \frac{C_{x,y(k)}}{S_x S_y}$$

With

$$C_{x,y(k)} = n^{-1} \sum_{i=1}^{n-k} \left(y_i - \overline{y} \right)\left(x_{i+k} - \overline{x} \right) \quad (2)$$

with x_1, x_2, $\cdots$, x_n = hourly values of surface water temperature or temperatures, at shallower depths; y_1, y_2, $\cdots$, y_n = hourly values of hyporheic water temperature or temperatures at deeper depths; $k = 0$, 1, 2, $\cdots$, m where k is equal to the lag, and m is equal to the maximum number of lags. $\overline{x}$ and $\overline{y}$ are means and S_x and S_y are the means and the standard deviations of the respective x and y series.

For the evaluation of cross-correlation, the dataset was reduced to hourly data to decrease the number of data and to reduce the possibility of over-fitting the statistical model. For the comparison of seasonal trends of both surface water and hyporheic water temperatures a 24-hour moving filter was applied to hourly data prior to cross-correlation, removing diel temperature fluctuations. Each filtered temperature at time t equaled the average temperature from 12 hours prior to and 12 hours after time t (including the temperature at time t in the ave-

raging).

For computation of between-site comparisons, gradients (the difference between surface water temperatures and temperatures at 140 cm depth) were used for cross-correlation in substitution of actual recorded temperatures. This eliminated the influence of differing surface stream temperatures, and allowed instead a comparison of the degree of temperature change with depth between sites.

First-order differencing was applied to all time series prior to cross-correlation, removing the data's temporal trend component and reducing autocorrelation. First-order differencing is achieved by subtracting from each term of the original series the preceding term. The transformation generates a new series defined by: $\hat{X}_t = (X_t - X_{t-1})$ where $\hat{X}_t$ = term of the filtered time series, and X_t = term of the original time series. All cross-correlations were computed using a lag (k) of 1 hr, and a maximum number of lags (m) of 125 determined so that $m \times k$ is less than or equal to n/3 as recommended in the literature [20].

For the evaluation of cross-correlation results, correlation coefficients (r) equal to or greater than 0.2 were treated as statistically significant. This was determined based on the number of observations used, and assuming rejection of the null hypothesis (there is no difference) if a > 0.01.

4. Results

4.1. Summary Statistics

A distinct difference in temperature patterns is seen when comparing summer and autumn results (**Figure 4**). In summer, mean streambed temperatures fall at or below mean surface stream temperatures, and pronounced cooling is witnessed with depth into the streambed at both sites. In autumn, these patterns are reversed with mean surface stream temperatures at or below mean streambed temperatures. A slight warming trend is also observed in mean streambed temperatures with depth. Additionally, temperatures appear more homogenized top to bottom, where mean temperatures at increasing depths are not distinctly different. It can be projected that the degree of difference of autumn to summer temperature patterns would increase in winter, and decrease again in spring with the next reversal. Irrespective of the differences observed between summer and autumn temperatures, a decrease in temperature ranges with streambed depth is experienced universally to varying degrees. In general, the observations above show the data from this study to be in line with general patterns witnessed in other HZ temperature studies, such as by Dogwiler and Wicks [15], in a karst environment featuring similar stream sediments as at Site 1, and by White et al. [16] in a Michigan river.

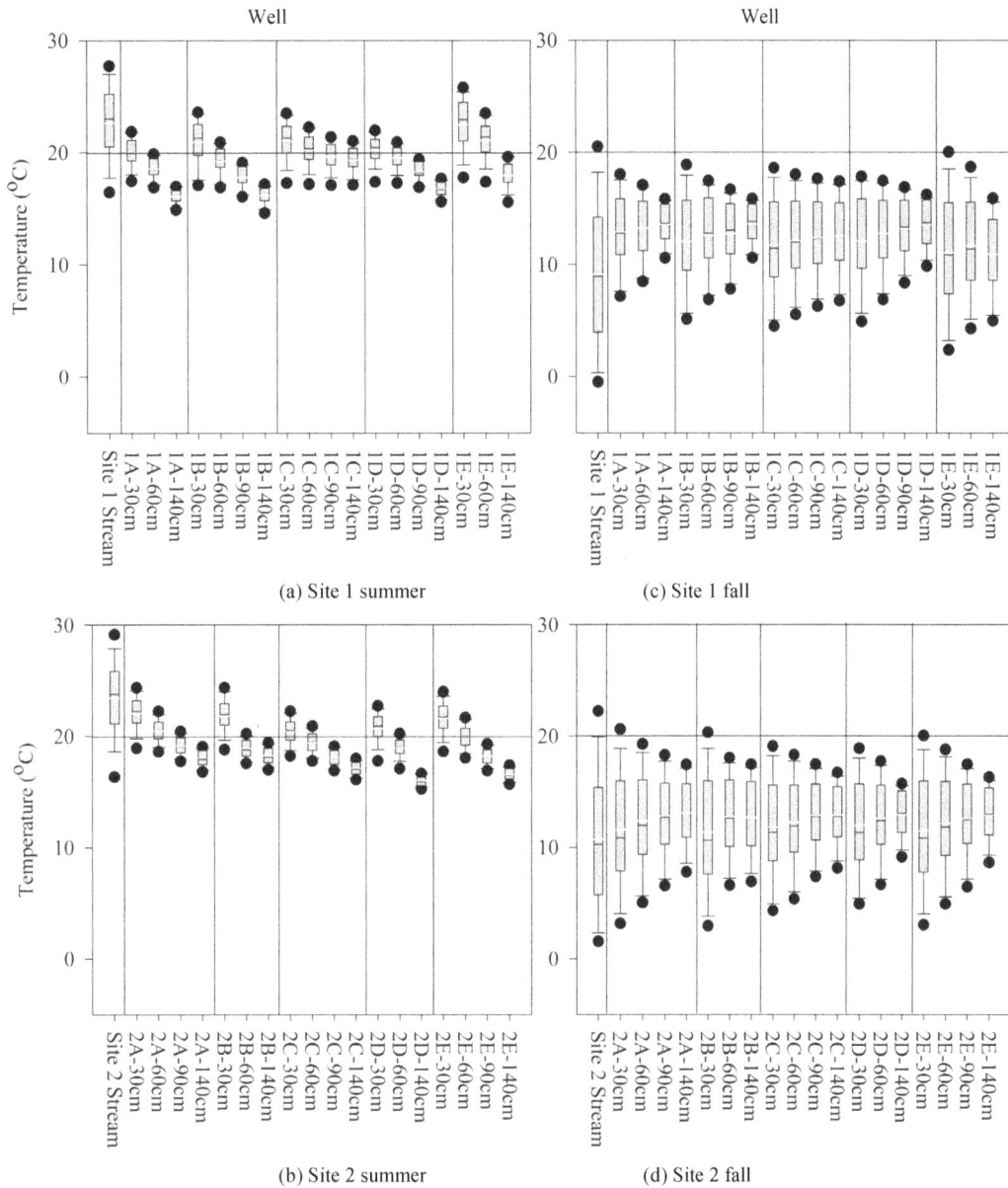

Figure 4. Box plots of temperature data with reference lines at 20°C. The edges of the boxes represent the 25th and 75th percentiles with the black line at the median and the white line at the mean; the whisker bars depict the 10th and 90th percentiles and the dots represent the 5th and 95th percentiles. (a) Site 1 summer (June 21, 2007 to September 23, 2007); (b) Site 2 summer (June 21, 2007 to September 23, 2007); (c) Site 1 autumn (September 24, 2007 to December 22, 2007); (d) Site 2 autumn (September 24, 2007 to December 22, 2007).

A site comparison of summer box plots reveals more uniform temperature decreases with increasing streambed depth in each well at Site 2. At Site 1, wells 1C and 1E have greater temperature ranges persisting at depth, suggesting that wells 1C and 1E maintain effective temperature transmission at depth. Additionally, Site 2 surface stream temperatures vary over a wider temperature range than at Site 1 (t (3903) = −67.98, $p < 0.01$), experiencing more days when temperatures are warmer, (up to a maximum temperature of 38°C). Interestingly however, Site 2 streambed temperatures do not noticeably reflect this

increased temperature range.

A site comparison of autumn box plots reinforces summer box plot observations. Temperatures in wells 1C and 1E again maintain larger temperature ranges at depth than do other wells at Site 1 (t (1851) = −92.84, $p < 0.01$). Surface stream temperatures at Site 2 again experience warmer temperatures, presumably due to the remainder of the low-flow period as well as to generally warmer temperatures in late autumn due to increased insolation. Surprisingly, Site 2 wells experience smaller temperature ranges than do equivalent Site 1 wells, suggesting slower transmission of surface temperatures into the streambed.

4.2. Seasonal Cross-Correlation

Results of the 24-hour averaging filter applied to well 1E and 2E (**Figure 5**) are representative of filter applications to all other wells. The greatest impact is on time series that feature strong diel components, such as surface stream temperatures. Temperatures at depth within the streambed were only mildly affected by the filter, due to their already dampened diel signals. Both a seasonal trend and short-term, 1 to 3 day, thermal fluctuations are observed in the filtered time series, closely matching the findings of Malard *et al.* [6].

All streambed temperatures show significant correlation to the seasonal trends in stream water (**Figure 6**). As expected, correlations between temperatures at 30 cm depth and stream water are highest within each well. The correlation coefficient generally decreased with depth into the streambed, as distance from the stream increases, and temperature signals become dampened through the mixing with groundwater. These results are as expected, based on research by Stonestrom and Constantz [4] amongst others, though not evaluated by cross-correlation.

With increasing depth in the streambed, temperature signals continue to be significantly correlated with seasonal trends in stream water over longer lag periods. This is likely due to the greater thermal homogeneity at depth, as illustrated in the filtered data (**Figure 5**). The filtered data at depth 140 cm is relatively insensitive to short-

term surface thermal fluctuations, resulting in lower correlation coefficients. However, temperatures remain more constant at 140 cm depth. Thus, a significant yet low correlation value persists for longer periods.

Lag times (the point where the highest correlation co-efficient along a single curve is obtained) of seasonal trends increase with depth to varying degrees, differing between sites as well as among individual wells. Seasonal lag times at 30 cm depth at Site 1 range from 5 hrs ($r = 0.941$) to 23 hrs ($r = 0.41$), and at Site 2 from 8 hrs ($r = 0.721$) to 18 hrs ($r = 0.575$). At 140 cm depth, seasonal lag times at Sites 1 and 2 ranged from 32 hrs ($r = 0.633$) to 109 hrs ($r = 0.279$) and 56 hrs ($r = 0.29$) to 68 hrs ($r = 0.312$), respectively. At 30 cm, relative heterogeneity of lag times is observed at both sites. However, at 140 cm depth, lag time heterogeneity persists only at Site 1, while Site 2 displays relatively uniform seasonal lags.

To further the understanding of subsurface connections, while also providing a means for lateral and lon-

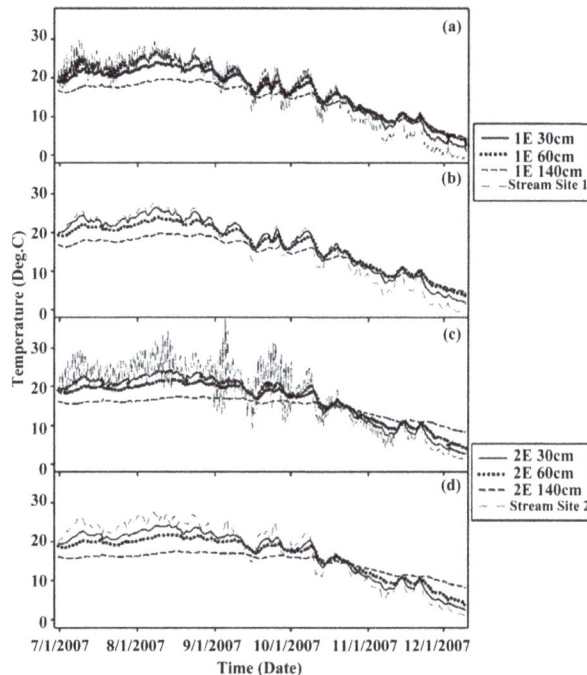

Figure 5. Comparison of unfiltered hourly time series (a) and (c) and filtered hourly time series (b) and (d) of wells 1E and 2E, respectively.

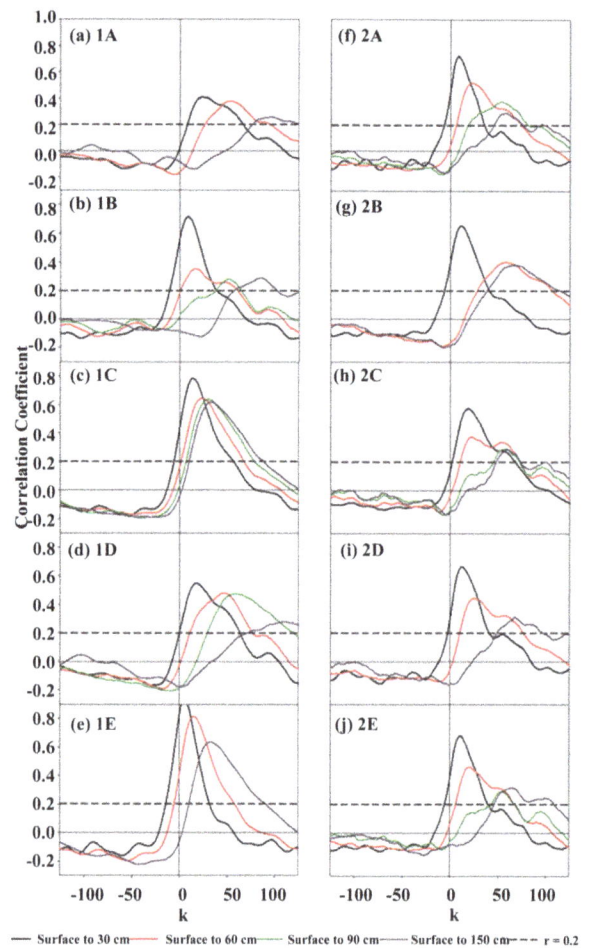

Figure 6. Cross-correlograms per well, showing correlation between hourly, filtered (24 hrs averaging filter and first order differencing) time series of surface stream temperatures and depths 30, 60, 90 and 140 cm within the streambed at Site 1 (a)-(e) and Site 2 (f)-(j).

gitudinal profile comparison, seasonal temperature trends were compared at each site by cross-correlation at equal depths along both profile lines (**Figure 7**). In general, as depth within the streambed increases, the correlation coefficient between temperatures at each depth decreases, regardless of profile type or site (**Figure 7**). One exception exists, between temperatures at wells 1C and 1E. Previously identified as featuring unique temperature patterns.

A second generalization can be made when comparing lateral and longitudinal profiles of Sites 1 and 2. Site 1 correlograms show great variation in peak r values, both between depths and at the same depth. When referring back to **Figures 4(a)** and **(b)**, both wells 1C and 1E showed wider temperature ranges than wells 1A, 1B, and 1D at 140 cm depth, indicating a greater influence of surface water temperatures within the streambed at these locations. Additionally, in well 1C the 90 cm and 140 cm depths have almost equal mean temperatures throughout the summer season. In contrast, wells 1A, 1B, and 1D show more regularly decreasing temperature ranges and mean temperatures with depth. Laterally at depth 140 cm,

seasonal trends in wells 1B and 1D lag behind well 1C, while longitudinally only seasonal trends in well 1A lag behind well 1C. At 140 cm seasonal trends in wells 1C and 1E are highly correlated. In contrast, Site 2 correlograms (**Figures 7(d)-(f)**) consistently peak at or very near $k = 0$. This is supported by the patterns seen in **Figures 4(c)** and **(d)**, where summer temperature ranges and mean temperature patterns change relatively uniformly across Site 2.

As for a comparison between lateral and longitudinal profiles within a single site, no universal patterns were detected. Local variability in streambed flow patterns and materials likely causes observed differences, with a high degree of unpredictability.

4.3. Diel Cross-Correlation

Significant correlation between diel stream and streambed temperatures is seen at 2 wells at Site 1, and at 4 wells at Site 2 (**Figure 8**). Additionally, with the exception of well 1E, significant correlation is seen only between stream and 30 cm depth temperatures. In well 1E, significant correlation is also seen between stream and 60 cm depth temperatures. Lag times of diel temperatures at Sites 1 and 2 range from 3 hrs ($r = 0.3110$) to 9 hrs ($r = 0.3650$) and 6 hrs ($r = 0.5030$) to 8 hrs ($r = 0.3260$) respectively. The trend of greater thermal variability at Site 1 persists in the diel temperatures.

As with seasonal temperature trends, diel temperature trends were analyzed along lateral and longitudinal profile lines across each site (**Figure 9**). At Site 1 significant correlation occurs at both 30 cm and 60 cm depth at $k = 0$. Correlation between wells 1C and 1E at 30 cm depth is unique in that it shows significant 24-hour fluctuations. This is likely the effect of their diel temperature trends correlating. Interestingly, diel patterns in well 1C lag behind those experienced in well 1E by 5 hours, despite well 1C being situated 1 meter upstream of well 1E. This pattern of temperature change opposing the direction of stream flow could indicate preferential flow paths at Site 1.

At Site 2 all wells show significant correlation between diel temperature patterns at 30 cm depth, displaying the unique 24-hour cycle. At depths 60 cm and 140 cm however, significant correlation exists only at $k = 0$.

5. Discussion

5.1. Role of Surface Waters

Surface waters are the source of increased temperature ranges and variability within the HZ, as both diel and seasonal temperature patterns are transmitted [13,29]. In contrast, groundwater, when mixed with surface water, has a dampening effect on diel temperature patterns

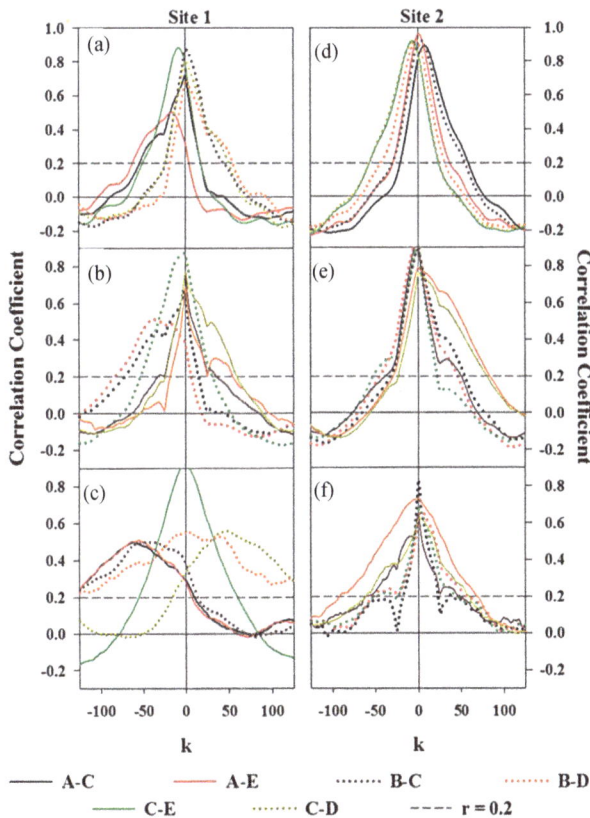

Figure 7. Cross-correlograms showing correlation between hourly filtered (24 hrs averaging filter and first order differencing) time series between wells along longitudinal (solid lines) and lateral (dotted lines) profiles. (a) Site 1, 30 cm; (b) Site 1, 60 cm; (c) Site 1, 140 cm; (d) Site 2, 30 cm; (e) Site 2, 60 cm; and (f) Site 2, 140 cm.

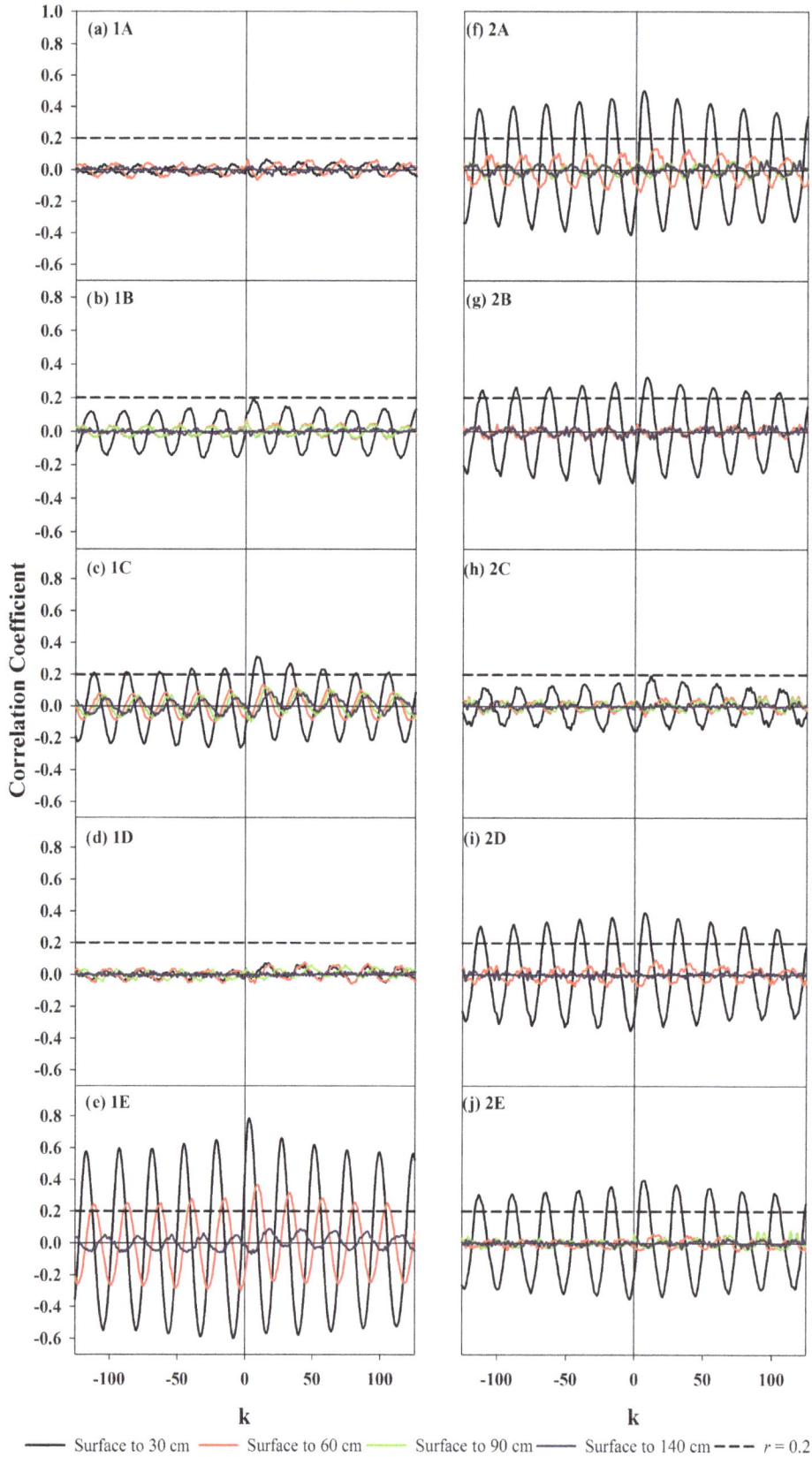

Figure 8. Cross-correlograms per well (indicated by letters A through E), showing correlation between hourly transformed (first order differencing) time series of surface stream temperatures and depths 30, 60, 90 and 140 cm within the streambed at Site 1 (a)-(e) and Site 2 (a)-(e).

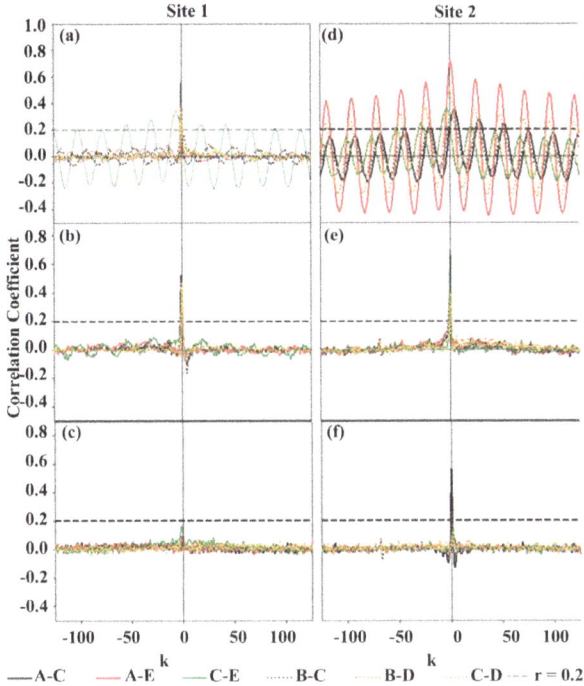

Figure 9. Cross-correlograms at 30 cm (a) and (d), 60 cm (b) and (e), and 140 cm (c) and (e), showing correlation between hourly transformed (first order differencing) time series between wells along longitudinal (solid lines) and lateral (dotted lines) profiles at Site 1 and Site 2, respectively.

within the HZ, as it imparts only seasonal temperature trends [4,13]. The decreasing temperature ranges and mean temperatures, as seen in box plots (**Figure 4**), can be attributed to the mixing of surface and groundwater and the increasing influence of groundwater with depth.

5.2. Differences between Site 1 & Site 2

Box plots reveal that while the stream water at Site 2 experienced greater temperature extremes than at Site 1, the extremes were not observed in the streambed temperatures at Site 2. This suggests that hyporheic exchange is lower at Site 2 than at Site 1. Possible explanations include the allowance of less surface water infiltration, the presence of increased groundwater discharge, the presence of meander flow-through, or perhaps the retardation of infiltration velocities associated with the finer grain sizes of the substrate at Site 2 as suggested by Ringler and Hall [25]. The hypothesis of less surface water infiltration is tied to the discharge of a greater groundwater component, dampening diel surface water signals. This possible explanation is consistent with the establishment of Site 2 as a gaining reach. The second possibility of retarded infiltration velocities is supported by Ringer and Hall [25], who found larger temperature gradients between stream and hyporheic waters at heavily silted sites, due to slower inter-gravel

flows. Based on grain size analyses, we believe that while silt size particles are not prevalent at Site 2, the small particles sizes and lower hydraulic conductivity, as compared to Site 1, may exert a similar effect. Though dampened in amplitude, diel surface water signals are still transmitted down to a 30 cm depth almost universally across Site 2 (**Figure 9**).

In addition to increased dampening of surface thermal trends, greater uniformity in thermal trends is seen in box plot and cross-correlation results at Site 2. Thus, Site 2 has more uniform HZ flow path patterns associated with more homogeneous sediment distribution. Vaux [23] and Cooper [24] observed that larger objects in or on the streambed surface respectively, cause significant disruptions to HZ flow paths and thereby thermal patterns as well, which does not appear to be the case at Site 2. Site 1 however, displays distinct thermal heterogeneity when comparing thermal trends in wells 1A, 1B, and 1D, to those in wells 1C and 1E, making the presence of sediment variations a possibility. While no large particles were observed on the streambed surface, Buyck [30] documented the presence of till anomalies in the stream bed up to depths of 60 cm.

Numerical modeling of the area indicated that Site 1 is a downwelling zone, while Site 2 is an upwelling zone [31]. While the site-specific details are not addressed, these flow dynamics potentially explain many of the trends observed in the statistical results of this study, as outlined below.Advection, as involved in a losing reach or downwelling zone, is commonly considered the most effective means of thermal transport, as fluid movement is typically faster and more efficient at heat transmission than the process of conduction. Therefore, the effective transmission of diel temperature signals into the substrate is likely due to advection of stream water into the HZ. This goes hand-in-hand with the established temperature-based method of defining losing and gaining reaches of a stream, where the presence of increased diel signal transmission into the HZ is an indication of a losing reach [4].

Lags between unfiltered hourly temperatures of the stream and at 30 cm depth (showing diel temperature variations) ranged from 3 to 9 hours. The smallest lag of 3 hours was experienced at Site 1, where sediments are coarser and feature a higher hydraulic conductivity. Site 2, with a lower hydraulic conductivity, experienced lags of 6 to 8 hours.

The persistent penetrations of diel surface water temperature patterns to depths of 30 cm in wells 1C and 1E, and to a depth of 60 cm (**Figure 8**) in well 1E, suggest the influence of a strong vertical advective component at Site 1. Additionally, these trends reinforce the identification of Site 1 as a downwelling zone, and pinpoint wells 1C and 1E as the point of most focused downwelling of surface water to a minimum depth of 30 cm in both wells, and to a minimum depth of 60 cm in well 1E. This is further reinforced by seasonal correlation coefficients of surface water to 140 cm depth remaining above 0.6 at both wells 1C and 1E, suggesting that surface water seasonal variations are responsible for 60% of the seasonal variability witnessed at this depth. It is therefore likely that advection penetrates deeper than the minimum values stated above, yet based on the available data no conclusive statement can be made.

Though both wells 1C and 1E appear to be the location of deepest surface water penetration, well 1E is the location of fastest surface water penetration to a depth of 30 cm, as shown by correlation results between 30 cm temperatures in wells 1C and 1E. Flow paths within the HZ and streambed can be controlled by a large number of factors. However, from what is known of Site 1 regarding sediment particle size and thermal heterogeneity, it is very likely that both flow paths and thermal regimes are impacted by sediment heterogeneities in the HZ and streambed. Buyck [30] found gray clay in the streambed, originating possibly from collapsed cut banks, or from underlying till layers. Such clay in the HZ would act as barriers to advection, and increase the chance of preferential flow path development, which could in turn lead to uncharacteristic flow patterns, as supported by research conducted by Vaux [23] and Becker et al. [26].

Site 2 flow path delineation is somewhat less precise than at Site 1. The comparison of lateral and longitudinal profiles at 30 cm depth reveals that the strongest correlation exists in the longitudinal direction, following the direction of stream flow. However, correlation between lateral temperature patterns at 30 cm depth exists also. This correlation reflects similar degrees of surface diel signal penetration to 30 cm depth. At depths greater than 30 cm, correlation of diel patterns is only significant at $k = 0$, suggesting lateral homogeneity of temperatures across the site. Based on statistical results, we believe flow paths at Site 2 are mostly in the longitudinal direction, at low velocities, and active surface water infiltration is limited to the upper 30 cm of the streambed. The influence of lateral flow is supported by the numerical modeling of Van der Hoven [31] showing the area to be an outlet for flow from underneath a meander lobe.

5.3. Controls on Thermal Transport

The process of conduction, while in part dependent on the thermal properties of a medium, is driven by temperature gradients, where steeper temperature gradients increase the effectiveness of conduction. Steepest temperature gradients appear to exist laterally at Site 1, between vertically down-welling warmer temperatures in well 1C and 1E, and cooler temperatures in wells 1A, 1B, and 1D. Subsequently, conduction may be an important

mode of heat transport in the lateral direction at Site 1. At Site 2 thermal gradients appear more gradual, suggesting conduction will be kept to a minimum. However, during the low-flow period at Site 2, the heating of streambed sediments by solar radiation may have provided a steep thermal gradient, allowing conduction an active role in the transport of heat into the HZ. A similar proposition was put forward by Shepherd et al. [32]. However, there are situations where advection and vertical conduction are of similar magnitude [22].

Quantitative delineation of the HZ based on thermal trends has not been possible. Though statements can be made as to where the HZ definitely persists, such as at 30 cm depth in the locations of wells 1C, 1E, 2A, 2B, 2D, and 2E, where significant correlation to surface stream diel temperature patterns was found, the exact cut-off point between the HZ and groundwater environments is difficult to pinpoint quantitatively without a thermal groundwater signature for the study location.

It is also possible that the maximum logger installation depth managed for this study was not deep enough to penetrate beyond the HZ. Even at 140 cm depth, seasonal temperature trends vary more than by the expected $\pm 3°C$ [10] range from the annual mean air temperature of 11.2°C. A likely alternative explanation to lacking penetration depth is the impact of conduction on temperatures at depth [22]. While the presence of advecting surface water defines the extent of the HZ, the presence of conduction may alter temperatures beyond the extent of the HZ, effectively masking the true groundwater thermal signature. Seasonal cross-correlation results between surface water and 140 cm depth at both sites (**Figure 6**) suggest at least 20% of the variability witnessed can still be explained by surface water variability. This may be coincidence, based on the large number of observations used in the correlation, as well as the small degree of change in the temperatures and the fact that groundwater also has a seasonal signal. However, if not coincidence, it seems possible that conduction could transmit 20% of the surface thermal signature to a depth of 140 cm below the streambed [22], especially considering that the seasonal trends are transmitted into the upper 30 cm by advection, leaving approximately 110 cm distance to be spanned by conduction.

6. Conclusions

Stream-groundwater interaction and HZ sediment physical and thermal properties are the major determining factors for temperature patterns within the HZ, simultaneously defining HZ flow paths of surface and groundwater, and the effectiveness of temperature transmission into the subsurface. Consequently, differences in one or all of these properties must exist between Sites 1 and 2 to explain the differences in temperature behavior, for although both site comparisons (**Figures** 7 and **8**) show little difference between thermal gradients, local differences were observed in all other statistical results.

Overall, distinct differences were identified in the thermal profiles of Sites 1 and 2. Site 1 appears as a downwelling zone with surface water penetrating deepest into the HZ at the location of wells 1C and 1E. Site 2 was characterized as a gaining reach, where the balancing between down-welling surface water and upwelling groundwater temperatures resulted in a more homogenized thermal environment. Additionally, a dampening of diel surface stream temperature ranges was noticed in upper HZ temperatures at Site 2. This dampening was attributed to a variety of possible causes, including a significant discharging groundwater component, which would produce a dampening effect on diel temperatures as previously outlined. This explanation is in line with Site 2 being recognized as a gaining reach. Additionally, the possibility of an increased percentage of finer sediments at the site was considered, resulting in slightly retarded inter-gravel flows causing dampening associated with the longer thermal transmission times.

A correlation between increased sediment homogeneity and more homogeneous thermal profiles was noted, though the lack of multiple sites makes definitive interpretation difficult. However, it has been established in the literature that larger sediment particles as well as possible low permeability zones can disrupt HZ flow paths and thermal regimes by altering the flowpaths[23,26].

The transmission of diel signals is limited by the efficiency of advection and diel thermal transfer requires higher transmission speeds than seasonal temperature signals. Supporting this, the deepest penetration depth of diel temperature patterns was 60 cm in well 1E, while seasonal surface temperature patterns were detected universally to a depth of 140 cm.

Thermal differences in lateral and longitudinal profiles were detected, and were attributed to variations in factors affecting thermal transport, such as the presence of preferential flow paths. The longitudinal profile exhibited a greater tendency for progressive transmission of thermal signals in the downstream direction, though a thermal transmission against the direction of stream flow was detected at Site 1.

Finally, only qualitative delineation of the HZ was possible in this study. The main limitation was the lack of a specific thermal groundwater signature for the study area. The persistence of surface seasonal temperature trends beyond the extent of surface diel temperature is likely due to the influence of conduction on temperatures below the reach of advection [33,34].

Both sediment particle size and degree of sorting impact thermal profiles. Site 1 (poorly sorted gravels) showed a high degree of thermal heterogeneity through preferential flow paths (local downwelling zone). Site 2 (moderately sorted sands) showed a vertically and laterally homogenized thermal environment with no defined preferential flow paths. Meander flow-through discharge can have a significant impact on streambed temperatures. The transmission of diel signals is limited by the efficiency of advection, requiring higher transmission speeds than seasonal temperature signals. The deepest penetration depth of diel temperature patterns was 60 cm in well 1E, where a local downwelling zone exists. Surface water temperatures influence the thermal regime not only of the hyporheic ecotone, but also of the shallow groundwater environment. Seasonal surface temperature patterns were detected universally to a depth of 140 cm.

7. Acknowledgements

The authors would like to extend a sincere thank to the following organizations: The Bloomington Normal Wastewater Reclamation District for access to the study site. The Geological Society of America (student research grant-Beach), Grant-In-Aid of Research from Sigma Xi (student research grant-Beach), and the PADI Foundation, for funding of this study.

REFERENCES

[1] H. B. N. Hynes, "Ecology of Running Waters," Liverpool University Press, Liverpool, 1970.

[2] G. C. Poole and C. H. Berman, "An Ecological Perspective on In-Stream Temperature: Natural Heat Dynamics and Mechanisms of Human-CausedThermal Degradation," *Environmental Management*, Vol. 27, No. 6, 2001, pp. 787-802.

[3] J. V. Ward, "THERMAL Characteristics of Running Waters," *Hydrobiologia*, Vol. 125, No. 1, 1985, pp. 31-46.

[4] D. A. Stonestrom and J. Constantz, "Heat as a Tool for Studying the Movement of Ground Water Near Streams," US Geological Survey, Denver, 2003.

[5] B. Conant, "Delineating and Quantifying Ground Water Discharge Zones Using Streambed Temperatures," *Ground Water*, Vol. 42, No. 2, 2004, pp. 243-257.

[6] F. Malard, A. Mangin, U. Uehlinger and J. V. Ward, "Thermal Heterogeneity in the Hyporheic Zone of a Glacial Floodplain," *Canadian Journal of Fisheries and Aquatic Sciences*, Vol. 58, No. 7, 2001, pp. 1319-1335.

[7] M. Hondzo and H. G. Stefan, "Riverbed Heat Conduction Prediction," *Water Resources Research*, Vol. 30, No. 5, 1994, pp. 1503-1513.

[8] S. E. Silliman, J. Ramirez and R. L. McCabe, "Quantifing Downflow through Creek Sediments Using Temperature Time Series: One-Dimensional Solution Incorporating Measured Surface Temperature," *Journal of Hydrology*, Vol. 167, No. 1-4, 1995, pp. 99-119.

[9] M. Hayashi and D. O. Rosenberry, "Effects of Ground Water Exchange on the Hydrology and Ecology of Surface Water," *Ground Water*, Vol. 40, No. 3, 2002, pp. 309-316.

[10] D. Deming, "Introduction to Hydrogeology," McGraw Hill, Boston, 2002.

[11] M. P. Anderson, "Heat as a Ground Water Tracer," Ground Water, Vol. 43, No. 6, 2005, pp. 951-968.

[12] E. W. Peterson and T. B. Sickbert, "Stream Water Bypass through a Meander Neck, Laterally Extending the Hyporheic Zone," *Hydrogeology Journal*, Vol. 14, No. 8, 2006, pp. 1443-1451.

[13] C. Schmidt, M. Bayer-Raich and M. Schirmer, "Characterization of Spatial Heterogeneity of Groundwater-Stream Water Interactions Using Multiple Depth Streambed Temperature Measurements at the Reach Scale," *Hydrology and Earth System Sciences Earth System Sciences Discussions*, Vol. 3, 2006, pp. 1419-1446.

[14] A. S. Arrigoni, *et al.*, "Buffered, Lagged, or Cooled? Disentangling Hyporheic Influences on Temperature Cycles in Stream Channels," *Water Resources Research*, Vol. 44, No. 9, 2008, Article ID: W09418.

[15] T. J. Dogwiler and C. M. Wicks, "Thermal Variations in the Hyporheic Zone of a Karst Stream," *Speleogenesis and Evolution of Karst Aquifers*, Vol. 3, No. 1, 2005, pp. 1-11.

[16] D. S. White, C. Elzinga, H. and S. P. Hendricks, "Temperature Patterns within the Hyporheic Zone of a Northern Michigan River," *Journal of North American Benthological Society*, Vol. 6, No. 2, 1987, pp. 85-91.

[17] C. E. Hatch, A. T. Fisher, J. S. Revenaugh, J. Constantz and C. Ruehl, "Quantifying Surface Water-Groundwater Interactions Using Time Series Analysis of Streambed Thermal Records: Method Development," *Water Resources Research*, Vol. 42, No. 10, 2006, Article ID: W10410.

[18] C. Schmidt, B. Conant Jr., M. Bayer-Raich and M. Schirmer, "Evaluation and Field-Scale Application of a Simple Analytical Method to Quantify Groundwater Discharge Using Mapped Streambed Temperatures," *Journal of Hydrology*, 2007.

[19] B. G. Fraser and D. D. Williams, "Seasonal Boundary Dynamics of a Groundwater/Surface-Water Ecotone," *Ecology*, Vol. 79, No. 6, 1998, pp. 2019-2031.

[20] A. Mangin, "For a Better Knowledge of Hydrological Systems from Correlogram and Variance Spectral Density," *Journal of Hydrology*, Vol. 67, No. 1-4, 1984, pp. 25-43.

[21] E. C. Evans, M. T. Greenwood and G. E. Petts, "Thermal Profiles within River Beds," *Hydrological Processes*, Vol. 9, No. 1, 1995, pp. 19-25.

[22] E. T. Hester, M. W. Doyle and G. C. Poole, "The Influence of In-Stream Structures on Summer Water Temperatures via Induced Hyporheic Exchange," *Limnology and Oceanography*, Vol. 54, No. 1, 2009, pp. 355-367.

[23] W. G. Vaux, "Intergravel Flow and Interchange of Water in a Streambed," *Fishery Bulletin*, Vol. 66, No. 3, 1968, pp. 479-489.

[24] A. C. Cooper, "The Effect of Transported Stream Sediments on the Survival of Sockeye and Pink Salmon Eggs and Alevins," *International Pacific Salmon Fishery Commission Bulletin*, Vol. 18, 1965, p. 75.

[25] N. H. Ringler and J. D. Hall, "Effects of Logging on Temperature, and Dissolved Oxygen in Spawning Beds," *Transactions of the American Fisheries Society*, Vol. 104, No. 1, 1975, pp. 111-121.

[26] M. W. Becker, T. Georgianb, H. Ambrosea, J. Siniscalchia and K. Fredricka, "Estimating Flow and Flux of Ground Water Discharge Using Water Temperature and Velocity," *Journal of Hydrology*, Vol. 296, No. 1-4, 2004, pp. 221-233.

[27] "SPSS for Windows Rel. 16.0.1.," SPSS Inc., Chicago, 2007.

[28] G. M. Jenkins and D. G. Watts, "Spectral Analysis and Its Applications," Holden-Day, San Francisco, 1968.

[29] M. Brunke and T. Gonser, "The Ecological Significance of Exchange Processes between Rivers and Groundwater (Special Review)," *Freshwater Biology*, Vol. 37, No. 1, 1997, pp. 1-33.

[30] M. S. Buyck, "Tracking Nitrate Loss and Modeling Flow through the Hyporheic Zone of a Low Gradient Stream through the Use of Conservative Tracers," Illinois State University, Normal, 2005.

[31] S. van der Hoven, N. Fromm and E. Peterson, "Quantifing Nitrogen Cycling Beneath a Meander of a Low Gradient, N-Impacted, Agricultural Stream Using Tracers and Numerical Modelling," *Hydrological Processes*, Vol. 22, No. 8, 2008, pp. 1206-1215.

[32] B. G. Shepherd, G. F. Hartman and W. J. Wilson, "Relationships between Stream and Intragravel Temperatures in Coastal Drainages and Some Implications for Fisheries Workers," *Canadian Journal of Fisheries and Aquatic Sciences*, Vol. 43, No. 9, 1986, pp. 1818-1822.

[33] E. Hoehn and O. A. Cirpka, "Assessing Hyporheic Zone Dynamics in Two Alluvial Flood Plains of the Southern Alps Using Water Temperature and Tracers," *Hydrology and Earth System Sciences Discussions*, Vol. 3, No. 2, 2006, pp. 335-364.

[34] W. W. Lapham, "Use of Temperature Profiles Beneath Streams to Determine Rates of Vertical Ground-Water Flow and Vertical Hydraulic Conductivity," United States Geological Survey, Reston, 1989, p. 35.

18

Combined Effect of Infiltration, Capillary Barrier and Sloping Layered Soil on Flow and Solute Transfer in a Heterogeneous Lysimeter

Le Binh Bien, Dieuseul Predelus, Laurent Lassabatere, Thierry Winiarski, Rafael Angulo-Jaramillo[*]

Laboratoire d'Ecologie des Hydrosystèmes Naturels et Anthropisés, UMR 5023 CNRS-ENTPE-UCBL, Université de Lyon, Lyon, France.

ABSTRACT

This aim of this paper is to describe a study of the combined effect of infiltration, capillary barrier and sloping layered soil on both flow and solute transport processes in a large, physical model ($1 \times 1 \times 1.6$ m^3) called LUGH (Lysimeter for Urban Groundwater Hydrology) and a 3D numerical flow model. Sand and a soil composed of a bimodal sand-gravel mixture were placed in the lysimeter to simulate one of the basic structural and textural elements of the heterogeneity observed in the vadose zone under an infiltration basin of Lyon (France). Water and an inert tracer (KBr) were injected from the top of the lysimeter using a specific water sprinkler system and collected at 15 different outlets at the bottom. The outlet flows and the 15 breakthrough curves obtained presented high heterogeneity, emphasising the establishment of preferential flows resulting from both capillary barrier and soil layer dip effects. Numerical modelling led to better understanding of the mechanisms responsible for these heterogeneous transfers and it was also used to perform a sensitivity analysis of the effects of water velocity (water and solute flux fed by the sprinkler) and the slope interface. The results show that decreasing velocity and increasing the slope of the interface can lead to the development of preferential flows. In addition, the offset of the centre of gravity of the flow distribution at the output increases linearly as a function of the slope angle of the layered soil. This paper provides relevant information on the coupling between hydrodynamic processes and pollutant transfer in unsaturated heterogeneous soil and emphasizes the role of the geometry of the interfaces between materials and boundary conditions as key factors for preferential flow.

Keywords: Capillary Barrier; Lysimeter; Preferential Flow; Unsaturated Soil

1. Introduction

Stormwater runoff is loaded with different contaminants (heavy metals, pesticides, fertilisers, etc.) of agricultural origin [1] and urban origin [2]. Consequently, when infiltrating the soil, the runoff water loaded with significant quantities of contaminants reaches the vadose zone and migrates to the groundwater [3] representing a major environmental issue.

The vadose zone plays a predominant role in the transfer of water and solutes as it occupies a central position for exchanges and interactions with the other compartments (atmosphere, biosphere, groundwater, etc.). The question raised is always that of how the contaminants that spread on the surface of the soil are transferred to this zone, and when, where and in what proportion they

reach the groundwater. Many authors have emphasized that this process is closely associated with preferential flows that participate in accelerating the transfer to the groundwater [4]. The evolution of these preferential flows depend on the heterogeneity of the texture [5-7] and the structure [8-10] of the soil and the soil moisture regime (*i.e.* moisture history, intensity and volume of precipitations) [11,12].

Many studies focused on coupled water-solute transfers have been carried out in the vadose zone by in-situ tracing [13-15]. Field tests have the advantage of being carried out under real conditions and highlight the influence of the vadose zone on the transfer of the solutes. However, the interpretation of results remains difficult given the myriad non controlled parameters involved: initial and boundary conditions, the geochemical quality of the water and materials, and the spatial variability of

[*]Corresponding author.

the lithology and structure of the vadose zone.

One of the alternatives to field tests is to perform laboratory studies using leaching columns. This technique, although accurate regarding the identification of certain transfer mechanisms [16-19], allows estimating the key parameters of water flow and the transport of solutes in unsaturated soils according to one dimensional geometry only. Since leaching columns are one-dimensional devices, their results are difficult to use for studying preferential flows in the field with two and three dimensional geometry.

Consequently, laboratory 3D pilot devices have been developed to study preferential flows under controlled conditions. Metric scale laboratory lysimeters are widely used [20] to better understand the water and solute transfer process in porous media [11,12,21,22]. They provide an intermediate approach between the two scales, *i.e.* the laboratory leaching column and a plot of land used for field studies. In these studies, no attempt was made to determine the effect of heterogeneity on coupled water and solute transfer mechanisms in an unsaturated medium. Abdou and Flury [23] focused on the role of heterogeneous structures and the impact of scale between a lysimeter and a test in the field. Nonetheless, their studies only dealt with numerical works performed in two dimensions. In all the studies mentioned above, the lysimeters were supplied with water uniformly over the entire surface of the soil. Thus the water tended to flow vertically while lateral flows were limited.

In this article, we present a methodology designed to improve and validate a conceptual and numerical model of the hydrodynamics in heterogeneous soil using a laboratory pilot rig. The purpose of this rig, known as LUGH (Lysimeter for Urban Groundwater Hydrology) [24], is to provide a 3D representation of the structural and textural heterogeneities observed in the fluvioglacial formations in the east Lyon region (France). This model permits studying in particular the initial and boundary conditions imposed and their role in preferential flows. The LUGH lysimeter is supplied only on one part of its surface to permit lateral flows, free drainage and the collection of the effluents distributed at several different outlets. This subdivision of effluents provides an extended view of the spatial and temporal distribution of solutes at the outlet of the lysimeter. The comparison of the numerical and experimental data makes it possible to test and validate transfer models coupling different processes in 3D and thus take into account the effect of the medium's heterogeneity and the effect of the boundary condition at the bottom of the lysimeter. The numerical resolution of different scenarios then permits testing the influence of infiltration speed and the slope at the interface between two different materials on the establishment

of heterogeneous flows.

2. Materials and Methods

2.1. Water Flow and Solute Transport

Modelling the transfer of water and solutes is based on the Darcy approach. Darcy's 3D flow is assumed to occur in the unsaturated porous medium studied with the LUGH lysimeter. This flow is characterised by the Richards equation [25] in the following way:

$$C(h)\frac{\partial h}{\partial t} = \vec{\nabla}\left[K(h)\vec{\nabla}H\right] \tag{1}$$

where the capillary capacity, $C(h)$ [L^{-1}], is the variation of the volumetric water content θ [L^3L^{-3}] b pressure head h [L], H [L] is the hydraulic head and $K(\theta)$ [LT^{-1}] is the unsaturated hydraulic conductivity, which depends on θ or h, and ∇ refers to the Nabla operator.

The functions of the usual water retention curve $\theta(h)$ and the hydraulic conductivity $K(h)$ to describe the flows in the unsaturated zone are given by van Genuchten equations with the Mualem condition [26,27] :

$$\theta(h) = \begin{cases} \theta_r + (\theta_s - \theta_r)\left(1 + |\alpha h|^n\right)^{-m} & h < 0 \\ \theta_s & h \geq 0 \end{cases} \tag{2}$$

with α (m^{-1}), m and n being parameters such that $m = 1 - 1/n$.

$$K(S_e) = \begin{cases} K_s S_e^l \left[1 - \left(1 - S_e^{1/m}\right)^m\right]^2 & h < 0 \\ K_s & h \geq 0 \end{cases} \tag{3}$$

with S_e [-] being the effective saturation:

$$S_e = \frac{\theta - \theta_r}{\theta_s - \theta_r} \tag{4}$$

where θ_r [L^3L^{-3}] and θ_s [L^3L^{-3}] denote the residual and saturated water contents, K_s [LT^{-1}] is the saturated hydraulic conductivity, l [-] is the connection coefficient of the pores estimated by Mualem [27] at an average of 0.5.

The transport of the non-reactive solute in the porous medium can be modelled by the convection-dispersion equation in an initial approach [28]:

$$\frac{\partial \theta C}{\partial t} = \vec{\nabla}\left(\theta D \vec{\nabla} C\right) - \vec{\nabla}(qC) \tag{5}$$

where C [ML^{-3}] is the concentration of the solute in the liquid, D [L^2T^{-1}] is the hydrodynamic dispersion coefficient.

D groups the molecular diffusion D_o and the kinematic dispersion: $D = \tau_L \cdot D_o + \lambda \cdot v$
where D_o [L^2T^{-1}] is the molecular dispersion, λ [L] is the dispersivity, τ_L [-] is the tortuosity, v [LT^{-1}] is the pore

Combined Effect of Infiltration, Capillary Barrier and Sloping Layered Soil on Flow and Solute
Transfer in a Heterogeneous Lysimeter

169

velocity, $v = q/\theta$.

These characteristics are assumed to be homogenous and invariant with humidity. The molecular diffusion D_o was obtained from the literature. The longitudinal dispersivity, α_1, was preselected according to the maximum grain size of the material as the initial value. Then, all the longitudinal α_1 and lateral dispersivities α_2 and α_3 were optimised by fitting experimental data. In steady state flow, the tortuosity τ_L was calculated using the volumetric water content as follows [29]:

$$\tau_L = \frac{\theta^{7/3}}{\theta_s^2} \qquad (6)$$

2.2. Materials

We used fine sand (0 to 2 mm in diameter) and gravel (4 to 10 mm in diameter). A third bimodal material was a mixture composed of sand and gravel, each making up 50% in weight. This proportion was chosen to ensure the bimodal grain size distribution of the third material. The grain sizes of the three materials were similar to those observed on the most common litho-facies found in the east Lyon region [30]. The hydrodynamic characteristics of the sand and the bimodal material were estimated individually using the BEST method [31], validated analytically [32,33] and experimentally to characterise the comparable matrices resulting from the fluvioglacial deposits of the east Lyon region [34] and those of other types of coarse material [7,35,36]. Then the hydrodynamic characteristics obtained were optimised using the RETC software [37] to adapt them to Equations (2) and (3) by the method proposed by Mubarak *et al.* [38]. The results are given in **Table 1**. The properties of the gravel are without importance given that only the sand and the bimodal material were used in this study.

The LUGH lysimeter (**Figure 1(a)**) is composed of a PVC (polyvinyl chloride) tank 1.6 m long, 1 m wide and

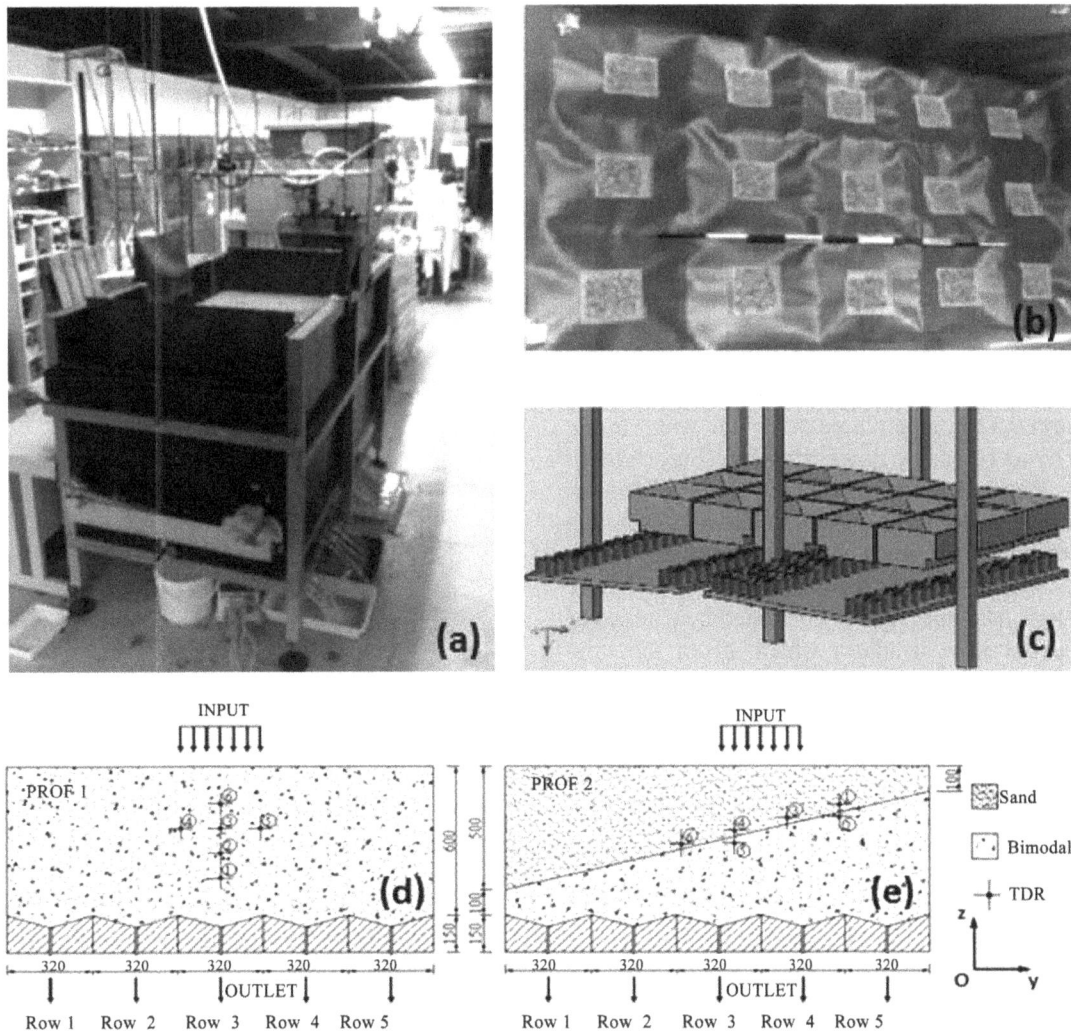

Figure 1. The LUGH lysimeter and drainage system (at top) and profiles used with positions of TDR sensors (at bottom).

Table 1. Hydrodynamic and hydrodispersive parameters used for modelling the LUGH lysimeter with COMSOL.

Hydrodynamic properties			Hydrodispersive properties		
Parameter	Sand	Bimodal	Parameter	Sand	Bimodal
θ_r (m^3·m^{-3})	0.023	0.019	α_1 (m)	2.0e−3	5.0e−3
θ_s (m^3·m^{-3})	0.377	0.377	α_2 (m)	5.0e−4	5.0e−4
n (-)	3.28	3.30	α_3 (m)	5.0e−4	5.0e−4
α (m^{-1})	5.05	10.15	D_o (m^2·s^{-1})	2.0e−7	2.0e−7
K_s (m·s^{-1})	1.5e−4	9.2e−5			

1 m deep. Fifteen concrete blocks (0.32 × 0.32 × 0.15 m^3) are arranged at the bottom of the lysimeter (**Figures 1(b)** and **(c)**) in the form of a funnel to recover the eluents in the lower boundary conditions. These blocks are labelled in the form of a matrix in 3 lines (A, B, and C) and 5 rows (1 to 5). The walls of the tank and the surface of the concrete blocks are covered by a watertight and non-reactive geomembrane that ensures the system is impermeable. A hole is pierced at the centre of each concrete block to permit free drainage and the collection of the effluent. Six TDR sensors (Time Domain Reflectometry, model CS616, Campbell Scientific, Logan, UT) with two rods 0.3 m long are placed at strategic points of the lysimeter to obtain a profile of the volumetric water content in the soil and next to the interfaces (**Figures 1(d)** and **(e)**). They are linked by an automatic acquisition system used to record the measurements every minute during the test.

The water (groundwater with an electric conductivity of 522 µS·m^{-1}) and the tracer solution were supplied along a strip 0.32 m long placed across the width of the lysimeter by an automatic spraying system (**Figure 1(a)**). Sprinklers are used to obtain relatively uniform humidification when calibrated at a given height and hydraulic supply pressure. Two solenoid valves upstream of the sprinklers, linked to PLCs, permit pulsed regulation of the discharge. The inlet flow is controlled by a series of closely spaced time windows which, given the time frame of the test, allow considering the supply as continuous. The hydraulic supply circuit is equipped with a three-way valve to permit switching between supplying the water and the tracer solution (stored in a plastic tank).

Two soil profiles were compacted in the LUGH lysimeter (**Figures 1(d)** and **(e)**). The first profile (PROF1) was produced only with the bimodal material (ρ_d = 1794 kg·m^{-3}). This mixture is analogous to the material found in large proportions in the fluvioglacial deposits of the Django Reinhardt site [30] and is used as a control (homogenous soil). The water and solute supply zone is placed vertical to row 3.

The second profile (PROF2) is composed of a layer of sand (ρ_d = 1634 kg·m^{-3}) placed above a layer of the bimodal mixture (ρ_d = 1794 kg·m^{-3}) forming an interface with a slope of 25% (angle of the interface in relation to the horizontal Φ = 14°). This profile represents the usual case observed in the field of a fine material deposited on a coarse material, implying the development of a capillary barrier [9,10,13,39]. In this case, the supply zone is slightly off-centre to the right, vertical to rows 3 and 4, to highlight the capillary barrier effect on the part downstream of the supply zone.

2.3. Protocols of the Infiltration test and Flow Tracing

Two experimental tests performed on the two profiles PROF1 and PROF2 are called E1 and E2 respectively. In each test, the lysimeter is sprayed at Darcian velocity q_1= 3.62e−5 ms^{-1} until the establishment of a steady state flow (*i.e.* constant flow at the outlet, and measurements of constant water content). Then, by switching the supply, a solution of potassium bromide (KBr) at a concentration of 10^{-2} moll^{-1} is supplied by pulse injection. The total volume of the solute supplied is 0.03 m^3 corresponding to a half pore volume, V_p, of the material placed directly under the infiltration zone, *i.e.*:

$$V_p = S_{inf} \times e_{ZNS} \times \varepsilon \qquad (7)$$

where S_{inf} [L^2] is the surface area of the infiltration, S_{inf} = 0.96 × 0.32 m^2, e_{ZNS} [L] is the total thickness of the profile, e_{ZNS}= 0.6 m, ε [-] is the total porosity, ε = 0.329 for PROF1 and ε = 0.353 for PROF2, *i.e.* 0.0606 m^3 and 0.0651 m^3 for PROF1 and PROF2, respectively.

Once the tracer pulse has been applied, the lysimeter continues to be supplied with water to "wash the system" and recover the tracer at the outlet, while keeping the Darcy velocity constant (steady state flow).

The quantity of water at the outlet of the 15 sampling blocks is recorded through time to characterise the flow rates leaving the system. For each outlet, we define w_i, the ratio between the volume of water eluted locally and the total volume injected to quantify the water balance.

Then, the concentrations in bromide are determined by electric conductimetry (electric conductimeter LF 318/SET) and ionic chromatography (DIONEX DX-100 Ion Chromatograph).

The elutions at the outlet of the 15 sampling blocks are processed classically by a dynamic systems approach by considering the flows of water measured at the outlet. The inlet signal is a pulse of solute at concentration C_0 and duration δt. The moments of order N are calculated at the outlet according to the following expression:

$$\mu_N = \int_0^{+\infty} \frac{C(t)}{C_0} t^N \mathrm{d}t \qquad (8)$$

For each elution curve, we calculate the mass balance, the mean residence time and the variance.

The mass balance of the solute, MB [-], of each outlet is calculated from the moment of order 0 and from the quantity injected at the inlet [40]:

$$MB = \frac{\mu_0}{\delta t} \qquad (9)$$

A global mass balance is also calculated at the scale of the lysimeter, by the following relation to verify the conservative character of the tracer:

$$MB_{\text{total}} = \sum_1^{15} w_i MB_i \qquad (10)$$

where w_i is the proportion of the discharge at the outlet.

The average residence time of the solute corresponds to the difference between the mean time of the breakthrough of the solute at the outlet minus the mean time of the entry of the solute at the inlet. This is calculated on the basis of the moment of order 1 and the moment of order 0 by [40]:

$$t_{sj} = \frac{\mu_1}{\mu_0} - \frac{\delta t}{2} \qquad (11)$$

The variance is calculated with the moment of order 2; it permits evaluating the degree of spread of the elution curves:

$$VAR = \frac{\mu_2}{\mu_0} - \left(\frac{\mu_1}{\mu_0}\right)^2 \qquad (12)$$

2.4. Numerical Modelling

Equations (1) and (5) are solved using calculation codes implemented in COMSOL Multiphysics [41]. The flow domain is divided into a tetrahedral mesh. It is tighter around the supply zone, at the base of the lysimeter and at the interface between the two materials for profile PROF2. In our study, the flow domains are discretized by 29,300 elements for profile PROF1 and by 37,100 ele-

ments for profile PROF2, respectively. The highest number of cells for the second case results from finer meshing next to the interface between the two materials. The lower boundary condition represents the outlet of the effluents through squares measuring 0.1×0.1 m^2. These squares correspond to the area of the filtering layer placed at the bottom of each concrete block of the lysimeter. The lower boundary condition corresponds to a free drainage condition (unit hydraulic gradient) giving rise to the following expression for the outlet velocity:

$$I = -K_s k_r \qquad (13)$$

where [LT^{-1}] is the Darcian velocity of the effluent in each outlet, K_s [LT^{-1}] is the saturated hydraulic conductivity, k_r [-] is the relative permeability taking into account the partial saturation of the material.

The negative sign indicates a flow leaving the lysimeter. The upper supply zone is represented by a uniform flow condition, whereas the rest of the surfaces correspond to a null flux boundary condition.

The experimental data of tests E1 and E2 are used to validate the 3D numerical model as well as the choice of the hydrodynamic and hydrodispersive parameters of the materials. Once the model had been validated, a sensitivity test (25 different flow scenarios) based on the geometric configuration of test E2 (heterogeneous profile) was performed to quantify the impact of the supply velocity (5 different velocities) and the slope angle (5 different angles) on the transfer of the water and the solute in the lysimeter (**Table 2**). The supply zone in the sensitivity test was placed astride rows 3 and 4, as in the case of test E2. Moreover, the vertical distance from the centre of the supply surface until the interface between the two materials was fixed at $z = 0.34$ m as in test E2. These velocities guaranteed the flows in the domain of validity of the Darcy equation and corresponded to the velocities used classically for studies of solute transfer in lysimeters [11,12,22]. They ensured the different and contrasted unsaturated moisture conditions of the lysimeter susceptible to influence capillary barrier phenomena.

3. Results and Discussion

3.1. Offset of Centre of Gravity of Outlet Flows

Under the uniform profile condition, the distribution of outlet flows in steady state of test E1 was almost symmetrical. There was a negligible diversion of the centre of gravity of the volume of infiltrated water, and of the centres of gravity at the corresponding outlet and inlet (**Figure 2(a)**). Below the supply zone (at the vertical of row 3), the discharges at the outlet of row 3 were higher and reached 30% of the total (sum of discharges of outlets A, B, and C). However, the discharges at the outlet

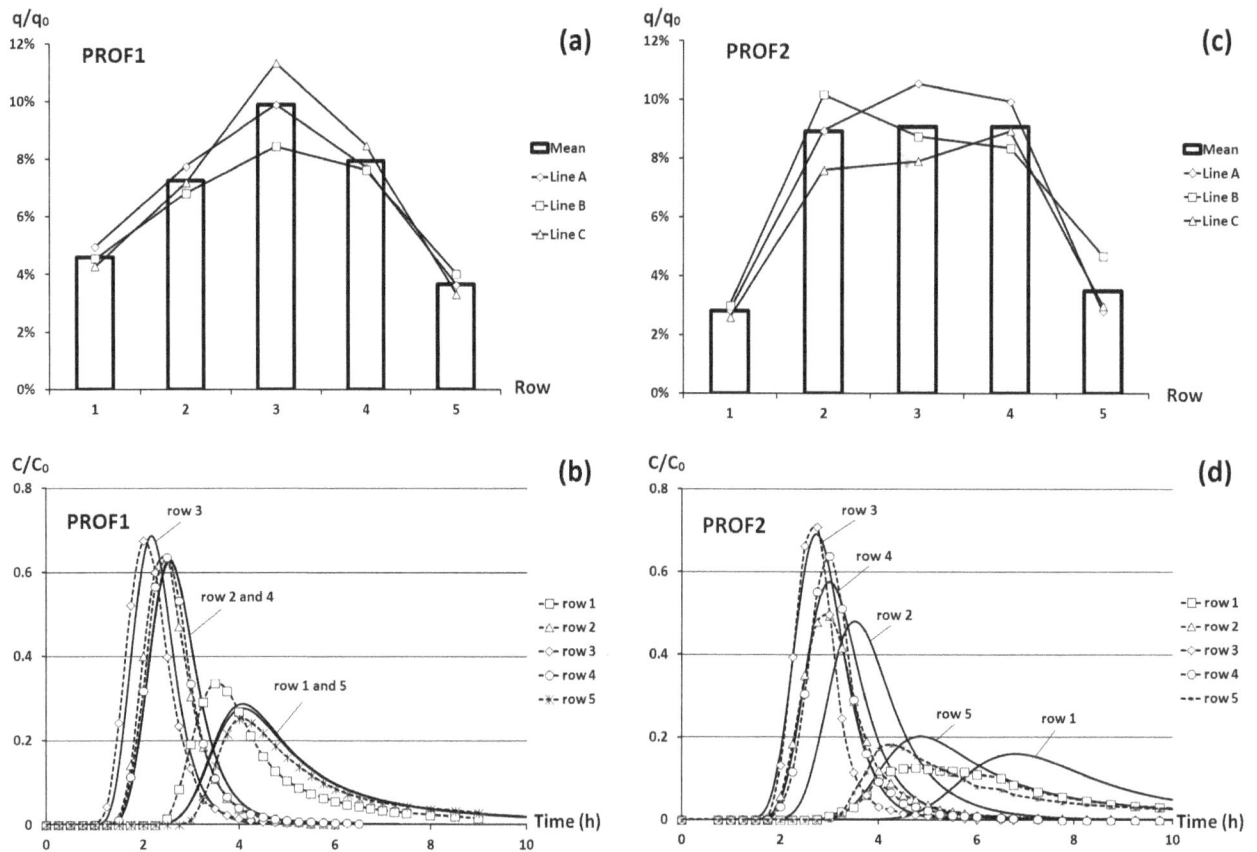

Figure 2. Top: relative discharge in steady state at each outlet (curves) and mean in each row (bars). Bottom: measured (dotted lines) and simulated (lines) elution curves; each curve corresponds to the mean of each row.

Table 2. Infiltration velocity at the surface and the slope angle of the sensitivity test.

Input q_i (m·s^{-1})	$q_1 = 3.62\text{e}-5$	$q_2 = 2.41\text{e}-5$	$q_3 = 1.21\text{e}-5$	$q_4 = 7.23\text{e}-6$	$q_5 = 3.62\text{e}-6$
Slope Φ_i (degrees)	$\Phi_1 = 14°$	$\Phi_2 = 10.5°$	$\Phi_3 = 7°$	$\Phi_4 = 3.5°$	$\Phi_5 = 0°$

of the lateral rows of the lysimeter (rows 1 and 5) also reached quite high values. The sum of the relative discharges of outlets 1A, 1B, and 1C was 14%, and that of outlets 5A, 5B and 5C was 11%. This shows that a large quantity of water flowed laterally. The TDR sensors indicated the possible existence of a zone in which water accumulated at the bottom of the lysimeter. It was also noted that these lateral flows were perfectly symmetrical: the sum of the discharges from rows 1 and 2 was equal to the sum of the discharges from rows 4 and 5.

Conversely, the distribution of the effluents of test E2 (supply zone centred on the vertical of rows 3 and 4) shows a large shift between the centres of gravity of the water at the outlet and the inlet (**Figure 2(c)**). The position of the latter was diverted by 13.5 cm downstream of the slope in comparison to that of the inlet. 62.3% of the discharge exited downstream (rows 1, 2, and 3) and the rest, 37.7%, exited upstream (rows 4 and 5). Furthermore,

the results of the TDR sensors showed that water accumulated in the sand along the interface between the two materials (**Figure 3**). Part of the flux of water arriving at the interface between the two materials therefore seems to have been diverted along the slope, simultaneously producing an increase in water content at the interface, a shift of the centre of gravity downstream and the uniformisation of the discharges of rows 2, 3, and 4. It was assumed that these preferential flows resulted from the effect of the capillary barrier at the interface between the two materials.

3.2. Elution Curves

The shapes of the bromide elution curves of the central rows (rows 2, 3, and 4) of test E1 (**Figure 2(b)**) were similar with peaks of the same order of magnitude, around the value $C/C_0 = 0.7$. Their tails decreased rapidly and almost simultaneously during the leaching phase

Combined Effect of Infiltration, Capillary Barrier and Sloping Layered Soil on Flow and Solute
Transfer in a Heterogeneous Lysimeter

173

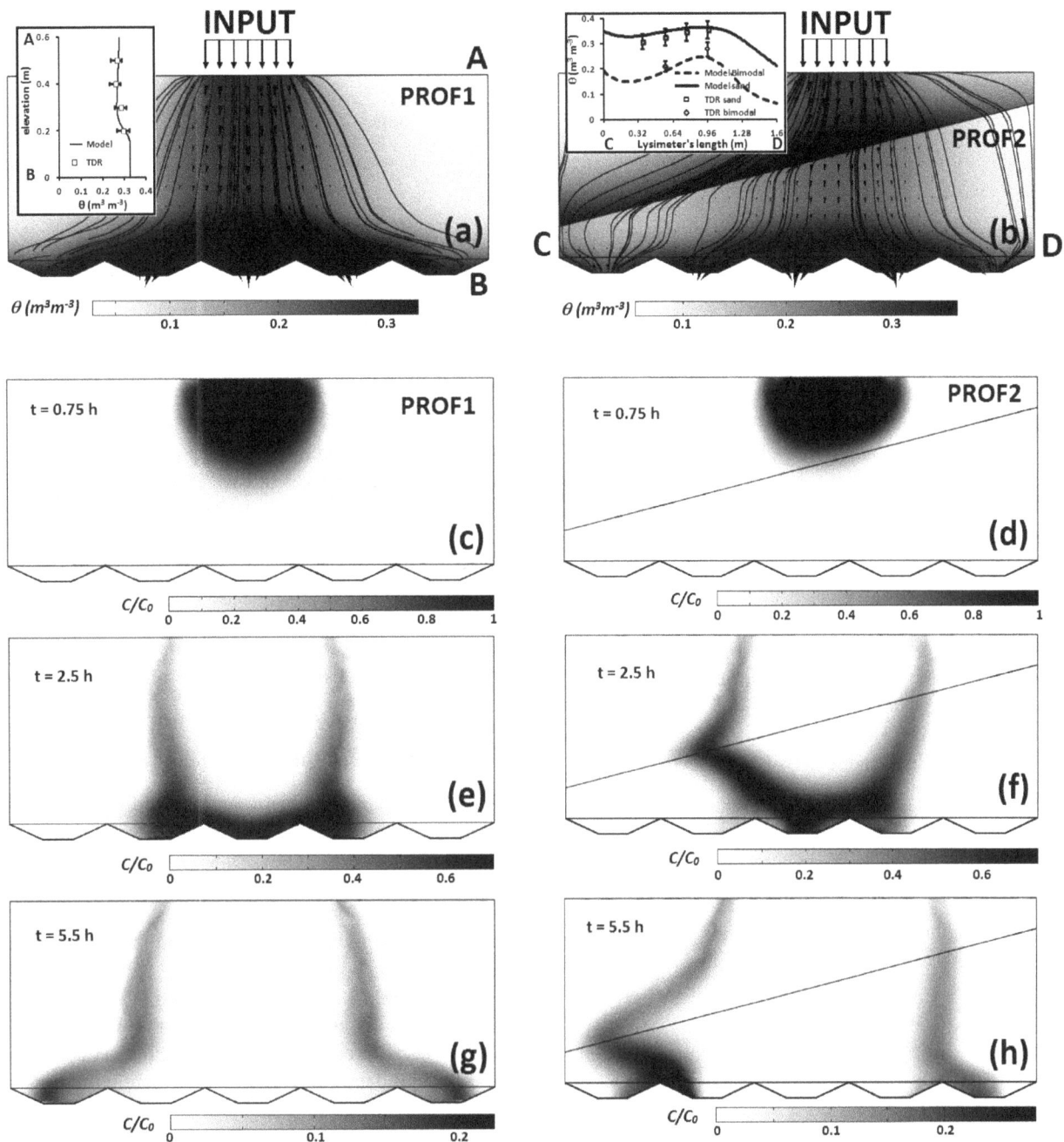

Figure 3. Distribution of volumetric water content, stream lines and vector field (a), (b); insert: comparison of the simulated (curves) and measured (dotted lines) water content. Bottom: transfer of solute as a function of time; t = 0.75 h corresponds to the moment of stopping the solute injection (c)-(h).

following the solute pulse. On the contrary, the solute appeared later in rows 1 and 5; the elution curves of these two rows were more spread out with lower peaks. The difference between the elution curves was representative of the spatial distribution of the solute at the bottom of the LUGH lysimeter. The solute arrived earlier at the centre and at higher concentrations, and then spread symmetrically on both sides.

At the scale of the lysimeter, the mass balance of the system was close to 1. Likewise, the mass balances were close to 1 for almost all the outlets (**Table 3**). We recorded values slightly lower than 1 for rows 1 and 5, showing incomplete elution linked to an insufficiently long experiment time (**Figure 2(b)**). The mass balance values demonstrated the conservative nature of the tracer, which followed the water perfectly.

Table 3. Characteristic parameters of the elution curves of tests E1 and E2; each value is the mean of lines A, B, and C of the same row (1 to 5).

	E1			E2		
	MB	T_{sj} (h)	σ (h)	MB	T_{sj} (h)	σ (h)
Row 1	0.91 ± 0.03	4.00 ± 0.26	1.24 ± 0.11	0.86 ± 0.08	7.21 ± 0.68	3.35 ± 0.23
Row 2	1.00 ± 0.00	2.23 ± 0.12	0.58 ± 0.02	0.98 ± 0.00	2.93 ± 0.35	0.76 ± 0.16
Row 3	1.00 ± 0.00	1.85 ± 0.10	0.54 ± 0.02	0.98 ± 0.02	2.44 ± 0.15	0.52 ± 0.01
Row 4	1.00 ± 0.00	2.31 ± 0.14	0.61 ± 0.02	0.98 ± 0.01	2.81 ± 0.14	0.65 ± 0.14
Row 5	0.82 ± 0.04	4.77 ± 0.31	1.34 ± 0.04	0.91 ± 0.06	7.16 ± 0.59	3.71 ± 0.08
Total	0.98	2.55	1.17	0.96	3.47	2.37

The residence times and standard deviations of rows 2, 3, and 4 were of the same order of magnitude. In combination with the distribution of discharges (**Figure 2(b)**), the flow zone formed by rows 2, 3, and 4 played a dominant role in solute transfer: up to 75% of the solute passed by these three rows, corresponding perfectly to the fraction of water eluted by them. Indeed, the solute transferred mostly vertically, following paths with short distances. Conversely, the residence times, variances and standard deviations of the elution curves of rows 1 and 5 were higher, thus indicating longer solute flow paths to these two outlets. This transfer characteristic resulted from the lateral flows induced by edge effects.

The elution curves of test E2 (**Figure 2(d)**) are also divided into two groups: the curves of the central rows (rows 2, 3, and 4) with high peaks and short tails and the curves of the sides (rows 1 and 5) that have low peaks and lags as well as very long tails. Analysis of the elution curves highlighted the diversion of the flow linked to the interface. Without the effect or diversion, the flow would have been perfectly symmetrical in comparison to the barycentre of the inlet positioned between rows 3 and 4. In this case, the elutions of rows 3 and 4 would have been the same. Likewise, the elutions of rows 2 and 5 would have been superposed. On the contrary, the data show that this was not the case. The solute transfer of row 3 was the fastest whereas the solute transfers of rows 1 and 2 were comparable to those of rows 5 and 4 respectively. These trends can be explained by a diversion of fluxes along the slope responsible for the shift of the solute downstream of the slope. Furthermore, the transfer of the solute in test E2 lasted longer and the elution curves were more spread out than for test E1, indicated by a longer residence time (**Table 3**).

The mass balance of the solute, calculated between 0 and 10 h for test E2, reached the value of 0.96 (**Table 3**). The solute injected was therefore recovered well at the outlet, mainly by the rows located at the centre of the

lysimeter (rows 2, 3 and 4). The mass balance was slightly underestimated due to the more significant spreading of the elution curves of rows 1 and 5 which had non null concentrations at $t = 10$ h (**Figure 2(d)**).

As with test E1, the results between the three series A, B and C were of the same order of magnitude but with larger differences in test E2 (**Table 3**). These differences were due to the local modification of the flow introduced by the interface between the two materials in test E2. These modifications were not present in test E1 in a homogeneous medium.

3.3. Validation of the Numerical Model

In steady state, the numerical model E1 was first validated by comparing it with the measurements of volumetric water content (TDR sensors). The results (**Figure 3(a)**) show that the simulated values are quite close to the experimental values. The calculated moisture profile predicted a saturated zone at the bottom of the LUGH lysimeter.

As with test E1, the volumetric water contents modelled for test E2 were compared to the TDR measurements. The difference between the simulated data and the experimental data was slightly larger than in the previous case. Nonetheless, the model and the measurements corresponded when taking the error margin relating to the measurements into account. The reduction of the correspondence between the model and the experiment can be explained by the sensitivity of the TDR sensors. Since the distance between the two rods of the TDR sensor was 5 cm, it can be assumed that the measurement was performed on the volume of influence of the sensor, in the order of 5 to 10 cm in diameter around the rods. Averaging the volume probed did not permit detecting the contrast of volumetric water content at the interface between the two materials, or the existence of a zone of water accumulating in the form of a more or less thick film of water on the sand side. Nonetheless, the results of the

Combined Effect of Infiltration, Capillary Barrier and Sloping Layered Soil on Flow and Solute
Transfer in a Heterogeneous Lysimeter

175

TDR sensors of test E2 clearly show the existence of a water accumulation zone at the interface between the two materials: the volumetric water content in the sand is higher than the volumetric water content in the bimodal material along the interface (**Figure 3(b)**). This increase in water content may stem from capillary retention capacities or to the diversion of flows due to the interface.

The numerical simulation of tests E1 and E2 was performed by assuming the symmetry of initial and boundary conditions through the width of the LUGH. In addition, the results of the simulation of the outlets positioned on series A, B and C are the same. The elution curves simulated on line B were calculated over 10 hours, counting from the application of the tracer. The eluted solute was obtained by integrating the distribution of the calculated concentrations crossing the surface of the squares at the bottom of the lysimeter. The calculated elution curves were then compared for the five rows (rows 1, 2, 3, 4 and 5) with the mean of the experimental elution curves (mean of lines A, B and C, **Figures 2(b)** and **(d)**).

In terms of transfer, the results of the model of test E1 represented the same trends as the curves measured on the five rows well (**Figure 2(b)**). This allowed us to validate coupling the Richards equation with the convection-dispersion equation for the study of water flow and solute transfer in model E1.

The numerical results provided the volumetric water content field and the vector field of the Darcy flux permitting us to study the water and solute transfer process in the LUGH lysimeter. We observed that, in test E1, the vertical flux corresponding to gravity flow mechanisms dominated with high velocities below the supply zone. The trajectory of the flow was almost vertical and the stream lines were diverted only near the bottom of the lysimeter (rows 2 to 4, **Figure 3(a)**). The fluxes of rows 1 and 5 were weak and dispersed. This explains the clear difference between the elution curves of rows 2 to 4, at the centre, and those of rows 1 and 5, at the sides. Indeed, the lower boundary condition is a condition of free drainage. However, the water is only drained when a layer of soil just above the bottom of the lysimeter reaches saturation [23]. This accumulation zone developed progressively, and at steady flow state, it was larger at the centre of the lysimeter and decreased towards the sides, thereby leading to the diversion of the flows at the bottom of the lysimeter. The presence of this accumulation and the induced lateral flows may have resulted from the finite geometry of the lysimeter and the proximity of the lower boundary condition. A numerical study that varied the dimension of the system tended to show that the lateral flows lessened at the same depth when the boundary condition was lowered (data not illustrated). These lateral flows are responsible for the output of the fluxes of water and solutes in rows 1 and 5 with a lag due to a trajectory longer than that for rows 2 to 4. The model therefore made it possible to demonstrate that the finite geometry of the lysimeter led to lateral flows and explain the outputs observed at the edge of the system (rows 1 and 5). In the field, where water can flow freely at depth, there is no such lower limit and transfers are essentially vertical.

The hydrodynamic and hydrodispersive parameters of the bimodal material were conserved following the optimisation step of model E1 and applied to the water and solute transfer conditions specific to case E2. The results of E2 showed that the outputs upstream of the slope (rows 3, 4 and 5) were simulated well and a little more lag (residence times) for the two outputs downstream (rows 1 and 2). As with the tortuosity parameter, the uncertainty obtained on dispersivities α_1, α_2, and α_3 could be relatively large. Theoretically, an uncertainty on the dispersivity and tortuosity parameters should affect the dispersion of the elution peak though not its position [40]. This uncertainty is overlapped by that on the other parameters of the model $(\theta_r, \theta_s, \alpha, n,$ and $K_s)$ for the two different materials which may explain the slight differences between the models and the measurements. Nonetheless, in our study, the differences between simulation and observation were considered acceptable and confirmed the model's capacity to reproduce the hydrodynamic hydrodispersive behaviour of model PROF2 of the LUGH lysimeter.

As with the case of profile PROF1, studying the distribution of volumetric water content and the vector field of the Darcy flux provided valuable information on the water and solute transfer process in the PROF2 system. The distribution of volumetric water content in E2 demonstrated the presence of a capillary barrier which was the cause underlying the accumulation of water at the interface between the two materials (**Figure 3(b)**). It resulted from the contrast between the hydrodynamic characteristics of these materials. For the same capillary pressure along the interface, the hydraulic conductivity in the sand was always greater than that in the bimodal material. From the hydrological standpoint, the lower layer, in this case bimodal, was less permeable, thus impeding the entire transfer of the flux through the interface, since part of the flux was diverted along it.

When the water arrived at the interface between the two materials, it accumulated in the sand, increasing capillary pressure locally before penetrating the bimodal medium. The form of the accumulation zone on an inclined plane developed progressively upstream and downstream from the supply zone asymmetrically due to gravity and capillarity. Below the interface, the hydrodynamic contrast between the two materials limited the volumetric water content in the bimodal medium and the flow returned to the vertical direction. Moreover, the flow velocity, indicated by the vector field of the flux, was quite low.

The water and the solute were then dispersed and distributed almost throughout the volume of the bimodal zone. The trajectory of the flow from the source (supply) was therefore extended by the effect of the slope and the residence time of the solute increased in comparison to test E1 (**Figures 3(c)-(h)**). Water accumulated at the bottom of the lysimeter in a similar way to that of test PROF1. However, the effect of the lower boundary condition appeared less pronounced.

Modelling the flow clarified understanding of solute transfers insofar as they followed the same path as the water. As with test E1, in test E2, the solute first passed through the sand symmetrically in relation to the supply surface. Then, the plume of solute was deformed by the capillary barrier and the slope of the interface. At the end of the solute pulse ($t = 0.75$ h), the quantity of solute present in the sand was pushed by the water ($t > 0.75$ h). The preferential transfer in the lysimeter was illustrated by the evolution of the shape of the plume corresponding to the zones of strongest concentrations. It is clear that the plume shifted along the interface. The volumes of solute injected and the input velocities of water and solute were the same for E1 and E2. The evolution through time of the relative concentration profile, C/C_0, of profile E2 presented a slight lag in comparison to that of profile E1. As a function of time, the horizontal dispersion in E2 was greater and penetration was less deep than in case E1. The upstream/downstream shift also appeared clearly. The elutions modelled at the outlet were characterised by this shift and were fully consistent with the experimental data.

3.4. Apparent Capillary Barrier and Moisture History

In this part, we used the validated model to predict the impact of the interface angle and the imposed velocity under upper boundary conditions on preferential flows and several related metrics (diversion of the barycentre of the leached volume, volume of water in the sand and bimodal gravel (associated with the accumulation at the interface), the vertical and horizontal components of the velocity at the interface). Afterwards, we studied the impacts of preferential flows on transfers in terms of residence time.

One of the most obvious consequences of the capillary barrier effect was the diversion of flows of water and solute at the interface between the two materials. This diversion led to a lateral shift of the centre of gravity at the outlet of the lysimeter in comparison to the inlet. Determining this shift is an important characteristic used to quantify the role of the capillary barrier.

The results of the sensitivity test showed that the shift of the centre of gravity of the water between the inlet and the outlet, ΔD [L], all velocities confounded, depended linearly on the slope of the interface (**Figure 4(a)**). For each velocity q_i, this relation is represented as $\Delta D_i = a_i \Phi_i + b_i$ with a_i and b_i being constants. The adjustment of these relations, indicated by the determination coefficient, R^2, is high for all five velocities ($R^2 > 0.998$).

The value of b_i [L] corresponds to the diversion for the case of a null slope with velocity q_i. This value is equal to 1 to 3 cm due to the bias introduced by the offset position of the supply in relation to boundary conditions of the lysimeter. These values remain negligible in comparison to the shifts caused by the inclination of the interface.

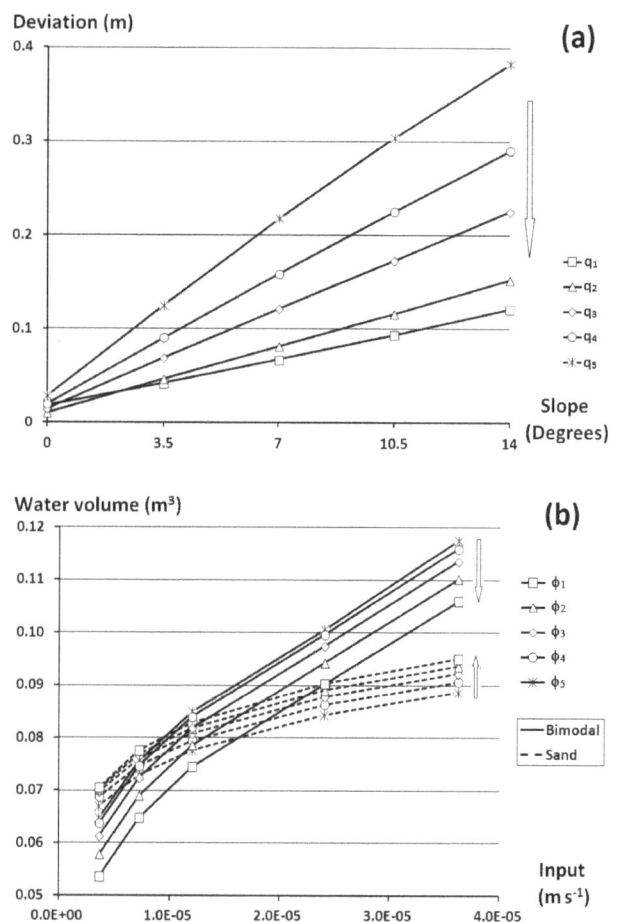

Figure 4. Diversion of the centre of gravity of the water discharges at the outlet as a function of the slope (a) and volume of water in the sand (dotted lines) and in the bimodal medium (lines) as a function of infiltration velocity (b) (the arrows show the trend of increase of angle).

The value of a_i [L·rad^{-1}] corresponding to the tangent of line $\Delta D_i = f(\Phi_i)$ represents the acceleration of the diversion as a function of the slope. It increases rapidly when the supply velocity decreases. This case shows how the moisture history of the soil and the boundary conditions (represented here by the velocity) can favour the occurrence of a preferential flow. Therefore, at low velocities, the water can flow laterally and preferentially further from the supply zone. The effect of the capillary barrier is more considerable. At a higher velocity, the initial kinetic energy is higher, resulting in significant inertia. The flows therefore deviate less. The value of a_i can be calculated using the value of q_i as in:

$$a_i = -0.448 \ln(q_i) - 4.1856 \qquad (14)$$

Finally, the diversion of the centre of gravity can be deduced as follows:

$$\Delta D_i = b_i - \Phi_i \left[0.448 \ln(q_i) + 4.1856 \right] \qquad (15)$$

This relation permits us to calculate the diversion of the centre of gravity of the water as a function of the infiltration velocity and the slope. This shows that the diversion results from the combined effect of the two factors of slope and imposed velocity at the upper boundary.

The volume of water in the lysimeter is also an important parameter for studying solute transfer. In particular, the presence of the interface results in the accumulation of water in the sand and has an umbrella effect, in turn resulting in a "loss" of water in the underlying gravel. Here, the aim is to link these volumes to the geometry of the system (angle of the interface) and to the boundary conditions.

For the same infiltration velocity, the volume of water contained in the sand increased slightly as a function of the angle of the interface, whereas that in the bimodal material decreased (**Figure 4(b)**). The variations of the volume of water in the sand and in the bimodal material as a function of the supply velocity were almost the same for the different input velocities. Indeed, curves $V_e(\Phi)$ are parallel.

Contrary to the effect of the slope, the velocity affected the stock of water in the lysimeter by favouring the humidification of the overall system. Nonetheless, we observed a more significant change in the bimodal material than in the sand. The variation of the volume of water between the lowest velocity and the highest one in the sand was 33.0% ± 0.91% (average value for the five different slopes), whereas that of the bimodal material was 87.4% ± 5.46%. This difference shows that the accumulation of water at the bottom of the lysimeter (in the bimodal material) was driven more by the increase of the supply velocity than by increasing the angle of the interface. This main characteristic must be taken into account

in further studies using lysimeters. Indeed, this accumulation resulting from the lower condition of free drainage also "disturbed" the results.

In order to better understand the water and solute transfer mechanism at the interface between the two materials, the velocity field at the interface was analysed. The aim was to determine how the slope of the interface and the supply velocity acted on components V_y and V_z of the velocity. Component V_x was not counted in this study due to the symmetrical arrangement in the direction of axis x (direction of the length of the lysimeter). The numbering 1 and 2 corresponds to the components in the sand and in the bimodal material respectively.

The values of the horizontal components V_{y1} were significantly non null and highlight the horizontal diversion of the flow at the interface (**Figure 5**). On the contrary, component $V_{y2,}$ in the bimodal material, was very small. This shows that the part of the water penetrating the interface flowed vertically in the bimodal material. Its lateral diffusion was negligible. When the angle of the interface was null, the horizontal component V_{y1} was null at the centre of the injection zone, positive on the right and negative on the left. Thus the flow was diverted to the edges of the lysimeter except at the barycentre of the injection zone (by symmetry) and next to the walls of the lysimeter (tangential flows). The diversion was maximal at about a third of the distance between the barycentre of the injection zone and the sides of the lysimeter. Introducing a slope (relative to the angle) resulted in establishing tangential flows at the interface (directed leftwards) and thus accentuating the corresponding negative values for the horizontal component V_{y1}. The intersection of the curve of V_{y1} with axis $y = 0$ defines a point of division between the flow upstream and downstream along the slope. This position was strongly dependent on the angle of the interface (**Figures 5(a)-(c)**). To the left of this point, $V_{y1} = 0$, the water flowed downstream of the slope, and the water running on the interface infiltrated into the underlying layer, explaining why the vertical velocity (and thus the infiltrated flux) was higher, i.e. $|V_{z1}| \geq |V_{z2}|$. The value of V_{y1} and V_{z1} in proportion to the value V_{z2} increased as a function of the slope (**Figures 5(a)-(c)**) and decreased as a function of the velocity of infiltration (**Figures 5(d)-(f)**). This proportion represents the development of preferential flows as a function of the two parameters above.

3.5. Residence Time and the Peaks of Elution Curves of Rows 1 and 3

Rows 1 and 3 were analysed essentially as they are representative of transfers at the centre and at the sides of the lysimeter, respectively. The residence times of row 3 were practically independen of the slope (**Figure 6(a)**).

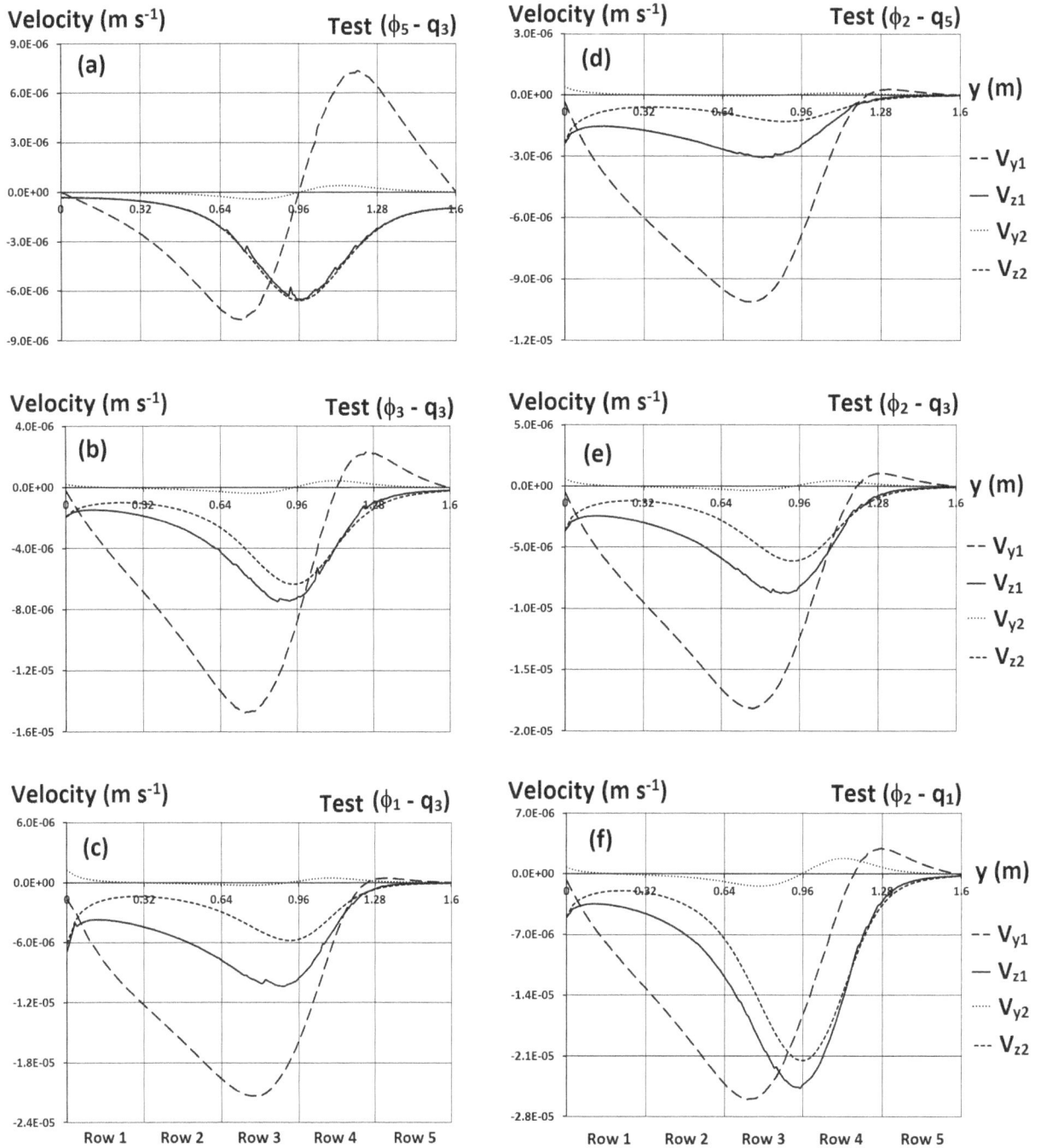

Figure 5. Velocity components in the sand (V_{y1} and V_{z1}) and in the bimodal medium (V_{y2} and V_{z2}) as a function of slope and the infiltration velocity.

They depended only on the infiltration velocity. Indeed, the water effluents of row 3 were formed by the fraction of water that penetrated the interface exactly in the direction of the supply. This flux was almost vertical. Likewise, the effect of the slope of the interface on the value of the peak of the elution curve of the solute of row 3 was slight. The effect of the supply velocity in this case is also slight (**Figure 6(d)**). The values obtained for the

relative concentration of the peak of the elution curve of row 3, C/C_0, were similar to those of the one-dimensional tests performed in the laboratory column. This can be explained by the transfer mechanism, which was almost one-dimensional close to row 3. The capillary barrier effect in this case is negligible.

Conversely, for row 1, the residence time depended on the slope of the interface and on the inverse of the supply

Combined Effect of Infiltration, Capillary Barrier and Sloping Layered Soil on Flow and Solute
Transfer in a Heterogeneous Lysimeter

179

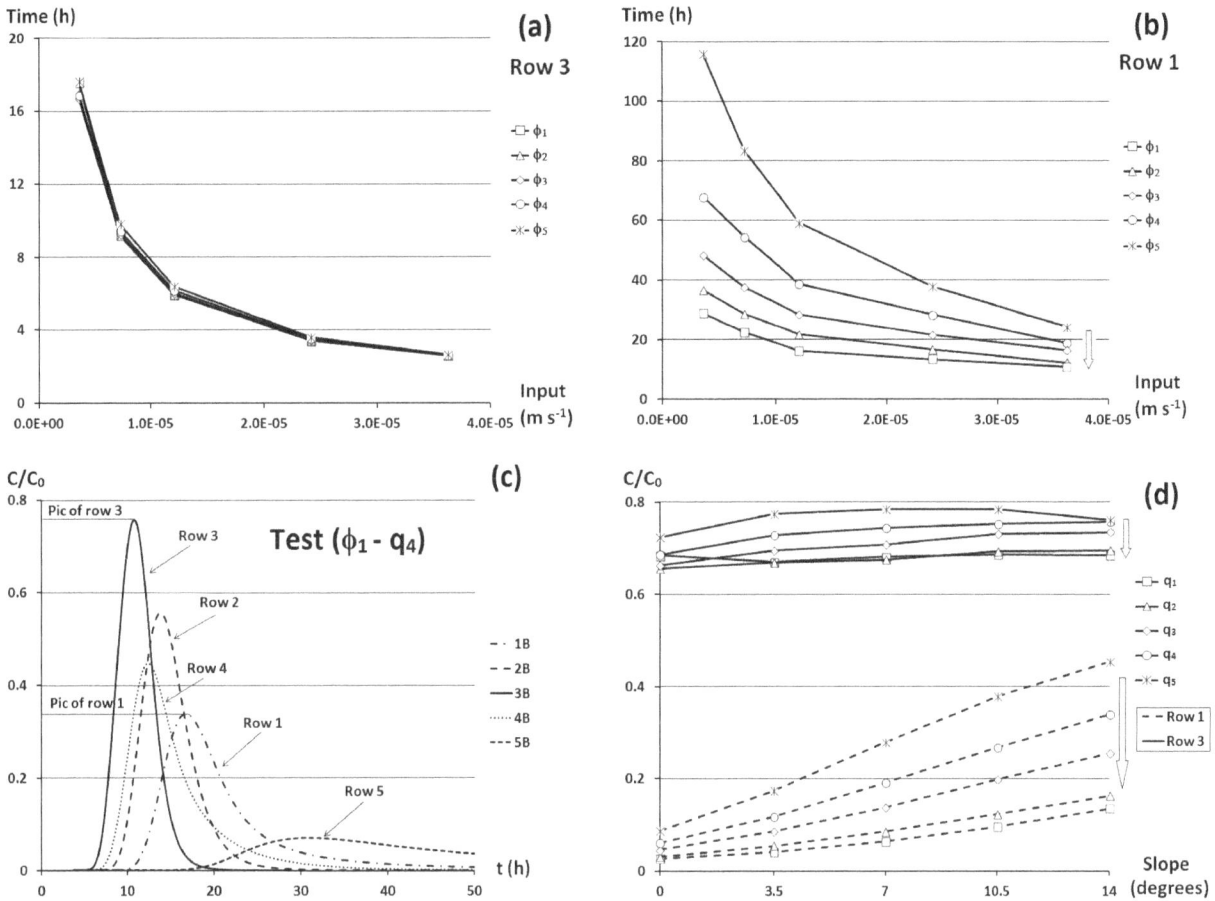

Figure 6. Comparison of residence time ((a) and (b)); The arrows show the trends of increase of angle and velocity and detail of elution curves for a specific case (c) and the peak of the elution curves of rows 1 and 3 (d).

velocity (**Figure 6(b)**). In addition, the capillary barrier effect is more obvious for the peak of the solute elution curves. The value of the peak of these curves increases considerably as a function of the slope (**Figure 6(d)**). This increase stands out more when the supply velocity decreases. This can be explained by the same argument proposed previously to explain the flow field: the fluxes crossing row 1 were generated by the part of the water diverted along the slope. The increase in the diversion of the accumulated water resulted in an increase in the velocity of the water in the direction downstream of the slope. The solute was therefore transferred more rapidly towards row 1. In other words, the preferential flows and their impacts on the solute transfers were favoured when the angle of the interface increased and the supply velocity decreased.

The combined effect of the supply velocity and the slope of the interface on the total elution curve of the lysimeter was also studied (**Figure 7**). Indeed, this information is often considered to the detriment of more precise sampling. We recall that the total curve was obtained by integrating all 15 elution curves. This curve represents the homogenised behaviour of the lysimeter. The shape of all the total elution curves takes that of a log-normal type of distribution with, however, a tail with a relatively wide spread.

Although the supply velocity was high enough (in this case $q_1 > q_2 > q_3$), the peak of the total curve was mainly determined by the effluents of the centre of the lysimeter (in particular row 3) which does not depend on the angle of the interface between the two materials. In this case, the curve obtained is slightly dissymmetric and mono-modal on which the influence of the angle is weak. The effect of, the angle starts becoming marked from $\Phi = 10.5°$ for flow rate q_4 and at $\Phi = 7°$ for flow rate q_5 (**Figure 7(b)**). In both the last two cases for which preferential flows were very developed, the elutions of the rows at the sides of the lysimeter (mainly row 1) in creased and contributed more to the total elution. In these cases (smallest velocities), increasing the angle resulted in increasing the dissymmetry of the global curve (**Figure 7(d)**) with a widening of the tail. This resulted in increasing the contribution of preferential flows at the sides of the lysimeter. We also observed that for each infiltra-

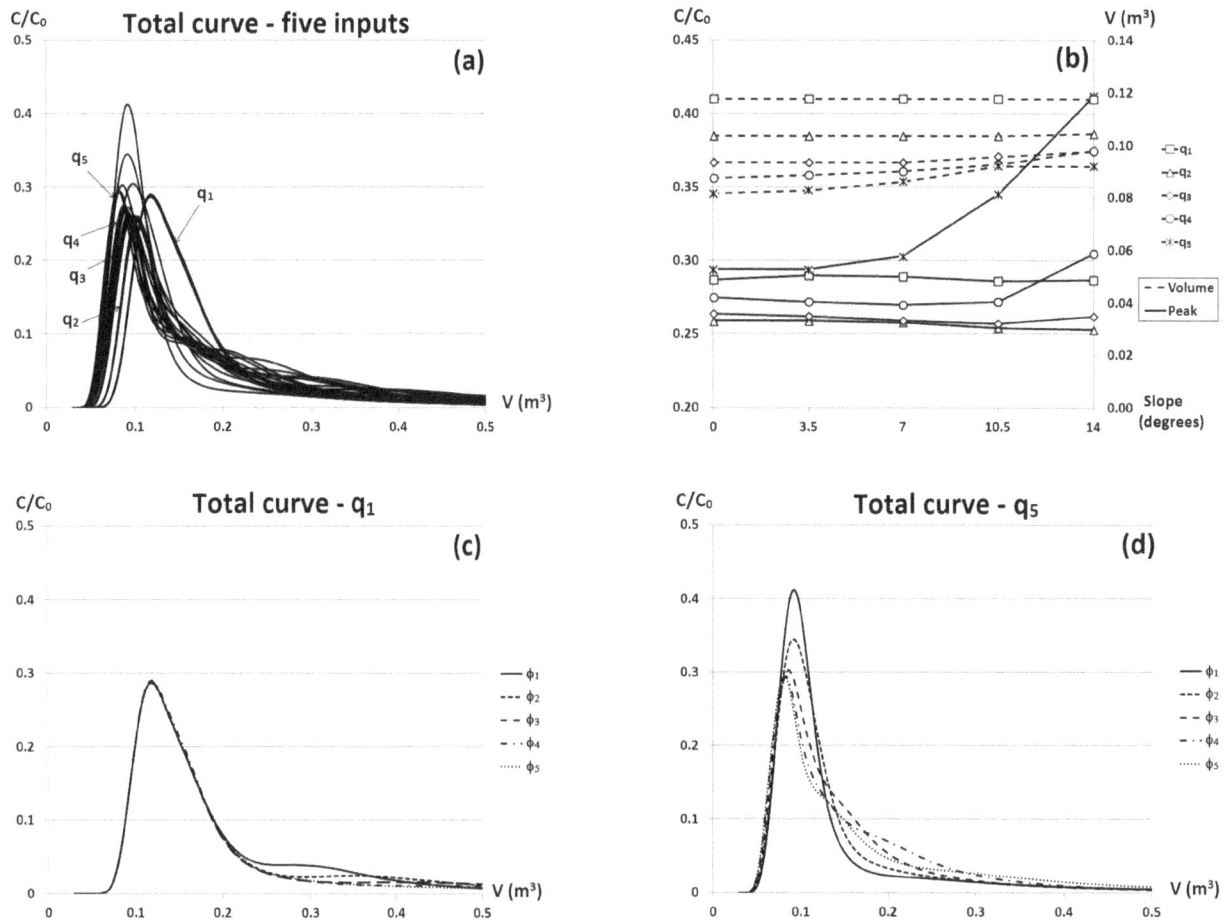

Figure 7. Peak and total volume of water injected as a function of slope and infiltration velocity.

tion velocity, the total volume of water injected necessary to push the solute to the total peak was quite similar, with a slight increase when the slope increased, above all for the lowest velocities. Conversely, this volume decreased and the elution curve was less spread when the velocity decreased. This shows that, for the total curve, the concentration of the solute was higher when velocities were lower.

4. Conclusions

This study focused on the development of a lysimeter for studying the transfer of heterogeneous flows of water and solute in a heterogeneous and unsaturated medium. To do this a pilot LUGH lysimeter was developed and used to observe the flow and transfer of a non-reactive solute in the injection of a steady state flow through the medium. The experimental and numerical results were demonstrated, explaining the occurrence of the capillary barrier phenomenon and the major role played by the initial and boundary conditions of the lysimeter in the establishment of this type of flow.

By using fifteen different outlets, the experimental

flow tracing data permitted studying the temporal and spatial evolution of the water discharges and fluxes of solute at the bottom of the lysimeter in comparison to the input at the surface and the configuration of the system. The heterogeneity of the discharges at the outlet, peaks, variances and residence times of the elution curves of these fifteen outlets provided detailed data on the hydraulic behaviour of the heterogeneous system and led to better understanding of the role played by the capillary barrier on the water and solute flows, by taking into account the effect induced by the finite volume of the lysimeter.

The model permitted representing the solute diffusion zone and understanding the effect cause by the presence of the slope between the two materials. Consequently, numerical modelling presents a considerable advantage in explaining the transfer process in the lysimeter. The capillary barrier between two materials in the lysimeter can be quantified by studying the vertical shift between the centres of gravity of the water supply and the distribution of the discharges at the outlet. Under the effect of the capillary barrier in a heterogeneous medium, the wa-

Combined Effect of Infiltration, Capillary Barrier and Sloping Layered Soil on Flow and Solute
Transfer in a Heterogeneous Lysimeter

181

ter and solute were dispersed, resulting in increased solute residence time in comparison to the homogeneous case.

A series of tests was performed to quantify the impact of supply velocity and slope angle between the two materials on water and solute transfer. It was shown that shift between the centres of gravity of the distribution of the water discharged at the outlet increased linearly as a function of the angle of the interface between the two materials. In addition, reducing the supply velocity and increasing the angle of the interface clearly determined the development of preferential flows. The effect of the bottom of the lysimeter played a very important role in the analysis of the experimental and numerical results. Indeed, the accumulation of water at the base of the lysimeter depended more on the supply velocity than on the capillary barrier effect. Furthermore, the transfer of the solute vertical to the supply zone was practically independent of the angle of the interface. The solute flux at the bottom boundary was impacted by the slope angle mainly for the lowest supply velocity.

From the methodological standpoint, the association of simple tests with a numerical model allowed us to refine the estimation of the key parameters involved in water flow and solute transport. The numerical model permitted both highlighting the behaviour of the coupled water-solute transfer in 3D as a function of space and time and completing the experiments with the sensitivity test in order to pre-dimension new tests.

REFERENCES

[1] EPA, "Protecting Water Quality from Agricultural Runoff," US Environmental Protection Agency, 2005.

[2] EPA, "Protecting Water Quality from Urban Runoff," US Environmental Protection Agency, 2003.

[3] J. T. Smullen, A. L. Shallcross and K. A. Cave, "Updating the U.S. Nationwide Urban Runoff Quality Date Base," *Water Science and Technology*, Vol. 39, No. 12, 1999, pp. 9-16.

[4] S. E. Allaire, S. Roulier and A. J. Cessna, "Quantifying Preferential Flow in Soils: A Review of Different Techniques," *Journal of Hydrology*, Vol. 378, No. 1-2, 2009, pp. 179-204.

[5] J. Bouma, A. Jongerius, O. Boersma, A. Jager and D. Schoonderbeek, "The Function of Different Types of Macropores during Saturated Flow through Four Swelling Soil Horizons," *Soil Science Society of America Journal*, Vol. 41, No. 5, 1977, pp. 945-950.

[6] K. J. Beven and P. Germann, "Macropores and Water Flow in Soils," *Water Resources Research*, Vol. 18, No. 5, 1982, pp. 1311-1325.

[7] P. Cannavo, L. Vidal-Beaudet, B. Bechet, L. Lassabatère

and S. Charpentier, "Spatial Distribution of Sediments and Transfer Properties in Soils in a Stormwater Infiltration Basin," *Journal of Soils and Sediments*, Vol. 10, No. 8, 2010, pp. 1499-1509.

[8] D. E. Hill and J.-Y. Parlange, "Wetting Front Instability in Layered Soils," *Proceedings of the Soil Science Society of America*, Vol. 36, No. 5, 1972, pp. 697-702.

[9] T. Miyazaki, "Water Flow in Unsaturated Soil Layered Slopes," *Journal of Hydrology*, Vol. 102, No. 1-4, 1988, pp. 201-214.

[10] K.-J. S. Kung, "Preferential Flow in a Vadose Zone: 2. Mechanism and Implications," *Geoderma*, Vol. 46, No. 1-3, 1990, pp. 59-71.

[11] M. T. Walter, et al., "Funneled Flow Mechanisms in a Sloping Layered Soil: Laboratory Investigation," *Water Resources Research*, Vol. 36, No. 4, 2000, pp. 841-849.

[12] B. Bussière, S. A. Apithy, M. Aubertin and R. P. Chapuis, "Water Diversion Capacity of Inclined Capillary Barriers," *Proceedings of the 56th CGS-IAH Conference*, Winnipeg, 29 September-1 October 2003, pp. 192-200.

[13] A. Heilig, T. S. Steenhuis, M. T. Walter and S. J. Herbert, "Funneled Flow Mechanisms in Layered Soil: Field Investigations," *Journal of Hydrology*, Vol. 279, No. 1-4, 2003, pp. 210-223.

[14] S. Kaskassian, et al., "L'essai d'Infiltration Couplé à un Traçage Non-Réactif: Un Outil Pour Evaluer le Transfert des Polluants dans la Zone Non-Saturée des Sols," *L'eau, l'Industrie, les Nuisances*, Vol. 349, 2012, pp. 38-45.

[15] K.-J. S. Kung, "Preferential Flow in a Sandy Vadose Zone: 1. Field Observation," *Geoderma*, Vol. 46, 1990, pp. 51-58.

[16] L. Février, "Transfert d'un Mélange Zn-Cd-Pb dans un Dépôt Fluvio-Glaciaire Carbonate. Approche en Colonnes de Laboratoire," Ph.D. Thesis, Ecole Nationale des Travaux Publics de l'Etat, Lyon, 2001.

[17] L. Lassabatère, T. Winiarski and R. Galvez Cloutier, "Retention of Three Heavy Metals (Zn, Pb and Cd) in a Calcareous Soil Controlled by the Modification of Flow with Geotextiles," *Environmental Science and Technology*, Vol. 38, No. 15, 2004, pp. 4215-4221.

[18] E. Lamy, L. Lassabatère, B. Bechet and H. Andrieu, "Modeling the Influence of an Artificial Macropore in Sandy Columns on Flow and Solute Transfer," *Journal of Hydrology*, Vol. 376, No. 3-4, 2009, pp. 392-402.

[19] L. Lassabatère, et al., "Concomitant Zn-Cd and Pb Retention in a Carbonated Fluvio-Glacial Deposit under Both Static and Dynamic Conditions," *Chemosphere*, Vol. 69, No. 9, 2007, pp. 1499-1508.

[20] C. Lanthaler, "Lysimeter Stations and Soil Hydrology Measuring Sites in Europe—Purpose, Equipment, Research Results, Future Developments," *A Diploma Thesis for the Degree of Magistra der Naturwissenschaften, School of Natural Sciences, Karl-Franzens-University, Graz*, 2004.

[21] R. Schoen, J. P. Gaudet and T. Bariac, "Preferential Flow

and Solute Transport in a Large Lysimeter, under Controlled Boundary Conditions," *Journal of Hydrology*, Vol. 215, No. 1-4, 1999, pp. 70-81.

[22] T. P. Anguela, "Etude du Transfert d'Eau et de Solutés dans un sol à Nappe Superficielle Drainée Artificiellement," Ph.D. Thesis, Ecole Nationale du Génie Rural, des Eaux et Forêts, Paris, 2004.

[23] H. M. Abdou and M. Flury, "Simulation of Water Flow and Solute Transport in Free-Drainage Lysimeters and Field Soils with Heterogeneous Structures," *European Journal of Soil Science*, Vol. 55, No. 2, 2004, pp. 229-241.

[24] L. B. Bien, X. Peyrard, L. Lassabatère, T. Winiarski and R. Angulo-Jaramillo, "Transferts d'Eau et de Particules dans la Zone Non-Saturée Hétérogène: Développement du Pilote de Laboratoire LUGH," *Proceedings of the 35 Ième Journées du GFHN*, Louvain-la-Neuve, Belgique, 23-25 Novembre 2010, pp. 165-168.

[25] L. A. Richards, "Capillary Conduction of Liquids through Porous Mediums," *Physics*, Vol. 1, No. 5, 1931, pp. 318-333.

[26] M. T. van Genuchten, "A Closed form Equation for Predicting the Hydraulic Conductivity of Unsaturated Soils," *Soil Science Society of America Journal*, Vol. 44, No. 5, 1980, pp. 892-898.

[27] Y. Mualem, "A New Model for Predicting the Hydraulic Conductivity of Unsaturated Porous Media," *Water Resources Research*, Vol. 12, No. 3, 1976, pp. 513-522.

[28] J. Bear, "Dynamics of Fluids in Porous Media," American Elsevier, New York, 1972, p. 764.

[29] R. J. Millington and J. P. Quirk, "Permeability of Porous Media," *Transactions of the Faraday Society*, Vol. 57, 1961, pp. 1200-1207.

[30] D. Goutaland, *et al.*, "Hydrostratigraphic Characterization of Glaciofluvial Deposits Underlying an Infiltration Basin Using Ground Penetrating Radar," *Vadose Zone Journal*, Vol. 7, No. 1, 2008, pp. 194-207.

[31] L. Lassabatère, *et al.*, "Beerkan Estimation of Soil Transfer Parameters through Infiltration Experiments—BEST," *Soil Science Society of America Journal*, Vol. 70, 2006, pp. 521-532.

[32] L. Lassabatère, R. Angulo-Jaramillo, J. M. Soria Ugalde,

J. Simunek and R. Haverkamp, "Analytical and Numerical Modeling of Water Infiltration Experiments," *Water Resources Research*, Vol. 45, 2009, Article ID: W12415.

[33] P. Nasta, L. Lassabatère, M. Kandelous, J. Simunek and R. Angulo-Jaramillo, "Analysis of the Role of Tortuosity and Infiltration Constants in the Beerkan Method," *Soil Science Society of America Journal*, Vol. 76, No. 6, 2012, pp. 1999-2005.

[34] L. Lassabatère, *et al.*, "Effect of the Settlement of Sediments on Water Infiltration in Two Urban Infiltration Basins," *Geoderma*, Vol. 156, No. 3-4, 2010, pp. 316-325.

[35] E. Gonzalez-Sosa, *et al.*, "Impact of Land Use on the Hydraulic Properties of the Topsoil in a Small French Catchment," *Hydrological Processes*, Vol. 24, No. 17, 2010, pp. 2382-2399.

[36] D. Yilmaz, L. Lassabatère, R. Angulo-Jaramillo and M. Legret, "Hydrodynamic Characterization of Basic Oxygen Furnace Slags through Adapted BEST Method," *Vadoze Zone Journal*, Vol. 9, No. 1, 2010, pp. 107-116.

[37] M. T. Van Genuchten, F. J. Leij and S. R. Yates, "The RETC Code for Quantifying the Hydraulic Functions of Unsaturated Soils," US Environmental Protection Agency, Oklahoma, 1991.

[38] I. Mubarak, J. C. Mailhol, R. Angulo-Jaramillo, S. Bouarfa and P. Ruelle, "Effect of Temporal Variability in Soil Hydraulic Properties on Simulated Water Transfer under High-Frequency Drip Irrigation," *Agricultural Water Management*, Vol. 96, No. 11, 2009, pp. 1547-1559.

[39] L. B. Bien, R. Angulo-Jaramillo, D. Predelus, L. Lassabatère and T. Winiarski, "Preferential Flow and Mass Transport Modeling in a Heterogeneous Unsaturated Soil," *Proceedings of the First Pan-American Conference on Unsaturated Soils (Pam-Am UNSAT 2013)*, Cartagena de Indias, 20 - 22 February 2013, pp. 211-216.

[40] D. Schweich and M. Sardin, "Les Mécanismes D'interation Solide-Liquide et Leur Modélisation: Applications aux Etudes de Migration en Milieu Aqueux," 1986, pp. 59-107.

[41] COMSOL AB, "COMSOL Multiphysics User's Guide, Version 3.5a," *COMSOL AB, Grenoble*, 2008, p. 624.

19

Weather Radar Data and Distributed Hydrological Modelling: An Application for Mexico Valley

Baldemar Méndez-Antonio[1], Ernesto Caetano[2], Gabriel Soto-Cortés[3], Fabián G. Rivera-Trejo[4], Ricardo A. Carvajal Rodríguez[1], Christopher Watts[1]

[1]Sonora University, Hermosillo, Mexico; [2]National Autonomous University of Mexico, Mexico City, Mexico; [3]Metropolitan Autonomous University, Mexico City, Mexico; [4]Tabasco Autonomous Juárez University, Villahermosa, Mexico.

ABSTRACT

The frequent occurrence of exceptionally very heavy rainfall in Mexico during the summer causes flash floods in many areas and major economic losses. As a consequence, a significant part of the annual government budget is diverted to the reconstruction of the disasters caused by floods every year, resulting hold up in the country development. A key element to mitigate the flash flood hazards is the implementation of an early warning system with the ability to process the necessary information in the shortest possible time, in order to increase structural and non-structural resilience in flood prone regions. The real-time estimation of rainfall is essential for the implementation of such systems and the use of remote sensing instruments that feed the operational rainfall-runoff hydrological models is becoming of increasing importance worldwide. However, in some countries such as Mexico, the application of such technology for operational purposes is still in its infancy. Here the implementation of an operational hydrological model is described for the Mixcoac river basin as part of the non-structural measures that can be applied for intense precipitation events. The main goal is to examine the feasibility of the use of remote sensing instruments and establish a methodology to predict the runoff in real time in urban river basins with complex topography, to increase the resilience of the areas affected by annual floods. The study takes data from weather radar operated by the National Meteorological Service of Mexico, as input to a distributed hydrological model. The distributed unit hydrograph model methodology is used in order to assess its feasibility in urban experimental basin. The basic concepts underlying the model, as well as calibration and validation are discussed. The results demonstrate the feasibility of using weather radar data for modeling rainfall-runoff process with distributed parameter models for urban watersheds. A product resulting from this study was the development of software Runoff Forecast Model (ASM), for application in distributed hydrological models with rainfall data in real time in watersheds with complex terrain, which are usually found in Mexico.

Keywords: Radar; Distributed Hydrologic Modeling; Resilience

1. Introduction

Soussan and Bourton [1] defined adaptation as the ability to respond and adapt to real or potential impacts under changing weather conditions in order to moderate damage. In the face of extreme hydrometeorological conditions the greater or lesser resilience or strength, defines vulnerability which a community is exposed. Lack of resilience is manifested at the structural, physical, economic, social, political, and institutional level. Thus, the human vulnerability and lack of resilience explain many of the large-scale disasters and y to decrease the risk of disaster, measures should be taken to modify those factors [2]. In order to build resilience in areas vulnerable to flooding, structural and nonstructural measures must be taken. Early warning systems (EWS) supported by operational hydrological runoff prediction models are a fundamental part of the non-structural measures. The disaster risk reduction and increased preparedness to natural hazards in different development sectors have multiplier effects and accelerate the achievement of the Millennium Development Goals [3].

The use of distributed hydrological models has increased in the last three decades [4,5]. These models are developed in order to physically represent the hydrological processes occurring in a watershed through analogies and mathematical simplifications. Early attempts to develop distributed models were limited by computing capacity required for processing and storage of information, but the digital revolution, responsible for the huge growth in the quantity of geospatial data available today, has allowed the development of more robust distributed hydrological models [6]. It is noteworthy that the spatial and temporal variability of precipitation has a strong impact on the outcome of distributed hydrological modeling, while the density of the rain gauge network also plays an important role [7-9].

Additionally, EWS have been greatly improved with the development of geographic information systems (GIS) and the Internet [10]. Remote sensing instruments, such as radar and weather satellites, due to their ability to estimate the spatial variability of precipitation in real-time, are ideal for use in distributed hydrological modeling. The combination of remote sensing, GIS and the Internet, allows the collection and transmission of real-time data more efficiently. The data generated by these systems can be used as input to operational hydrological models and so estimate the basin response in the presence of extreme precipitation events. In Mexico, although efforts have been made to use this technology by both federal local governments and academic institutions, the results are still only at the diagnosis level [11].

This paper examines the use of weather radar data for operational hydrological modeling in an experimental catchment located in the metropolitan area of Mexico City. The methodology proposed here could easily be applied to others basins anywhere that radar data is available. Additionally, EWS can be integrated into contingency plans allowing countries to build resilience and prevent large losses from annual flooding, thus improving regional and national development.

2. Study Area

The selected case study area is the Mixcoac River experimental basin, located in Mexico City, (**Figure 1**). This

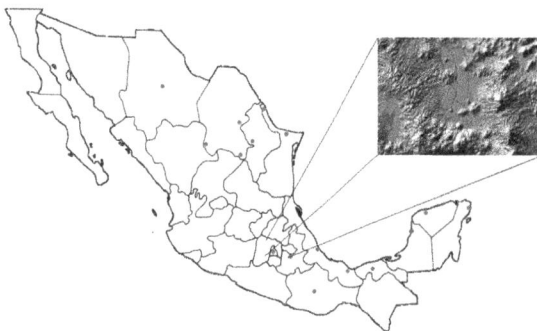

Figure 1. Basin study location and Mixcoac River Basin topography.

is a mountain river, with low runoff in dry period and sudden and intense floods in the rainy periods. Intense precipitation occurs often due to topographically enhanced convection over this area [12]. This basin was selected for this study because of the availability of input data from: 1) the Cerro Catedral weather radar, located 35 km from the study basin (**Figure 1**); 2) runoff and precipitation estimates from 78 rain gauges network within the radar coverage area. This data set allows us to establish a robust relationship between rainfall and runoff that eventually will help calibrate the hydrological model.

3. Hydrological Modeling

In traditional hydrological models, the runoff is produced by the fraction of precipitation not absorbed by the soil, this flow component is referred as direct or surface runoff, and the volume portion of precipitation which has produced it, is called excess or effective precipitation [13]. This type of runoff is known as Horton excess infiltration mechanism and usually occurs in rivers with steep slopes or low permeability soils. Another mechanism (Dunne) occurs when soil moisture is saturated, known as over-saturation, is more common in flat areas with permeable soils or near wetlands [14]. The river Mixcoac is a typical mountain stream, so the Horton mechanism is employed for obtain the infiltration.

By using a transfer function, the precipitation surpluses are converted into direct runoff, and are added to the base flow to obtain the total runoff hydrograph. This scheme corresponds to lumped parameter hydrologic models, which use spatial averages for physiographic features and precipitation and reproduce the temporal variability of the output basin response [15]. Moreover, the distributed hydrological modeling considers the spatial variability of the physical properties and precipitation dividing the basin into sub-basin or cells. Naturally, the development in remote sensing and geographic information systems has facilitated spatially distributed information management. An advantage of distributed models is that they allow the analysis of different elements that influence the hydrological response and can be modified by human intervention, such as the vegetation and land use, and with appropriate calibration. Distributed models can estimate changes in the hydrological response of the basin to extreme precipitation events, caused by these interventions.

Distributed models obtain the flows in each of the sub-basins simultaneously, which can indicate the state of the system at any point of the drainage network and improve flood risk assessment. Namely in distributed models the spatial variation of the characteristics and processes are explicitly considered, while in aggregate models, spatial variations are averaged or ignored.

Although it is believed that implementing Geographic

Information Systems (GIS), the problem is solved, this is just a tool to facilitate the determination of watershed hydrologic parameters, but no improvements due to spatial data scarcity and/or temporal components of the hydrological cycle. If the input precipitation data is obtained from a sparse density/spatial distribution rain gauge network, these data cannot characterize adequately the rain fields at the same level of spatial detail as the terrain described by the models. This occurs because the rain gauge network does not necessarily detect the most intense storm so often interpolated data or extrapolated are used, resulting in misrepresentation of observed rain fields. In this sense, modelers have great interest in estimating precipitation from data from remote sensing instruments, such as weather radar, which potentially provide estimated rainfall at the spatial detail required by distributed hydrological models. In this regard, a variety of models such as HEC-HMS, SWAT, CEQUEAU, MIKE-SHE, TOP MODEL, MERCEDEZ, Topkapi, etc., consider the use of such data [16].

It is also important to note that distributed models give good results without an excessive amount of parameters. From the operational point of view, the model must be simple and of fast execution, without losing the physical representation of phenomena. The simplicity and agility of the model operation are key factors for applications in operational forecasting, since otherwise there may be too little time for decision making and taking action to mitigate the effects of flash floods. Models with many parameters are only used in experimental watersheds instrumented for this purpose [17]; for this reason, they are not attractive for distributed modeling in countries like Mexico where measurements are scarce.

4. Methodology

4.1. Data Grid Generation

In distributed hydrological modeling, the Mixcoac river basin is represented by a set of square cells, where each cell is considered as the basic runoff production unit (**Figure 2**), with a spatial resolution of 1×1 km. The georeferenced grid containing the basin physiographic

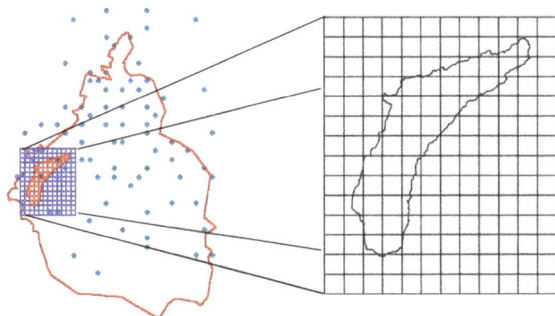

Figure 2. Weather radar grid data area.

parameters, such as: the land use, soil type and curve number (CN) obtained from the digital elevation model (DEM). The digital precipitation arrays were estimated from a calibration equation for this radar survey area [18].

Two storms were selected where radar rainfall (every 15 minutes) and runoff data were available. These storms occurred on July 28 and August 23 1998.

4.2. Distributed Hydrological Model

The distributed hydrological model comprises two conceptual sub-models: one for the runoff production in each of the components of the distributed system, and another, which represents travel runoff to be added, from each cell to reach the basin outlet. The following describes each of the sub-models.

4.3. Runoff Production Sub-Model

The runoff production in each hydrological unit (cell) is obtained from an infiltration or loss model. These losses consist of initial abstraction and the ground infiltrated water during the storm. Initial losses include water intercepted by the vegetation, the water stored in the depressions of the surface (puddles) and water infiltrated into the ground until it is saturated.

The Natural Resources Soil Conservation Service (SCS), or Curve Number (CN) was used to determine runoff because of its simplicity [19]. This is one of most widely used methods for estimating runoff volumes estimates from a single parameter, the main physical characteristics of the watershed such as slope and use and soil type are considered to produce runoff. This method has the advantage of high predictability and stability; in addition, it is a conceptual method which estimates runoff directly from precipitation [20].

Once the basin has been divided into rectangular cells, the value of the curve number (CN) is set to each of the soil properties (type and land use) and then the direct runoff in each of the cells is estimated. The water volume not converted into runoff is infiltrated into the ground, where part is stored as soil humidity and the rest passes to form deeper underground storage.

The SCS method does not explicitly include any infiltration scheme, so this was estimated directly from accumulated runoff, accumulated precipitation, soil storage capacity and initial losses. The conversion of rainfall to runoff, essential to surface hydrological modeling, is based on the conservation of mass or water balance;

$$P = P_e + I_a + F_a \qquad (1)$$

where P = total precipitation (cm); P_e = effective precipitation; I_a = initial infiltration i (cm); F_a = accumulated infiltration (cm). I_a and F_a represent losses and their quantification is based on two fundamental assumptions [19]. The first states that the relationship between the effective volume of precipitation (P_e), or direct runoff, and the maximum potential runoff ($P - I_a$), correspondent to an impervious surface, is equal to the ratio between the real infiltration F_a and the maximum potential infiltration S (Equation (2)). The second hypothesis assumes that the initial infiltration is directly proportional to the potential retention (Equation (3)):

$$\frac{P_e}{P - I_a} = \frac{F_a}{S} \qquad (2)$$

$$I_a = \lambda S \qquad (3)$$

Emerging Equations (1) and (2):

$$P_e = \frac{(P - I_a)^2}{P - I_a + S} \qquad (4)$$

From Equation (3), assuming $\lambda = 0.2$,

$$P_e = \frac{(P - 0.2S)^2}{P + 0.8S} \qquad (5)$$

S (cm) is given:

$$S = \frac{(2540 - 25.4CN)}{CN} \qquad (6)$$

And P_e (cm) is obtained [21] as follows:

$$P_e = \frac{\left(P - \dfrac{508}{CN} + 5.08\right)^2}{P + \dfrac{2032}{CN} + 20.32} \qquad (7)$$

The curve number (CN) is determined from the land use and soil type defined by the US Soil Conservation Service, and the values of P and P_e are expressed in cm. Equations (6) and (7) are valid for $P \geq I_a$. The parameter I_a depends on regional geological and climatic factors.

The main hydrological interest in land use maps lies in the infiltration modeling as a function of soil properties, thereby capturing spatial variability. To determine infiltration parameters from soil properties require some re-classification of representative soil parameters units for the hydrological model. The SCS proposed a criterion for to estimate the effective precipitation in terms of total precipitation and soil characteristics from a table of values for the curve number according to soil type [22]. For the particular case of Mexico, the classification hydrological soil texture and curve number given by [23] were used (**Tables 1** and **2**). Soil properties (**Table 1**), depending on their permeability, is used to define the hydrological soil type. **Table 2** describes the use of soil, which together with the soil type defined in **Table 1**, and the slope, enables allocation of the curve numbers. Geo-referenced soils, vegetation and topography maps

Table 1. Hydrological soil type classification (source [23]).

Hydrological Soil Group	Properties	Permeability
A	Sands with little slime and clay (minimum runoff)	Very High
B	Fine sands and slime	Good
C	Very fine sands, slime and quite clay	Medium
D	Clays in large quantities, shallow soils almost impermeable (maximum runoff)	Low

(DEM) facilitate the processing and spatial allocation of curve numbers. Once curve numbers maps were obtained (CN), then Equation (7) is applied to obtain the runoff generated by each storm.

4.4. Runoff Routing Sub-Model

The routing of effective precipitation (P_e) at the watershed outlet is an interdependent component in the hydrological cycle, while a proportion of the rainfall is lost to infiltration, excess rain generates runoff, which accumulates and drains through the stream network to the basin outlet. The widely used transfer hydrological method the unit hydrograph [24]—is applied to this experimental watershed.

The basin was divided into cells, which represent a set of elements where the continuity equation is applied. The change in the volume V_s stored in the drain network element during a time interval expresses the difference between the stored volume at the end of the previous period and the stored volume in the end of the next period. That is, the change in storage V_s is equal to the difference between the volume of water entering V_i and leaving the soil volume V_o during the time interval Δt:

$$V_s = V_i - V_o \left(\overline{I_t} - \overline{O_t} \right) \Delta t \overline{O_t} \qquad (8)$$

where $\overline{I_t}$ and $\overline{O_t}$ are the mean input and output flood discharges, respectively, during the time interval Δt. The previous equation can also be represented as:

$$V_s = \int_{t_0}^{t_1} (I - O)\, dt \qquad (9)$$

or in finite differences as:

$$V_s = \left(\frac{I_{t_0} + I_{t_1}}{2} - \frac{O_{t_0} + O_{t_1}}{2} \right) \Delta t \qquad (10)$$

The stream flow routing from any point to the basin outlet can be modeled by a simple aggregation through the distributed unit hydrograph or the Clark modified unit hydrograph [25] also called isochrones distributed unit hydrograph [26], (**Figure 3**).

Table 2. Soilcover (source [23]).

Cover type	Slope %	Hydrologic soil group			
		A	B	C	D
No cultiveted	-	77	86	91	94
Row crops					
Straight row	>1	72	81	88	91
Straight row	<1	67	78	85	89
Contoured	>1	70	79	84	88
Contoured	<1	65	75	82	86
Terraced	>1	66	74	70	82
Terraced	<1	62	71	78	81
Small grain					
Straight row	>1	65	76	84	88
Straight row	<1	63	75	83	87
Contoured	>1	63	74	82	85
Contoured	<1	61	73	81	84
Terraced	>1	61	72	79	82
Terraced	<1	59	70	78	81
Rotation meadow					
Straight row	>1	66	77	85	89
Straight row	<1	58	72	81	85
Contoured	>1	64	75	83	85
Contoured	<1	55	69	78	83
Terraced	>1	63	73	80	83
Terraced	<1	51	67	76	80
Grassland					
-	>1	68	79	86	89
-	<1	39	61	74	80
Contoured	>1	47	67	81	88
Contoured	<1	6	35	70	79
Permanent grassland		30	58	71	78
Forest					
Very sparse	-	56	75	86	91
Sparse	-	46	68	78	84
Normal	-	36	60	70	77
Dense	-	26	52	62	69
Very dense	-	15	44	54	61
Roads					
Dirt	-	72	82	87	89
Paved	-	74	84	90	92
Residential area	-	77	85	90	92
Cemeteries	-	68	79	86	89
Brush	-	48	67	77	83
Parks	-	53	72	83	88

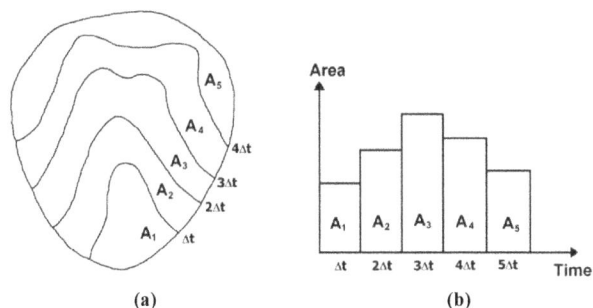

Figure 3. (a) Isochrones; and (b) Time-area histogram for a watershed (adapted from [27]).

The runoff routing to the basin outlet was performed using the Muskingum method. This method employs the continuity equation (Equation (10)) and a relationship between storage V and inputs and outputs of the analysis section (Equation (11)).

$$V = K\left[xI + (I-x)O\right] \quad (11)$$

where I is the inflow, O is outflow, K is the attenuation coefficient and x storage weight factor that relates the input and output storages in the current segment.

4.5. Distributed Unit Hydrograph

In order to use weather rainfall data, the method of Clark unit hydrograph must be modified to apply in distributed hydrological models and hydrological forecasting [25]. The conceptual model of this approach to distributed models is shown in **Figure 4**.

This type of unit hydrograph is interpreted as the result of the combination of a pure translation process, followed by a routing in a linear storage. According to this scheme, the actual travel time of a water particle is given by the time-area diagram plus the retaining time in the linear reservoir [17].

This method requires the estimation of four parameters for determining the hydrograph of the basin: the time of concentration t_c; the Muskingum storage attenuation coefficient K; the basic flow recession constant R and a time-area histogram that is used to obtain the initial infiltration I_a and infiltration potential maximum S. The concentration time t_c is defined as the time which the precipitation takes to reach the basin outlet from most remote point. This is a measure of pure delay, regardless of the effect of storage. In the literature there are several equations to calculate the concentration time t_c [28], in this study the Kirpich equation was used:

$$t_c = 0.000325\left(\frac{L^{0.77}}{S^{0.385}}\right) \quad (12)$$

with,

t_c—concentration time (hours),

L—river bed length (m),

S—basin average slope (dimensionless).

The storage attenuation coefficient K is the second parameter and is a measure of the delay caused by natural storage. For calibration of this parameter an initial value of $K = 0.6t_c$ is assumed [23]. The recession constant R is a measure of the flow rate decrease in the recession curve between Q_i and Q_{i-1} interval. That is,

$$R = Q_i / Q_{i-1} \qquad (13)$$

The fourth parameter (the time-area histogram) represents the watershed area contributing to runoff at the basin outlet in a given time and transforms the effective rainfall hyetograph into a runoff hydrograph, regardless of time storage. This area is obtained by constructing a map of isochrones, defined with travel time from each cell to the basin outlet (Equation (12)). By relating the areas between isochrones with corresponding time interval, the time-area histogram of the basin is obtained. This parameter is very important in this methodology because, together with the storage constant K, it determines the runoff response of the basin to its outlet (**Figure 4**).

5. Hydrological Model Calibration

Hydrologic model calibration was performed with the storms of July 28 and August 23, 1998 with precipitation mapped every 15 minutes. The study period was from 18:00 to 00:00 h (local time) on 28 July and from 16:45 at 18:45 h (local time) on 23 August.

The rain period matrices for storm August 23, 1998 are shown in **Figure 5**. They show the evolution of the storm approaching the basin until it finally dissipates. This allowed obtaining the rainfall-runoff model to experimental river basin Mixcoac.

The grid parameters represent the cells as sub-basins, in this way, from the current length and slope of each cell the travel time to the basin outlet is estimated to create

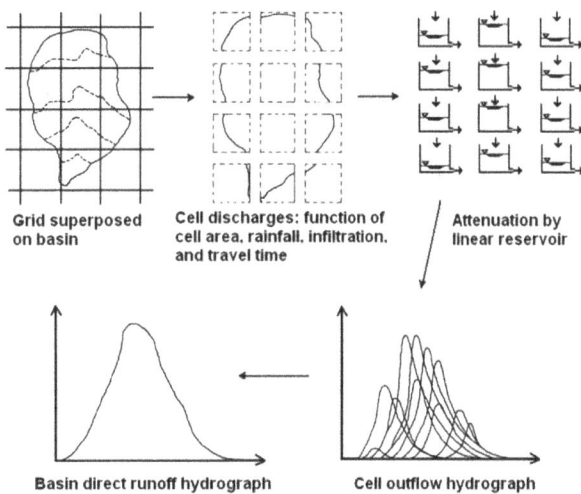

Grid superposed on basin | Cell discharges: function of cell area, rainfall, infiltration, and travel time | Attenuation by linear reservoir

Basin direct runoff hydrograph | Cell outflow hydrograph

Figure 4. Conceptual model of clark method for distributed parameters (adapted from [25]).

the isochrones (**Figure 6**). The spatial variation of the curve number (CN) was determined using a Geographic Information System (GIS) in raster format, ensuring that the study area and the data format matching the radar rainfall grid. For this grid, land use and soil type maps (**Figure 7**) of the study area were used. This format allows the CN of each of the cells in the runoff generation model to be included as input (**Figure 8**).

Finally using the modified Clark method, this uses the Histogram Time-Area (HTA) defined with subareas built between consecutive isochrones from the remote areas to the basin outlet. This HTA is the basis to transform rainfall into runoff and is determined from the convolution equation. The time interval used in the hydrograph basin response defines the travel time between two adjacent isochrones defined by [27] as:

$$Q_j = \sum_{i=1}^{j} E_i A_{j-i+1} \qquad (13)$$

where: j is the number of time intervals, Q is the basin outlet runoff, E is the excess rainfall intensity and A is the area enclosed between isochrones. This method calibrates the model until the hydrograph estimate is comparable to the observed hydrograph of selected storms. The rainfall and runoff data were used as observed data to calibrate the hydrological model.

Results

With the proposed methodology, the hydrological modeling was performed and the results obtained are shown in the **Figures 9** and **10**. The parameters obtained for the two storms are shown in **Table 3**.

Where I_a is the initial infiltration, S is the retention potential, t_c is the concentration time of the watershed, K is the Muskingum storage coefficient, Q_{bi} is the starting base runoff, R is the recession constant and Q_u is the threshold base runoff.

Figures 9 and **10** show, for the event on July 28, the difference in volume between the measured and the observed hydrograph is 5%, and for the event of August 23 the difference is 1%, while the peak occurs in the first case with a delay of 30 minutes, and the second with an advance of 30 minutes. These differences in volumes and the time in the peak flow could be associated to the limited information available to calibrate the model.

Table 3 lists the seven parameters required for model

Table 3. Parameters resulting for each storm.

Date	Basin Model Parameters						
	I_a	S	t_c	K	Q_{bi}	R	Q_u
July 28	0.6	0.125	0.25	1.0	0.6	0.8	0.1
August 23	0.6	0.41	0.25	1.0	0.6	0.8	0.1

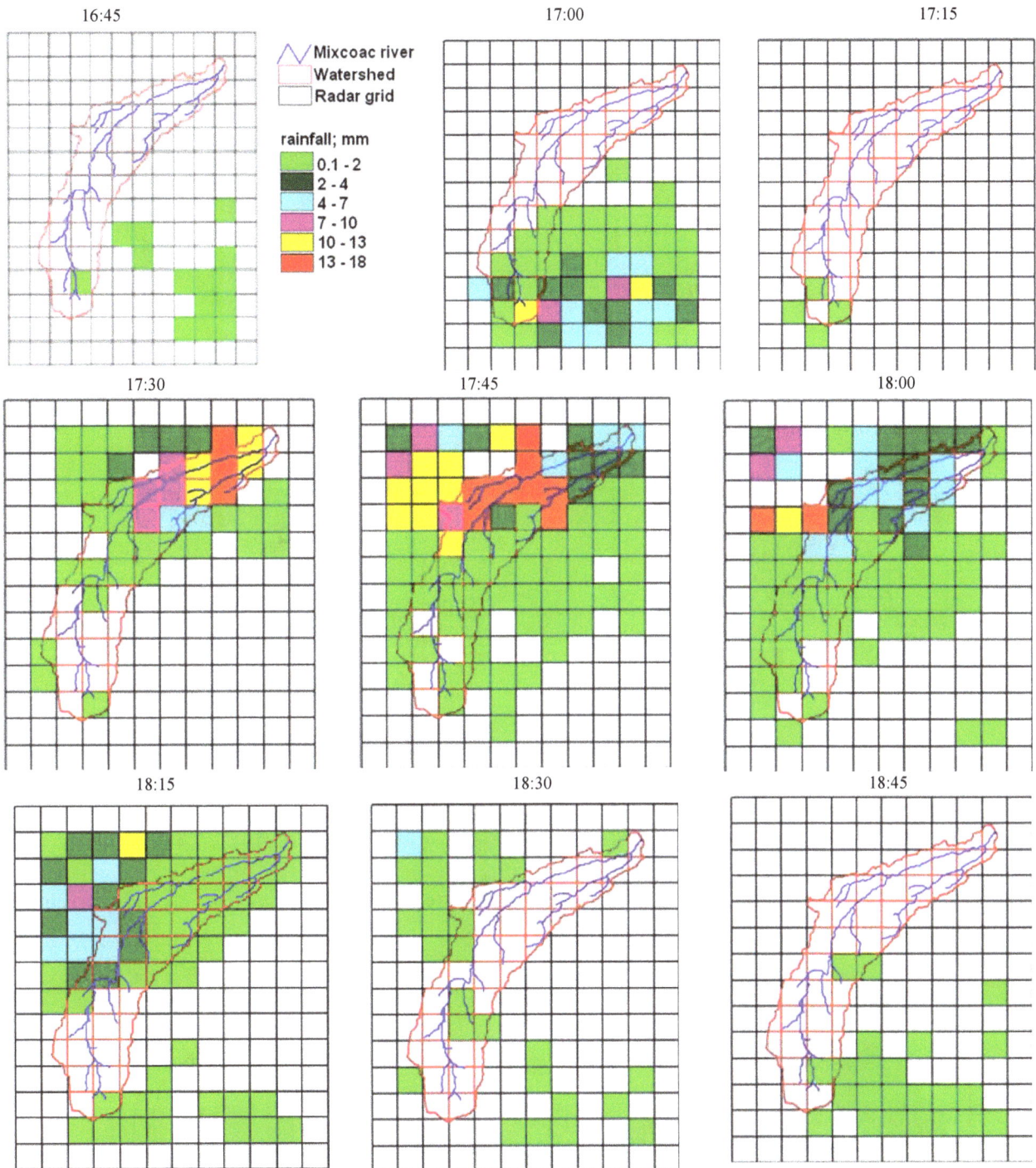

Figure 5. Precipitation matrix every 15 minutes for 23 August 1998 (16:45 - 18:45, local time).

calibration: initial infiltration (I_a), and potential retention (S) are influenced by the antecedent soil moisture. The time of concentration (t_c) and storage coefficient (K) affect the shape of the hydrograph. Meanwhile, initial base runoff (Q_{bi}), the recession constant (R) and threshold based runoff (Q_u) are affected by the observed historical base flow parameters.

The physical parameters of the basin, which can be

considered constant, were obtained with the software HEC-GEOHMS. However, thinking of a future operational hydrological model and with the intent to encourage the use of radars and distributed hydrological models, as an additional product of this study, the software Runoff Forecast Model (RFM) was created, for calculation of variable parameters required by the model's hydrological basin Mixcoac [29]. This software is available com-

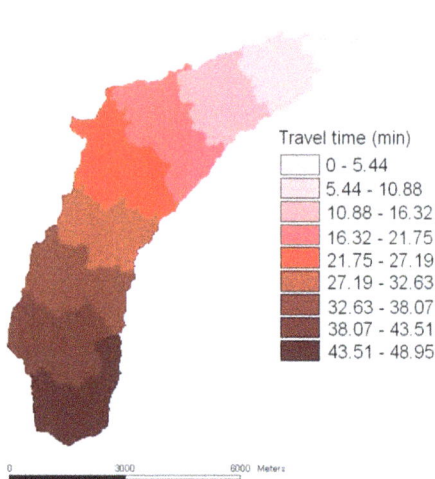

Figure 6. Mixcoac basin isochrones maps.

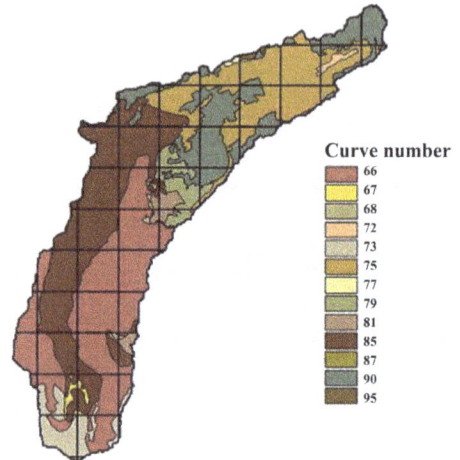

(a)

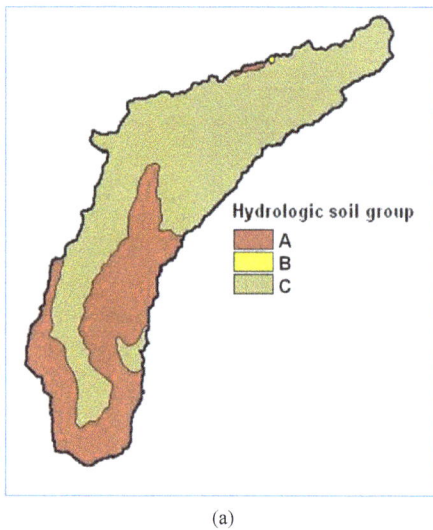

(b)

Figure 7. Map of hydrologic soil type (a) and use land (b).

pletely free of all hydrologists interested in distributed hydrological models. The RFM software, along with its manual (in Spanish) and application example can be

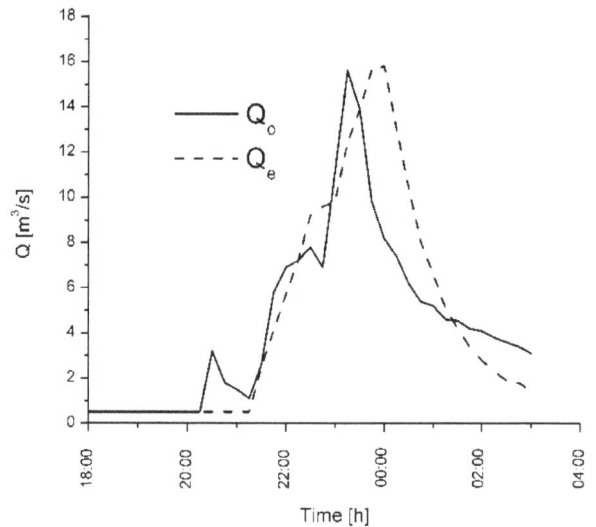

Figure 8. Soils curve number in the Mixcoac Basin.

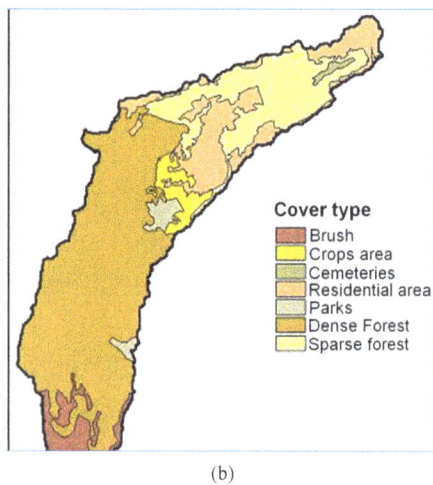

Figure 9. Outflow hydrograph for the storm of July 28, 1998.

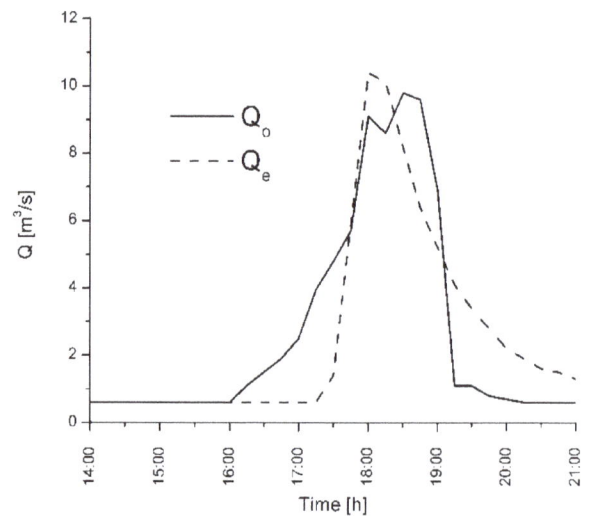

Figure 10. Outflow hydrograph for the storm of August 23, 1998.

downloaded from the Engineering Institute of the National Autonomous University of Mexico [30].

The results show that radar data input in operational hydrological models can be of great help for flood forecasting in real time and useful to the Meteorological Service of any country. Currently, the weather radar is not seen as the only solution to flood problems, but as part of a whole observation system that includes automatic weather stations, radar, satellite and also regional weather forecast models. The estimation of rainfall from satellite data has been developed extensively in recent years and the proposed Global Precipitation Measurement (GPM) is giving great support to researchers, developers and operational hydrologists. Moreover, mesoscale forecast models, such as WRF provide a great advantage in urban hydrology and other fast response watershed, as the forecast window is important in helping to anticipate sites of high flood risk and issue warning notices for this population. Additionally, radars and satellites can be used to generate immediate forecasts (nowcasting), a very useful product in urban hydrology and mountain river basins. All of these rainfall product can be used as input into the operational hydrological models developed here.

6. Conclusions

The major physical significance of distributed models, as used here, is to consider temporal and spatial variability of storms and the spatial variability of soil characteristics of the basin in order to accurately reproduce the hydrological processes inside the watershed, thereby generating more realistic and accurate hydrographs. In this sense, the weather radar is an excellent option to estimate the spatial variability of precipitation affecting hydrological processes within a watershed. In addition, as the response of the basin is non linear, the distributed model allows more basin accurate integration.

Concerning model parameters considered, the initial infiltration and potential retention of soil moisture were calibrated considering runoff volumes shown in observed hydrographs and it was seen that only in the case of the second parameter there is a variation from one storm to another, mainly due to changes in the humidity. In the case of the two selected storms, infiltration parameters showed that, for the storm of July 28, the infiltration is less than in the case of the August 23 storm. This is an indication that the second storm occurs when the soil moisture is greater than the first one.

The difference between measured and estimated volumes are 5% and 1% for storms in July and August, respectively, providing an accurate estimation of the peak discharge volume, although its timing is less accurate. However, if a larger number of events are available for calibration, the results can be considerably improved. Moreover, of the seven parameters used in the calibration

model, six remain constant and only one is variable—the antecedent moisture. This is a problem not yet solved by hydrologists, since it is difficult to estimate the soil moisture conditions when a rainfall event occurs. Except for this problem, the model is operationally simple and produces valuable information for decision-making in high-risk area. In addition, of the seven parameters required by the model, three of them, relating to base flow, can be obtained directly from the historical analysis, leaving only four for calibration.

This effort also aims to establish the basis for real time storm monitoring systems, in order to integrate them into an early warning system in Mexico river basins with major flooding risks, as part of the non-structural measures which should be implemented to increase resilience in flood prone areas. The investment in this type of measure is far less than the economic losses which occur each year in these areas of the country [31]. The methodology shown can be used in other basins. Additionally, this type of hydrological modeling is also useful to assess the uncertainty in runoff forecasting models in real time.

In an attempt to encourage the use of remote sensing instruments in operational distributed hydrological models, the Runoff Forecast Model (RFM) was developed. This model can be downloaded from the website of the Institute of Engineering of UNAM.

7. Acknowledgements

The authors would like thank the National Weather Service of Mexico and the Water System of the City of Mexico, for the data provided and PAPIIT, under contract IT100712, and CONACYT-SEMANART, under contract 107997, for the financial support in the development of the distributed hydrological model using weather radar data.

REFERENCES

[1] J. Soussan and I. Burton, "Adapt and Thrive: Combining Adaptation to Climate Change, Disaster Mitigation, and Natural Resources Management in a New Approach to the Reduction of Vulnerability and Poverty," *UNDP Expert Group Meeting, Integrating Disaster Reduction with Adaptation to Climate Change*, Havana, 17-19 June 2002, pp. 28-44.

[2] Undp Expert Group Meeting, "A Climate Risk Management Approach to Disaster Reduction and Adaptation to Climate Change. Integrating Disaster Reduction with Adaptation to Climate Change," *Undp Expert Group Meeting*, Havana, 17-19 June 2002, 234 p.

[3] UIP and UNISDR, "Disaster Risk Reduction: An Instrument for Achieving the Millennium Development Goals," Advocacy Kit for Parliamentarians, Inter-Parliamentary Union, Geneva, 2010.

[4] K. J. Beven, "Distributed Models," In: M. G. Anderson

and T. P. Burt, Eds., *Hydrological Forecasting*, John Wiley & Sons Ltd., Chichester, 1985, pp. 405-435.

[5]   M. Smith, "NOAA Technical Report NWS 45," National Oceanic and Atmospheric Administration, Boulder, 2004, p. 62.

[6]   V. P. Singh and D. A. Woolhiser, "Mathematical Modelling of Watershed Hydrology," *Journal of Hydrologic Engineering*, Vol. 7, No. 4, 2002, pp. 270-292.

[7]   J. M. Faures, D. C. Goodrich, D. A Woolhiser and S. Sorooshian, "Impact of Small-Scale Rainfall Variability on Runoff Modeling," *Journal of Hydrology*, Vol. 173, No. 1-4, 1995, pp. 309-326.

[8]   J. Morin, D. Rosenfeld and E. Amitai, "Radar Rain Field Evaluation and Possible Use of Its High Temporal and Spatial Resolution for Hydrological Purposes," *Journal of Hydrology*, Vol. 172, No. 1-4, 1995, pp. 275-292.

[9]   D. Guichard, R. García, F. Francés and R. Domínguez, "Influencia de la Variabilidad Espacio-Temporal de la Lluvia Mediterránea en la Respuesta Hidrológica en Cuencas Pequeñas y Medianas," XXI Congreso Latinoamericano de Hidráulica, Sao Paulo, 2004.

[10]   E. B. Vieux, "Distributed Hydrologic Model Using GIS," Kluwer Academic Publisher, Norwell, Vol. 38, 2001, p. 293.

[11]   F. Rivera-Trejo, G. Soto-Cortés and B. Méndez-Antonio, "The 2007 Flood in Tabasco, Mexico: An Integral Analysis of a Devastating Phenomenon," *International Journal of River Basin Management*, Vol. 8, No. 3-4, 2010, pp. 255-267.

[12]   V. Magaña, J. Pérez and M. Méndez, "Diagnosis and Prognosis of Extreme Precipitation Events in the Mexico City Basin," *Geofísica Internacional*, Vol. 41, No. 2, 2003, pp. 247-259.

[13]   R. E. Horton, "The Role of Infiltration in the Hydrologic Cycle," *Eos Transactions*, Vol. 14, No. 1, pp. 446-460.

[14]   M. Sivapalan, K. Beven and E. Wood, "On Hydrologic Similarity 2. A Scaled Model of Storm Runoff Production," *Water Resources Research*, Vol. 23, No. 12, 1987, pp. 2266-2278.

[15]   S. M. Jay, "Comparison of Distributed Versus Lumped Hydrologic Simulation Models Using Stationary and Moving Storm Events Applied to Small Synthetic Rectangular Basins and an Actual Watershed Basin," Ph.D. Dissertation, The University of Texas, Arlington, 2007, p. 419.

[16]   B. Lastoria, "Hydrological Processes on the Land Surface: A Survey of Modeling Approaches," Universidad de Trento, Trento, p. 60.

[17]   J. Vélez, "Desarrollo de un Modelo Hidrológico Conceptual y Distribuido Orientado a la Simulación de Creci-

das," Ph.D. Thesis Dissertation, Universidad Politécnica de Valencia, Valencia, 2001, p. 266.

[18]   B. Méndez-Antonio, R. Domínguez, G. Soto-Cortés, F. Rivera-Trejo, V. Magaña and E. Caetano, "Radars, an Alternative in Hydrological Modeling. Lumped Model," *Atmósfera*, Vol. 24, No. 2, 2011, pp. 157-171.

[19]   V. P. Singh and D. K. Frevert, "Mathematical Models of Large Watershed Hydrology," Water Resources Publications, Highlands Ranch, 2002.

[20]   V. M. Ponce and R. H. Hawkins, "Runoff Curve Number: Has It Reached Maturity?" *Journal of Hydrologic Engineering*, Vol. 1, No.1, 1996, pp. 11-19.

[21]   F. Aparicio, "Fundamentos de Hidrología de Superficie," Editor Limusa, México, 1994, p. 305.

[22]   US Army Corps of Engineers, "Hydrologic Modeling System HEC-HMS, User's Manual, V. 2.1," Hydrologic Engineering Center, 2001, p. 178.

[23]   R. M. Domínguez and S. J. Gracia, "Manual de Diseño de Obras Civiles, Pérdidas," Comisión Federal de Electricidad, México, 1981, p. 46.

[24]   L. K. Sherman, "Stream Flow from Rainfall by the Unit Graph Method," *Engineering News-Record*, Vol. 108, 1932, pp. 501-505.

[25]   D. W. Kull and A. D. Feldman, "Evolution of Clark's Unit Graphs Method to Spatially Distributed Runoff," *Journal of Hydrologic Engineering*, Vol. 3, No. 1, 1998, pp. 9-19.

[26]   D. R. Maidment, "Developing a Spatially Distributed Unit Hydrograph by Using GIS," *Proceeding of HydroGIS'93*, IAHS Publication, Wallingford, No. 211, 1993, pp. 181-192.

[27]   B. P. Saghafian, P. Julien and H. Rajaie, "Runoff Hydrograph Simulation Based on Time Variable Isochrone Technique," *Journal of Hydrology*, Vol. 261, No. 1-4, 2002, pp. 193-203.

[28]   M. V. Ponce, "Engineering Hydrology: Principles and Practices," Prentice Hall, Upper Saddle River, 1996, p. 640.

[29]   R. M. Domínguez, G. E. Garduño, B. Méndez-Antonio, R. Mendoza, J. M. L. Arganis and E. E. Carrizosa, "Manual del Modelo Para Pronóstico de Escurrimiento. Instituto de Ingeniería," Universidad Nacional Autónoma de México, México DF, 2008, p. 101.

[30]   http://aplicaciones.iingen.unam.mx/ConsultasSPII/Buscar publicacion.aspx

[31]   Cenapred, "Características e Impacto Socioeconómico de los Principales Desastre Ocurridos en la República Mexicana en el Año 2008," Secretaría de Gobernación, Sistema Nacional de Protección Civil y Cenapred, 2008, p. 368.

20

Flash Flood Risk Assessment Using Morphological Parameters in Sinai Peninsula

Ashraf M. Elmoustafa, Mona M. Mohamed

Irrigation and Hydraulic Department, Faculty of Engineering, Ain Shams University, Cairo, Egypt.

ABSTRACT

Flash floods are considered to be one of the worst weather-related natural disasters. They are sudden and highly unpredictable following brief spells of heavy rain. Egypt is subjected to flash floods, especially the eastern desert and Sinai Peninsula where floods from the mountains of Red Sea and Sinai are causing heavy damage to man-made features. This manuscript presents the methodology adopted to generate a weighted risk map for main watersheds located in Sinai according to main morphological parameters. Using digital elevation model (DEM) implemented into a Geographic Information System (GIS) the Sinai watersheds were delineated and morphological parameters calculated. The parameters where then used in a multi criteria analysis process to calculate a morphological risk factor. The resulted risk maps of this study can help initiating appropriate measures to mitigate the probable hazards in the area with prioritization.

Keywords: Flash Floods; Morphology; MCA; Sinai Peninsula; Arid Region

1. Introduction

The Sinai Peninsula is a triangular peninsula in Egypt about 60,000 km^2 (23,000 sq mi) in area. It is situated between the Mediterranean Sea to the north, and the Red Sea to the south, **Figure 1** , and is the only part of Egyptian territory located in Asia as opposed to Africa, effectively serving as a land bridge between two continents; also it is one of the largest mining fields in Egypt beside the potential of agriculture and industrial development which give it a great importance to the economy of developed country of Egypt. This part of Egypt is sometimes subjected to flash floods events resulting from heavy, short duration and sudden rainfall events. This work evaluates and tests a new criterion for flood risk assessment studies for Sinai Peninsula.

Many studies were carried out to assess the seriousness of the floods in places susceptible to floods. EL-Shamy (1992); established two relation graphs to classify basins flood risk based on the relations between bifurcation ratio (Rb) and the drainage density (D) and the relations between bifurcation ratio versus the drainage frequency (F) [1]. Elmoustafa (2012); used a Weighted Normalized Risk Factor (WNRF) for floods risk assessment. The four parameters used are Area, Slope, time of concentration

and runoff volume. A weight coefficient (W) was assumed constant for all factors and equal to 1/(No. of parameters). It was noticed during the analysis for a case study in the Eastern desert that the drainage basin area has a great effect on the floods generated at its outlet while other factors have less effect than the drainage area such as the slope and roughness [2]. Another study used nine morphological parameters to make risk map for watersheds that affect on area from Marsa Alam to Ras Banas The morphological parameters used were Area of watershed, Slope, Drainage Density, Drainage Frequency, R$_B$, Rt, Roughness factor, Shape factor, and Heights Factor. The risk value was calculated for every parameter for all watersheds [3].

$$\text{Risk value} = 4 \times \frac{\left(x - x_{\min}\right)}{\left(x_{\max} - x_{\min}\right)} + 1 \qquad (1)$$

Morphological parameters are the main factors affecting the flood hydrograph shape and hence its strength and should be addressed when studying watersheds flash floods risk assessment [4,5]. In this manuscript the following algorithm was followed:

A Geographic Information System (GIS) was used to calculate the required morphological parameters for all watersheds of the study area using the available Digital Elevation Model (DEM).

Main morphological parameters were used to estimate a risk factor representing each of them.

HEC-HMS was used to study the sensitivity of each morphological parameter of every main catchment in the study area to a pseudo storm applied to all the catchments.

New methodology was proposed to combine those risk factors through weights assigned to each of them based on the HEC-HMS results.

2. Objectives of the Study

The objective of this research is to come up with a risk map by developing a risk factor that reflects the watersheds morphological parameters effects on flood hydrographs and test it by studying the response of all catchment in the study area to a pseudo storm. These results are essential to define the higher risk locations in the study area through a resulted risk maps of the effecting watersheds that can help initiate appropriate measures to mitigate the probable hazards in the area with prioritization.

3. General Description of the Study Area

The Sinai Peninsula extends Longitude 32.25°E and 34.8°E; Latitude 27.8°N and 31.2°N. Geologically Sinai can be roughly divided into three areas. The northern region consists of sand dunes and fossil beaches formed by the changing levels of the Mediterranean Sea during the glacial periods two million years ago. The landscape is flat and uniform, interrupted only by some vast sand and limestone hills.

The scarcely inhabited Al Tih Plateau is the central geological area with limestone dating from the Tertiary Period. The highlands extend towards the south until it goes over into the third area consisting of granite and volcanic rock. Limestone and sandstone sediments are replaced by granite and basalt. Both rocks are produced by volcanic activity on the bottom of the ocean from the Precambrium.

4. Scope and Methodology

This study deals with the evaluation of different geomorphological impacts on flash floods risk degree. To achieve the goal of the present research, DEM from the NASA Shuttle Radar Topographic Mission (SRTM) provided digital elevation data of 90 m resolution, was exported to a Geographic Information System (GIS) environment to extract all possible morphological parameters of the catchment in the area.

The watershed, streams and sub-basins were delineated with a threshold of a 25 km^2 for the smallest stream definition, **Figure 2**.

Watershed characteristics, such as watershed area, perimeter, drainage line number, sum of drainage lines and longest flow path were extracted and a sample is presented in the next **Table 1**.

Then the geometric watershed characteristics were automatically calculated. From these characteristics other morphological parameters were calculated as follows:

1) **Drainage frequency** (F), **Figure 3**, defined as the ratio between Streams' Number to the total watershed area (Horton, 1932).
2) **Drainage density** (D), **Figure 4**, defined as the ratio between the summation of Streams' Lengths and total watershed area (Horton, 1945).
3) **Surface flow length** (L$_o$), **Figure 5**, it is the distance travelled by water before reaching any stream.
4) **Shape factor** (Ish), **Figure 6**, and is equal to 1.27* (A/P2) as per McCuen, 1989. Where P is the watershed perimeter and A is its corresponding area.
5) **Time of concentration** (T$_c$) and was estimated based on the Kirpich formula, (Soil Conservation Services (SCS), 1940).

$$T_c = \left[0.01944 \times \left(L^{0.77} \right) \middle/ S^{0.385} \right]$$

where (L) is longest flow path in meters and (S) is its corresponding slope.

Table 2 gives sample of the calculated parameters.

5. Standardization of Parameters

The hydro-morphological parameters obtained for each watershed are expressed in different units. It is therefore difficult to compare across criteria. For many of the arithmetic Multi Criteria Analysis (MCA) techniques, it is necessary to reduce the scores to the same unit, this is called standardization [5,6]. The difference between the

Figure 1. Location of the study area.

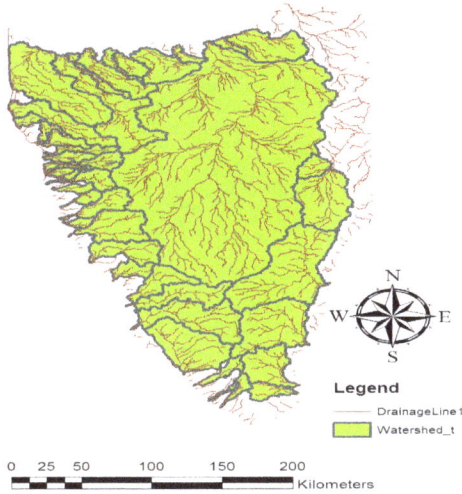

Figure 2. Watersheds of study area.

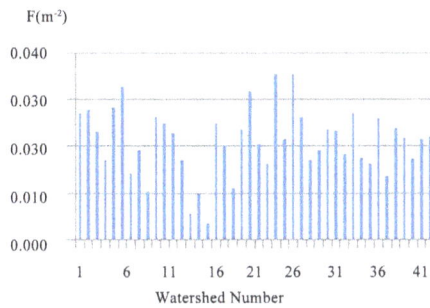

Figure 3. Drainage frequency of watersheds.

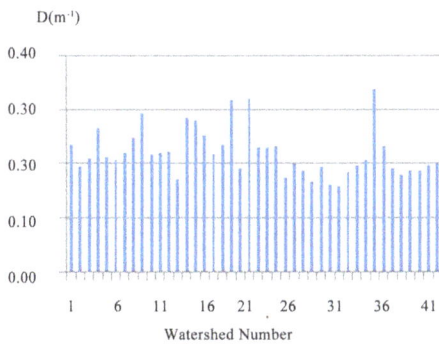

Figure 4. Drainage density of watersheds.

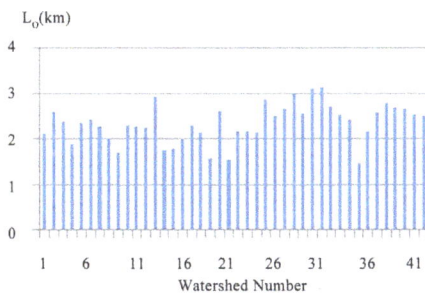

Figure 5. Surface flow length of watersheds.

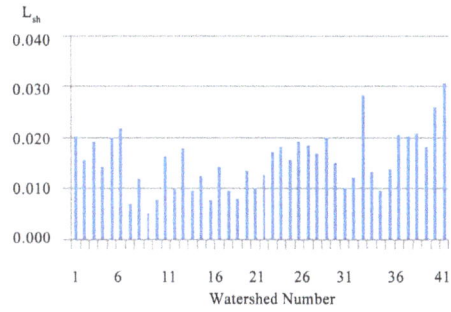

Figure 6. Shape factor of watersheds.

Table 1. Watershed characteristics sample.

Watershed	Length km	Area km²	D.L. no.	ΣD.L. lengths (Km)	Longest flow path (m)	Slope
1	276	1222	33	288	80,172	0.0045
2	108	144	4	28	33,193	0.0033
3	1256	23,810	550	4984	364,050	0.0038
4	124	176	3	47	42,487	0.0001
5	247	958	27	203	78,832	0.0070
6	140	336	11	69	47,331	0.0058

actual parameter and that of the lowest value is divided by the difference between the parameters of the highest value and that of the lowest value. This led to standardized factors that reflect the degree of risk for each parameter compared to the same parameter in the other sheds.

Area Standardized Risk Factor (ASRF)

$$= \frac{Area - Area\ Min.}{Area\ Max - Area\ Min} \quad (2)$$

Slope Standardized Risk Factor (SSRF)

$$= \frac{Slope - Slope\ Min.}{Slope\ Max - Slope\ Min} \quad (3)$$

Drainage Frequency Standardized Risk Factor (FSRF)

$$= \frac{F - F\ Min.}{F\ Max - F\ Min}$$

$$(4)$$

Drainage density Standardized Risk Factor (DSRF)

$$= \frac{D - D\ Min.}{D\ Max - D\ Min} \quad (5)$$

Surface flow length Standardized Risk Factor (L_oSRF)

$$= \frac{L_o - L_o Min.}{L_o Max - L_o Min}$$

$$(6)$$

Slope Standardized Risk Factor (ShSRF)

$$= \frac{\text{Ish} - \text{Ish Min.}}{\text{Ish Max} - \text{Ish Min}} \quad (7)$$

T_c Standardized Risk Factor (TSRF)

$$= \frac{T_c - T_c \text{ Min.}}{T_c \text{ Max} - T_c \text{ Min}} \quad (8)$$

where, Max. refers to the maximum value of the mentioned parameters and Min. refers to the minimum value of the mentioned parameters.

It was noticed that extreme high values may affect the results, one main drainage area is extremely high (23,810 km^2), **Figure 7**, that appear as extreme line on the graph, while all the other values fall below 1000 km^2. This will affect the risk factors calculated and should be reconsidered and risk factors carefully adjusted.

The box plot technique was applied to test all the data for values that are extremely high or an outlier. An outlier is an observation that is numerically distant from the rest of the data which may lead to biased results. Box plot technique is useful to display differences between populations without making any assumptions of the underlying statistical distribution. It is non-parametric and spacing between the different parts of the box helps indicate the degree of dispersion (spread) and deviation in the data, and identify outliers. **Table 3** represents the box plot test results for the morphological parameters that were previously discussed and will be used in the nest steps of the analysis. After the exclusion of the outlier,

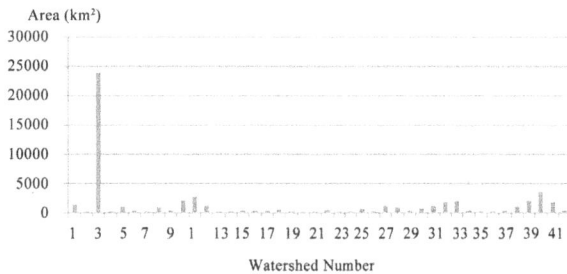

Figure 7. Extreme value of watersheds area.

Table 2. Sample of the calculated parameters.

Watershed	Drainage density (D)	Drainage frequency (F)	Surface flow length (L$_o$) (km)	Shape factor (Ish)	T_c (min.)
1	0.24	0.03	2.12	0.02	928
2	0.19	0.03	2.57	0.02	529
3	0.21	0.02	2.39	0.02	3181
4	0.27	0.02	1.89	0.01	3224
5	0.21	0.03	2.36	0.02	774
6	0.21	0.03	2.42	0.02	564

Table 3. Box plot test results.

	Area	S	D	F	L$_o$	Ish	T_c
Q1	195	0.01	0.19	0.020	2.14	0.01	426
Minimum	62	0.0001	0.16	0.004	1.47	0.01	209
Median	366	0.01	0.21	0.02	2.37	0.02	677
Maximum	23,810	0.05	0.34	0.04	3.12	0.03	3224
Q3	1091	0.01	0.23	0.03	2.61	0.02	831
Inter quartile range	896	0.01	0.04	0.01	0.47	0.01	406
Inter quartile range	3779	0.0416	0.36	0.05	4.01	0.04	2048

the Standardized Risk Factors (SRF) were then recalculated.

6. Sensitivity Analysis Using HEC-HMS

The sum of factors is called the Weighted Standardized Risk Factor (WSRF). A weight coefficient (W) will be assigned to each factor based on the results of watersheds sensitivity analysis on HEC-HMS, a computer programs developed and used for hydraulic modelling for the past 30 years (Chow *et al.*, 1988) [7].

All watersheds were subjected to the same condition by applying a typical pseudo storm with all input parameters (as curve number and precipitation) the same so that the more important the factor is the higher the W value will be assigned to it.

$$\text{WSRF} = \sum \left(W_i \times SRF_i \right) \quad (9)$$

where,

W_i is the weight to be assigned to each risk factor.

SRF_i each factor of those presented by Equations (1) to (7).

The outflow hydrographs for the given rainfall event were generated and main effective outputs of the hydrographs (*i.e.* peak discharge and time to peak) were tested against all morphological parameters. For example, it was noticed that correlation between area and both time to peak and peak discharge is directly proportional and correlation between slope and both time to peak and peak discharge is indirect proportional, **Figures 8** to **11** represent sample of this analysis results.

6.1. Sensitivity Analysis Results

Flash flood means high peak and short duration, therefore peak discharges and time to peak results were studied and their correlation with all morphological parameters were calculated. Flood risk is considered directly proportional with peak discharge value and indirect proportional with time to peak.

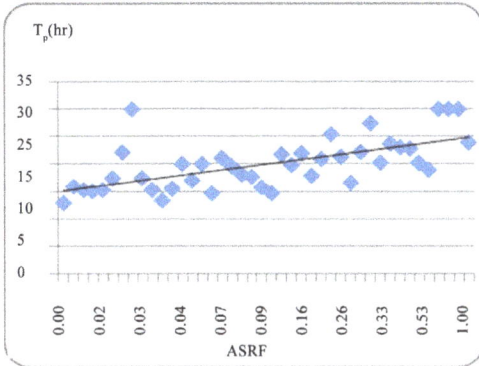

Figure 8. Area vs time to peak.

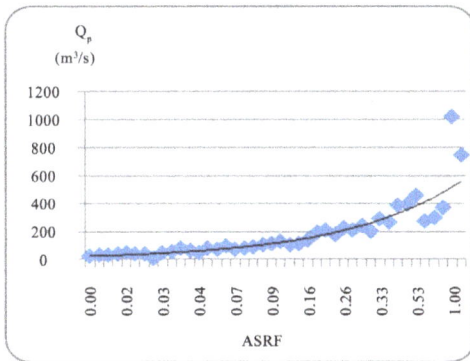

Figure 9. Area vs peak discharge.

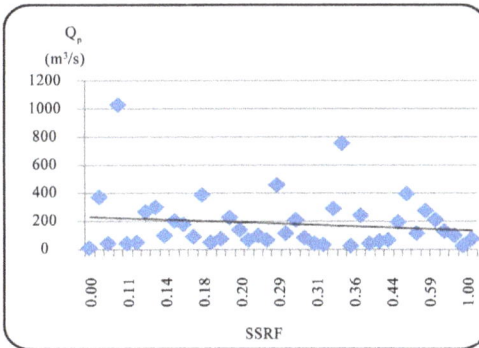

Figure 10. Slope vs peak discharge.

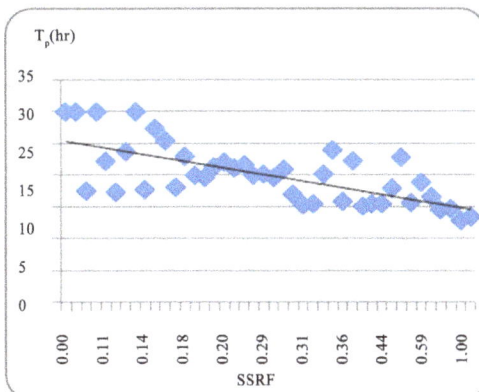

Figure 11. Slope vs time to peak.

6.1.1. Peak Discharge Risk

The correlation between each factor and the hydrograph parameters were computed to help estimating the weight of each risk factor, **Table 4**. It was noticed that correlations between SSRF, L_oSRF and ShSRF are negative therefore inverse values of slope and shape factor will used and A, D, F, T_c values will be directly used in SRF calculation as they were presented with positive correlations.

It was also noticed from the correlation factor results that the drainage basin area has a great effect on the generated floods followed by time of concentration and slope, other factors have less effect such as the shape factor, drainage frequency, Drainage density and surface runoff length. Correlations of $1/L_o$ and drainage density are found to be of negative value, which is opposite to what was expected, and this is because watersheds with small areas have higher drainage density, **Figure 12**, and to overcome this, streams threshold had to be reduced and this led to long processing time. Therefore $1/L_o$ and D SRFs were neglected in WSRF calculation in this work.

Based on the resulted correlation factors, the weight coefficient w values were then calculated as followed:

$$w = \frac{\text{factor's correlation}}{\Sigma \text{ standardized risk factors' correlations}} \quad (10)$$

Table 5 represents the final weights assigned to each risk factor with total summation of 1. Standardized Risk Factors were recalculated and WSRF were estimated using these weight coefficients. **Table 6** shows sample of SRF calculations results with respect to the peak flow value risk.

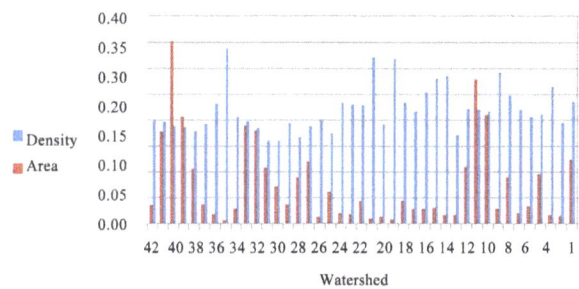

Figure 12. Area and drainage density for watersheds.

Table 4. Correlation between peak discharges and SRF values.

1/S SRF	ASRF	DSRF	F SRF	1/L$_o$SRF	ShSRF	T$_c$SRF
0.14	0.94	−0.34	0.13	−0.34	0.32	0.54

Table 5. The weight coefficient (W).

1/S SRF	ASRF	DSRF	FSRF	1/L$_o$SRF	ShSRF	T$_c$SRF
0.07	0.45	0.0	0.06	0.0	0.15	0.26

Figure 13 represents the main watersheds flowing through Sinai Peninsula, with different colours each representing the risk level of the watershed, WSRF, as computed by Equation (8).

The analysis showed that the flood risk factor for the main watersheds flowing through Sinai Peninsula could be classified into 5 categories according to their WSRF, **Figure 14**.

6.1.2. Time to Peak Risk

The correlation between each factor and the hydrograph parameters were computed to help estimating the weight of each risk factor. **Table 7** shows results for the correlation between the hydrograph time to peak and SRFs. It

was noticed that the drainage basin slope has a great effect on the floods generated peak time followed by shape factor, correlation of 1/D is positive, which is opposite to what was expected, and this is because watersheds with small areas have higher drainage density, **Figure 12**, and to overcome this, streams threshold had to be reduced and this led to long processing time. Therefore 1/D SRF is neglected in WSRF calculation in this work.

The weight coefficients were calculated as explained before; **Table 8** represents the final weights assigned to each risk factor with total summation of 1. Standardized Risk Factors were recalculated and WSRF were estimated using these weight coefficients. **Table 9** shows sample of SRF calculations results with respect to the peak flow time risk.

Figure 15 represents the main watersheds flowing through Sinai Peninsula, with different colours each representing the risk level of the watershed, WSRF, as computed by Equation (8).

The analysis showed that the flood risk factor for main watersheds flowing through Sinai Peninsula could be classified into 5 categories according to their WSRF, **Figure 16**.

6.1.3. Overall Risk

To combine both effects of peak discharge value and time to peak risk for flash flood risk assessment of the study area, the maximum value of WSRF due to peak discharge and WSRF due to time to peak is considered as the overall WSRF (**Table 10**).

Overall WSRF

$$= \text{Max}\left(\text{WSRF\_Peak Discharge}, \text{WSRF\_Time to Peak}\right) \quad (11)$$

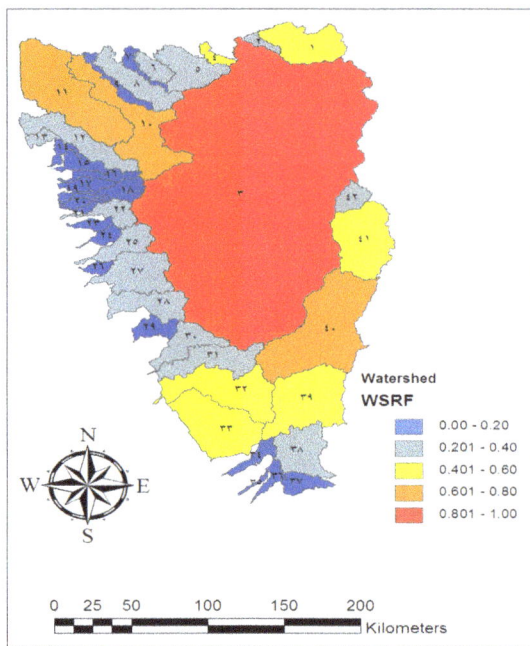

Figure 13. weighted risk map for main watersheds draining towards Sinai coast, with respect to the peak flow value.

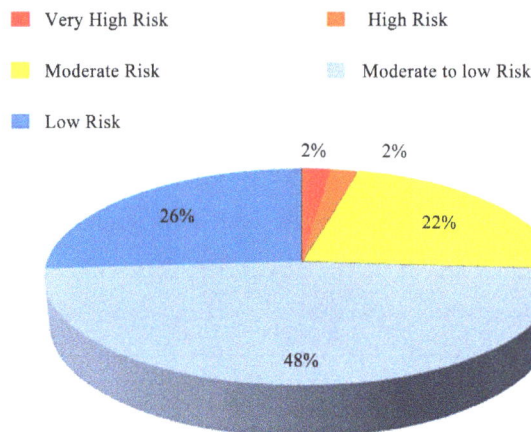

Figure 14. WSRF ranges for main watersheds draining towards Sinai coast taken peak discharge risk.

Table 6. The standardized risk factor for peak discharge risk (SRFQn).

Water-shed	ASRF	T_cSRF	1/SSRF	FSRF	ShSRF	WSRF_P
1	0.34	0.45	0.60	0.74	0.59	0.45
2	0.02	0.20	0.83	0.76	0.41	0.23
3	1.00	1.00	0.72	0.61	0.54	0.89
4	0.03	1.00	1.00	0.43	0.35	0.43
5	0.26	0.35	0.37	0.77	0.57	0.37
6	0.08	0.22	0.46	0.92	0.65	0.28

Table 7. Correlation between time to peak and SRF values.

SSRF	1/ASRF	1/DSRF	1/FSRF	1/L_oSRF	ShSRF	1/T_cSRF
−0.65	−0.56	+0.01	−0.05	−0.07	−0.1	−0.89

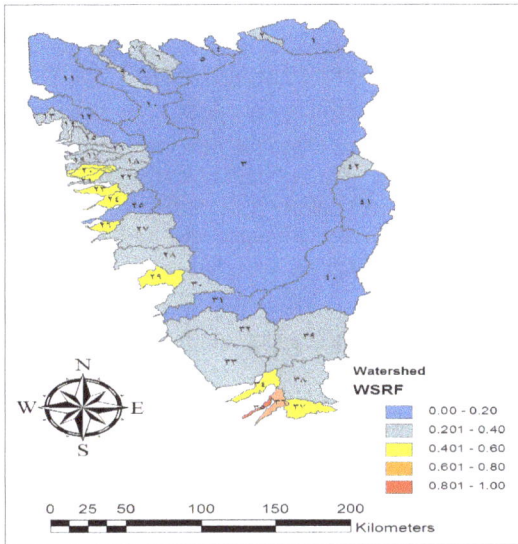

Figure 15. Weighted risk map for main watersheds draining towards Sinai coast, with respect to the peak flow time.

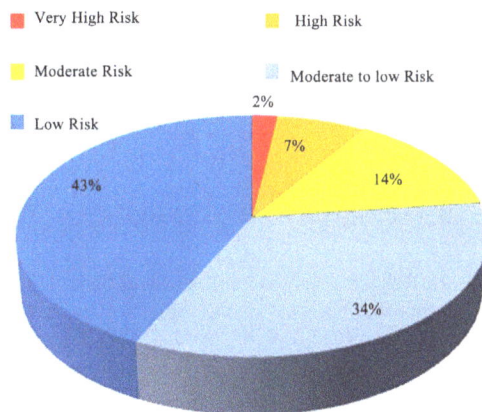

Figure 16. WSRF ranges for main watersheds draining towards Sinai coast.

Table 8. The weight coefficient (W).

SSRF	1/ASRF	1/DSRF	1/FSRF	$1/L_o$ SRF	ShSRF	$1/T_c$SRF
0.28	0.24	0.00	0.02	0.03	0.04	0.38

Table 9. The standardized risk factor due to time to peak risk (SRF).

Watershed	$1/A$ SRF	$1/T_c$SRF	SSRF	$1/F$ SRF	$1/L_o$ SRF	Sh SRF	WSRF_T
1	0.05	0.17	0.13	0.12	0.42	0.59	0.15
2	0.43	0.35	0.09	0.10	0.19	0.41	0.29
3	0.00	0.00	0.11	0.20	0.28	0.54	0.06
4	0.35	0.00	0.00	0.41	0.59	0.35	0.12
5	0.06	0.22	0.20	0.10	0.29	0.57	0.19
6	0.18	0.33	0.16	0.03	0.26	0.65	0.25

Figure 17 represents the main watersheds flowing through Sinai Peninsula, with different colours each representing the overall risk level of the watershed, WSRF, as computed by Equation (10). The analysis showed that the flood risk factor for main watersheds flowing through Sinai Peninsula could be classified into 4 categories according to their overall WSRF, **Figure 18**.

7. Conclusions and Recommendations

Flood protection measurements depending solely on recurrence interval have been adopted for long time without giving weight to the morphological parameters of the watersheds that cause such floods. The work presented the use of multi criteria analysis technique to develop a risk factor when defining flood events.

Based on the analysis results the following conclusions were obtained:

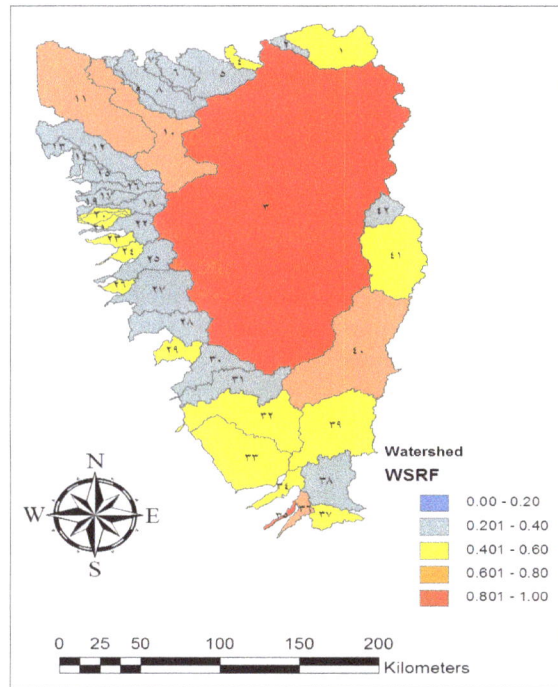

Figure 17. weighted risk map for main watersheds draining towards Sinai coast.

Table 10. Overall weighted standardized risk factors (WSRF) sample.

Watershed	WSRF_Peak Discharge	WSRF_Time to Peak	Overall WSRF
1	0.45	0.15	0.45
2	0.23	0.29	0.29
3	0.89	0.06	0.89
4	0.43	0.12	0.43
5	0.37	0.19	0.37
6	0.28	0.25	0.28

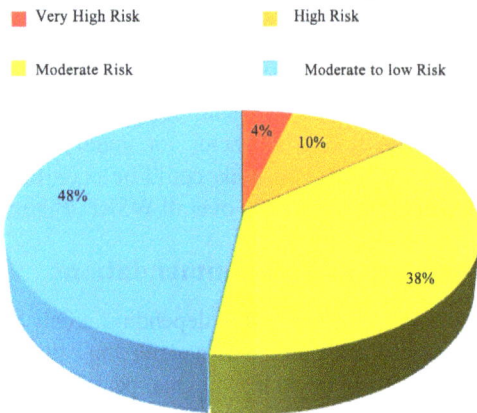

Figure 18. Overall WSRF ranges for main watersheds draining towards Sinai coast taken discharge and time risk.

1) A new criterion was used to evaluate the risk factor for the floods in Sinai. This criterion could be used in other places with similar characteristics.
2) The main watersheds flowing through Sinai Peninsula are classified into four categories where 4% of watersheds have very high risk, 10% has high risk, 38% has moderate risk and 48% has moderate to low risk.
3) The produced risk map is helpful to know the locations that have high flood risk in order to prevent loss of life and minimize damages to property.
4) The drainage basin area is the morphological parameter that has the highest effects on the peak floods generated followed by time of concentration and slope; other factors have less effect such as the shape factor, drainage frequency, drainage density and surface runoff length.
5) The drainage basin slope is the morphological parameter that has the highest effect on the time to peak

followed by the shape factor.

It is also recommended to use the Weighted Standardized Risk Factor (WSRF) obtained during the design of flood protection measurements and/or the calculation of design peak flows for crossing structures. This may lead to more economic design procedure that can be adopted in drainage design guidelines and manuals. Studies should be carried out to investigate how to implement these results in the design procedure.

REFERENCES

[1]   I. Z. El-Shamy, "New Approach for Hydrological Assessment of Hydrographic Basins of Recent Recharge and Flooding Possibilities," 10*th Symposium Quaternary and Development*, Egypt, 18 April 1992, p. 15.

[2]   A. M. Elmoustafa, "Weighted Normalized Risk Factor for Floods Risk Assessment," *Ain Shams Engineering Journal*, Vol. 3, No. 4, 2012, pp. 327-332

[3]   Resources Technology Company, "Flash Flood Risk Assessment for Area from Marsa Alam to Ras Banas Report," 2008.

[4]   A. M. Youssef and M. A. Hegab, "Using Geographic Information Systems and Remote Sensing Techniques for Investigation of New Proposed Sohag-Hurghada Highway across the Egyptian Desert," 2005.

[5]   M. G. El-Behiry, A. Shedid, A. Abu-Khadra and M. El-Huseiny, "Integrated GIS and Remote Sensing for Runoff Hazard Analysis in Ain Sukhna Industrial Area," Egypt, 2005.

[6]   M. L. El-Rakaiby, "Drainage Basins and Flash Flood Hazard in Selected Parts of Egypt," *Egyptian Journal of Geology*, Vol. 33, No. 1-2, 1989, pp. 307-323.

[7]   R. E. Horton, "Drainage Basin Characteristics," *Transactions—American Geophysical Union*, Vol. 13, 1932, pp. 350-361.

Permissions

The contributors of this book come from diverse backgrounds, making this book a truly international effort. This book will bring forth new frontiers with its revolutionizing research information and detailed analysis of the nascent developments around the world.

We would like to thank all the contributing authors for lending their expertise to make the book truly unique. They have played a crucial role in the development of this book. Without their invaluable contributions this book wouldn't have been possible. They have made vital efforts to compile up to date information on the varied aspects of this subject to make this book a valuable addition to the collection of many professionals and students.

This book was conceptualized with the vision of imparting up-to-date information and advanced data in this field. To ensure the same, a matchless editorial board was set up. Every individual on the board went through rigorous rounds of assessment to prove their worth. After which they invested a large part of their time researching and compiling the most relevant data for our readers. Conferences and sessions were held from time to time between the editorial board and the contributing authors to present the data in the most comprehensible form. The editorial team has worked tirelessly to provide valuable and valid information to help people across the globe.

Every chapter published in this book has been scrutinized by our experts. Their significance has been extensively debated. The topics covered herein carry significant findings which will fuel the growth of the discipline. They may even be implemented as practical applications or may be referred to as a beginning point for another development. Chapters in this book were first published by Scientific Research Publishing Inc.; hereby published with permission under the Creative Commons Attribution License or equivalent.

The editorial board has been involved in producing this book since its inception. They have spent rigorous hours researching and exploring the diverse topics which have resulted in the successful publishing of this book. They have passed on their knowledge of decades through this book. To expedite this challenging task, the publisher supported the team at every step. A small team of assistant editors was also appointed to further simplify the editing procedure and attain best results for the readers.

Our editorial team has been hand-picked from every corner of the world. Their multi-ethnicity adds dynamic inputs to the discussions which result in innovative outcomes. These outcomes are then further discussed with the researchers and contributors who give their valuable feedback and opinion regarding the same. The feedback is then collaborated with the researches and they are edited in a comprehensive manner to aid the understanding of the subject.

Apart from the editorial board, the designing team has also invested a significant amount of their time in understanding the subject and creating the most relevant covers. They scrutinized every image to scout for the most suitable representation of the subject and create an appropriate cover for the book.

The publishing team has been involved in this book since its early stages. They were actively engaged in every process, be it collecting the data, connecting with the contributors or procuring relevant information. The team has been an ardent support to the editorial, designing and production team. Their endless efforts to recruit the best for this project, has resulted in the accomplishment of this book. They are a veteran in the field of academics and their pool of knowledge is as vast as their experience in printing. Their expertise and guidance has proved useful at every step. Their uncompromising quality standards have made this book an exceptional effort. Their encouragement from time to time has been an inspiration for everyone.

The publisher and the editorial board hope that this book will prove to be a valuable piece of knowledge for researchers, students, practitioners and scholars across the globe.

List of Contributors

Kronenberg Rico
Meteorology, Technische Universität Dresden, Dresden, Germany
Mathematical Modelling, Bauman Moscow State Technical University, Moscow, Russia

Güttler Tino, Franke Johannes and Bernhofer Christian
Meteorology, Technische Universität Dresden, Dresden, Germany

Mohammed Bahir, Rachid El Moukhayar and Hamid Chamchati
Geodynamics Laboratory Magmatic Géoressources and Georisks, Université Cadi Ayyad, Marrakech, Morocco

Najiba Chkir
Geography Departement, Faculty of Letters and Humanities, Sfax University, Sfax, Tunisia

Paula Galego Fernandes and Paula Carreira
Nuclear Technology Institute, Sacavém, Portugal

Roberto Franco-Plata, Carlos Miranda-Vázquez, Héctor Solares-Hernández and Luis Ricardo Manzano-Solís
Faculty of Geography, Autonomous University of the State of Mexico, Toluca, Mexico

Khalidou M. Bâ and José L. Expósito-Castillo
Inter-American Center of Water Resources, Autonomous University of the State of Mexico, Toluca, Mexico

Haruna Garba and Folagbade Olusoga Peter Oriola
Department of Civil Engineering, Nigerian Defence Academy, Kaduna, Nigeria

Abubakar Ismail
Department of Water Resources and Environmental Engineering, Ahmadu Bello University, Zaria, Nigeria

Nikhil Bhatia, Laksha Sharma, Shreya Srivastava, Nidhish Katyal and Roshan Srivastav
School of Mechanical and Building Sciences, VIT University, Vellore, India

Chieko Gomi
Graduate School of Bioresources, Mie University, Tsu, Japan
Aichi Prefectural Government, Nagoya, Japan

Yasuhisa Kuzuha
Graduate School of Bioresources, Mie University, Tsu, Japan

Prabeer Kumar Parhi
Center for Water Engineering and Management, Central University of Jharkhand, Ranchi, India

Henrique G. Momm
Department of Geosciences, Middle Tennessee State University, Murfreesboro, USA

Ronald L. Bingner, Robert R. Wells and Seth M. Dabney
National Sedimentation Laboratory, United States Department of Agriculture – Agricultural Research Service, Oxford, USA

Lyle D. Frees
United States Department of Agriculture – Natural Resources Conservation Service, Salina, USA

David P. Groeneveld and David D. Barz
HydroBio Advanced Remote Sensing, Santa Fe, USA

John P. O. Obiero, Lawrence O. Gumbe, Christian T. Omuto and Januarius O. Agullo
Department of Environmental and Biosystems Engineering, University of Nairobi, Nairobi, Kenya

Mohammed A. Hassan
Department of Earth and Environmental Science and Technology, Technical University of Kenya, Nairobi, Kenya

Jozsef Szilagyi
Department of Hydraulic and Water Resources Engineering, Budapest University of Technology and Economics, Budapest, Hungary
School of Natural Resources, University of Nebraska-Lincoln, Lincoln, USA

Safouane Mouelhi
National Researches Institute of Water, Forests and Rural Engineering, Tunis, Tunisia

Khaoula Madani
National Water Distribution Utility, Tunis, Tunisia

Fethi Lebdi
Food and Agriculture Organisation, Addis-Abeba, Ethiopia

Reza Kabiri, Andrew Chan and Ramani Bai
Faculty of Engineering, University of Nottingham Malaysia Campus, Kajang, Malaysia

Fredrik Huthoff
Department of Water Engineering & Management, Faculty of Engineering Technology, University of Twente, Enschede, The Netherlands
HKV Consultants, Lelystad, The Netherlands

Menno W. Straatsma
Department of Earth System Analysis, Faculty of Geosciences, Utrecht University, Utrecht, The Netherlands

Denie C. M. Augustijn and Suzanne J. M. H. Hulscher
Department of Water Engineering & Management, Faculty of Engineering Technology, University of Twente, Enschede, The Netherlands

Germain Esquivel-Hernández, Rosa Alfaro-Solís and Juan Valdés-González
Escuela de Química, Universidad Nacional, Heredia, Costa Rica
Laboratorio de Química de la Atmósfera, Escuela de Química, Universidad Nacional, Heredia, Costa Rica

Erin S. Brooks
Department of Biological and Agricultural Engineering, University of Idaho, Moscow, USA

Ricardo Sánchez-Murillo
Escuela de Química, Universidad Nacional, Heredia, Costa Rica
Waters of the West-Water Resources Program, University of Idaho, Moscow, USA

Kristen Welsh
Waters of the West-Water Resources Program, University of Idaho, Moscow, USA
Division of Research and Development, Centro Agronómico Tropical de Investigación y Enseñanza (CATIE), Turrialba, Costa Rica

Jan Boll
Waters of the West-Water Resources Program, University of Idaho, Moscow, USA
Department of Biological and Agricultural Engineering, University of Idaho, Moscow, USA

Vanessa Beach and Eric W. Peterson
Department of Geography-Geology, Illinois State University, Normal, USA

Le Binh Bien, Dieuseul Predelus, Laurent Lassabatere, Thierry Winiarski and Rafael Angulo-Jaramillo
Laboratoire d'Ecologie des Hydrosystèmes Naturels et Anthropisés, UMR 5023 CNRS-ENTPE-UCBL, Université de Lyon, Lyon, France

Baldemar Méndez-Antonio, Ricardo A. Carvajal Rodríguez and Christopher Watts
Sonora University, Hermosillo, Mexico

Ernesto Caetano
National Autonomous University of Mexico, Mexico City, Mexico

Gabriel Soto-Cortés
Metropolitan Autonomous University, Mexico City, Mexico

Fabián G. Rivera-Trejo
Tabasco Autonomous Juárez University, Villahermosa, Mexico

Ashraf M. Elmoustafa and Mona M. Mohamed
Irrigation and Hydraulic Department, Faculty of Engineering, Ain Shams University, Cairo, Egypt

www.ingramcontent.com/pod-product-compliance
Lightning Source LLC
Chambersburg PA
CBHW080256230326

41458CB00097B/5016